Surveying

Surveying

Sixth edition

A Bannister
MC, MSc, CEng, FICE

S Raymond
MSc, PhD, DipTP (Manchester), CEng, FICE,
MRTPI, FIHT (*Consulting Engineer*)

Directors, GeoResearch Ltd, Stockport
(Formerly Readers in Civil Engineering,
University of Salford)

R Baker
MSc, CEng, MICE, MIWEM
(Lecturer in Civil Engineering,
University of Salford)

 LONGMAN

Addison Wesley Longman Limited
Edinburgh Gate, Harlow
Essex CM20 2JE, England
and Associated Companies throughout the world.

British Library Cataloguing in Publication Data
A catalogue record for this book is available from the British Library

ISBN 0-582-07688-9

Library of Congress Cataloging-in-Publication Data
Bannister, A. (Arthur)
    Surveying / A. Bannister, S. Raymond, R. Baker. -- 6th ed.
        p.   cm.
    Includes bibliographical references and index.
    ISBN 0-470-21845-2
    1. Surveying. I. Raymond, Stanley. II. Baker, R. (Raymond),
    1956–   III. Title
    TA545.B25   1992
526.9--dc20
                                                    91-43507
                                                    CIP

Set in Compugraphic Times 9/11

Produced through Longman Malaysia, PA

Contents

Preface to sixth edition

In this edition of *Surveying*, as in earlier editions, our aim has been to produce a text which gives an up-to-date and concise treatment of the subject. Developments in electronics since 1984, when the last edition was published, have caused us to make radical alterations to the format of the book. Accordingly we have sought to minimise our treatment of optical distance measurement and further emphasise electromagnetic distance measurement and introduce total station instruments, data loggers and the Global Positioning System. We have also taken the opportunity to consolidate aspects of setting out into a single chapter in order to highlight the importance of this part of construction engineering.

The Publishers contacted a number of lecturers in universities and polytechnics, home and abroad, asking them to appraise our proposals for this edition and we received much helpful advice. We have tried to blend this with our thoughts and we hope that our goal has been effectively achieved in that the text satisfies the requirements not only of students and others engaged in the construction industry but also of others with an interest in this subject such as mining engineers, geographers and archaeologists.

Arthur Bannister
Stanley Raymond
Raymond Baker
October 1991

Acknowledgements

When preparing the manuscript we have been indebted to the advice of our academic colleagues mentioned in the Preface, unfortunately their names were not disclosed and accordingly we trust, therefore, that they will accept our gratitude as a group. We wish to thank the following for permission to reproduce questions from their examination papers: Engineering Council, Institution of Civil Engineers, Institution of Structural Engineers, University of London, University of Salford. In addition we would like to thank the Editor of New Civil Engineer for permission to reproduce material from that journal. We have also appreciated the assistance of the many instrument manufacturers who supplied catalogues and other details about their products and photographs for inclusion in the text: Geotronics UK Ltd; Carl Zeiss (Oberkochen) Surveying Ltd; Leica UK Ltd; Hall and Watts Ltd; Sokkisha UK Ltd; Spectra Physics Ltd; Valeport Marine Scientific Ltd; Kelvin Hughes Ltd; Stanley Tools Optimal Solutions Ltd.

Cover photograph courtesy of Chris Tivy, was supplied by Longman.

We are grateful to the following for permission to reproduce copyright material:

British Standards Institute for Table 10.1 (Extracts from British Standards are reproduced with the permission of BSI. Complete copies can be obtained from BSI, Linford Wood, Milton Keynes, MK14 6LE); British Transport Commission for Fig. 13.9(a,b); Carl Zeiss (Oberkochen) Surveying Ltd for Figs 4.16, 5.14, 5.15, 5.17, 13.11, 13.12; Geotronics UK Ltd for Figs 5.8(a,b), 5.11, 5.12, 5.13, 7.21, 12.17; Hall & Watts Ltd for Figs 4.4, 9.3; Kevin Hughes Ltd for Fig. 12.7; Leica UK Ltd for Figs 3.20, 3.25, 3.26, 4.7, 4.11, 4.15, 4.16, 4.17, 4.23, 5.9, 5.16, 7.7, 7.8, 13.7, 13.10, 13.20; the editor, *New Civil Engineer* for Figs 6.18, 10.20; Optimal Solutions Ltd for Fig. 3.24; Stanley Tools for Figs 2.2, 2.3; Sokkisha UK Ltd for Figs 5.10, 10.11; Spectra-Physics Ltd for Fig. 10.12; Telumat UK Ltd for Fig. 5.6; Valeport Marine Scientific Ltd for Figs 12.4, 12.19.

Whilst every effort has been made to trace the owners of copyright material, in a few cases this may have proved impossible and we take this opportunity to offer our apologies to any copyright holders whose rights we may have unwittingly infringed.

Arthur Bannister
Stanley Raymond
Raymond Baker
October 1991

1 Introductory

Surveying may be defined as the art of making measurements of the relative positions of natural and man-made features on the earth's surface, and the presentation of this information either graphically or numerically. It dates back to antiquity. Heron, a Greek who lived in Alexandria in about the first century AD, provided the first serious account of surveying techniques. From this it is clear that the work of Euclid and other geometers was used in measuring up and setting out. Naturally, much has altered but, nevertheless, a few techniques have shown little change in principle over the centuries. For instance, the illustration at the beginning of this chapter, taken from Rathborne's *The Surveyor* (London, 1616), can be compared with modern plane-table equipment mentioned in Chapter 6.

The commonest method of presentation is by way of a plan, a true-to-scale representation of an area in the two dimensions which form the horizontal plane. The third dimension, height, is normal to the horizontal and can be shown on the plan in various ways. The term 'levelling' is encountered here and refers to operations connected with the representation of relative difference in altitude between various points on the earth's surface.

Plane surveying

Surveying is divided primarily into geodetic surveying and plane surveying. In geodetic surveying large areas of the earth's surface are involved and the curvature of the earth must be taken into account. In plane surveying relatively small areas are under consideration, and it is taken that the earth's surface is flat, i.e. it gives a horizontal plane. Measurements plotted will represent the projection on the horizontal plane of the actual field measurements. For example, if the distance between two points A and B on a hillside is l, the distance to be plotted will be $l \cos \alpha$, where α is the angle line AB makes with the horizontal, assuming a uniform slope.

A horizontal plane is one which is normal to the direction of gravity, as defined by a plumb bob at a point, but owing to the curvature of the earth such a plane will in fact be tangential to the earth's surface at the point. Thus, if a large enough area is considered on this basis, a discrepancy will become apparent between the area of the horizontal plane and the actual curved area of the earth's surface.

It can be shown that for surveys up to 250 km^2 in area this discrepancy is not serious, and it is obvious therefore that plane surveying will be adequate for all but the very largest surveys. However, precautions are required when connecting such surveys to control points established and co-ordinated by geodetic surveys.

Geodetic surveying

Geodetic surveying is actually a branch of surveying distinguished both by use and technique. As will be explained shortly, frameworks of angular and distance measurements between points are necessary to control all surveys and when surveying large areas, such as a whole country, these measurements must be taken to the highest possible standard. Modern methods for this task include global positioning systems which use transmissions from satellites to obtain the three dimensional co-ordinates of any point on the earth's surface to a high degree of accuracy. The study of the size and shape of the earth and its gravity field is known as geodesy, hence the name of this type of surveying.

Branches of surveying

Surveys are often classifed by purpose, as follows.

Topographic surveys

These produce maps and plans of the natural and man-made features. There is no clear distinction between a map and a plan but it is generally accepted that in a plan detail is drawn such that it is true to scale, whilst in a map many features have to be represented by symbols, the scale being too small. Height information can be added either as spot heights, which are individual height points, or contours, which give a less detailed, but more visual, representation of the area. Frequently spot heights only are shown on plans.

Plans tend to be used for engineering design and administration purposes only, but maps have a multitude of uses — navigational, recreational, geographical, geological, military, exploration — their scales ranging from 1:25 000 to, say, 1:1 000 000.

Engineering surveys

These embrace all the survey work required before, during and after any engineering works. Before any works are started large-scale topographical maps or plans are required as a basis for design. The proposed position of any new item of construction must then be marked out on the ground, both in plan and height, an operation generally termed setting out, and finally, 'as built' surveys are often required.

Especially for the design and construction of new routes, e.g. roads and railways, but in many other aspects of surveying, it is often required to calculate the areas and volumes of land and data for setting out curves for route alignment.

Typical scales are as follows:

Architectural work, building work, location drawings: 1/50, 1/100, 1/200.

Site plans, civil engineering works: 1/500, 1/1000, 1/1250, 1/2000, 1/2500.

Town surveys, highway surveys: 1/1250, 1/2000, 1/2500, 1/5000, 1/10 000, 1/20 000, 1/50 000.

Cadastral surveys

These are undertaken to produce plans of property boundaries for legal purposes. In many countries the registration of ownership of land is based on such plans.

Basic principles

The fundamental principles of surveying are few and simple in concept; for instance, on any area of land to be measured, it will always be possible to choose two points and to measure the distance between them. This line AB can be drawn to scale on paper. Other points can be located relative to the line by taking *two* other measurements which can of course be similarly drawn to scale on the paper, and in this way a map is constructed. The two measurements can consist of two distances, one distance and an angle, or two angles as illustrated in Fig. 1.1, A and B representing in each case the two original points, and C a point to be located.

In the simplest surveys the above approach could well suffice, C representing the points of detail to be located from line AB, which would be referred to as a base line. When the whole area to be surveyed cannot be seen from this one line, additional lines have to be defined, relative to the first, using such pairs of measurements. The points of the junctions of these lines are called control points and together with the lines they constitute a framework or control. Normally the control points (control stations) need to be intervisible with several other control stations and principles governing their location are given in subsequent chapters. It will be realized that in addition to locating points of detail by further angular and distance measurements, engineering features can be set out from the framework in the same way.

Because the rest of the mapping or setting out work is based on this framework it has to be surveyed to greater accuracy than the detail. The relative positions of points can be calculated far more accurately than they can be directly plotted, especially when very large areas are involved. Hence, except for the simplest surveys, an appreciable amount of calculation is involved and the positions of the control points can be given in terms of co-ordinates.

The discussion so far has related to the production of the plan of an area or the location of engineering features in plan position. Height information is often essential, and is typically derived by levelling methods described in Chapter 3.

The reliability of a survey

Since every technique of measurement is subject to unavoidable error, surveyors must be aware of all sources and types of error and how they combine. If the accuracy of a measurement is defined as the nearness of that value to its true value (a quantity we can never know) then a surveyor must ensure that the techniques he chooses will produce a result that is sufficiently accurate. He must know, therefore, how

Fig. 1.1

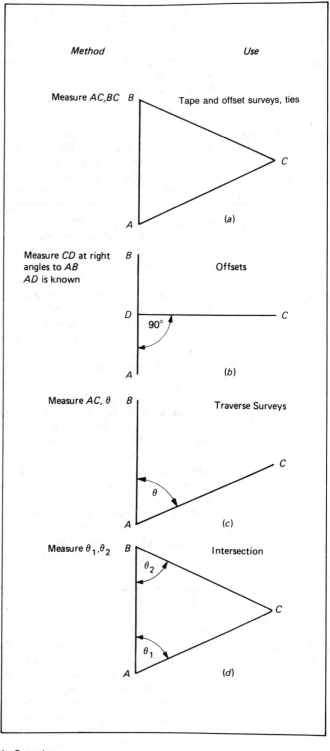

Method | Use

Measure AC,BC — Tape and offset surveys, ties

(a)

Measure CD at right angles to AB
AD is known — Offsets

90°

(b)

Measure AC, θ — Traverse Surveys

θ

(c)

Measure θ₁,θ₂ — Intersection

θ₂

θ₁

(d)

accurate he needs to be, how to achieve this accuracy and how to check that the required accuracy has been achieved.

Accuracy required

When surveying to produce a plan, the accuracy required is defined by the scale of the plot, since there should be no plottable error in the survey data. A good draughtsman can plot a length to within 0.25 mm and so, if a plan of an area is required at a scale of 1/1000, i.e. 1 mm on the plan represents 1 m on the ground, the smallest plottable distance is 0.25 m. Thus, for a survey at 1/1000 scale, all the measurements must be taken such that the relative positions of any point with respect to any other must be determined to 0.25 m or better.

The specifications of surveys for other purposes such as engineering works or property boundary definition might well be determined by engineering tolerances or legal standards.

Achieving the specification

Equipment and methods must be chosen to ensure that the specification is achieved, and this involves checking that the right sort of data, i.e. the correct combination of angles and distances, will be collected and that the data will be to the required accuracy. There are several types of error that occur and a knowledge of their importance and characteristics is essential in the understanding of the limitations of the techniques of measurement. The salient features will be stated now but further information is given in later chapters.

Mistakes

Blunders or mistakes are often inaccurately referred to as gross errors. Miscounting the number of tape lengths when measuring a long distance or transposing numbers when booking are two very simple, but all too common, examples of mistakes. These types of mistakes can occur at any stage of a survey, when observing, booking, computing or plotting, and they would obviously have a very damaging effect on the results if left uncorrected. However, by following strictly a well-planned observing procedure it is possible to reduce the number that occur and then independent checks at each stage should show up those that have been made. In practice, none should ever go undetected and uncorrected.

Systematic errors

Systematic errors arise from sources which act in a similar manner on observations. The method of measurement, the instruments used and the physical conditions at the time of measurement must all be considered in this respect. Expansion of steel tapes, frequency changes in electromagnetic distance measuring (EDM) instruments and collimation in a level are just a few examples of possible sources of systematic errors.

These errors are of vital importance in activities which consist of adding together a succession of individual observations (see sections on levelling and traversing). If all the individual measurements contain the same type of systematic error, which by their nature always act in the same direction, then the total effect is the sum of them all.

It must be ensured that measurements are as accurate as required by removing the effects of all factors that, if neglected, would result in a significant error. The errors caused by some factors can be eliminated with the correct observing procedure and others countered by applying corrections.

Systematic errors are not revealed by taking the same measurement again with the same instruments. The only way to check adequately for systematic error is to remeasure the quantity by an entirely different method using different instruments.

Random errors

Random errors are really all those discrepancies remaining once the blunders and systematic errors have been removed. Even if a quantity is measured many times with the same instrument in the same way, and if all sources of systematic error have been removed, it is still highly unlikely that all results will be identical. The differences, caused mainly by limitations of instruments and observers, are random errors.

It is found in practice that these errors, although called random, have the following characteristics:

(1) small errors occur more frequently than large ones
(2) positive and negative errors are equally likely to occur
(3) very large errors seldom occur.

These characteristics are typical of errors which are normally distributed and it is assumed that we can use the mathematical theory based on the normal distribution to deal with the errors met with in surveying. All the necessary properties and effects of random errors are described in detail in Chapter 8.

Understanding the errors that limit the accuracy of the measurement techniques is but one step to ensuring specifications are achieved, as will be seen when the methods of survey are described:

(a) The survey area is always totally covered with the simplest possible framework of high quality measurements. If the rest of the survey work is carried out within this control the possible damaging accumulation of errors can be contained. This is often termed 'working from the whole to the part'.
(b) Observing procedures are designed so that (i) most mistakes that occur are discovered immediately and (ii) possible sources of systematic errors eliminated.
(c) Additional, or redundant, observations are taken so that all data can be checked for the mistakes, systematic errors and random errors that do occur. For example, the three angles of a triangle would be observed although only two are required to define the shape as in Fig. 1.1(d). The third angle could be deduced but, when measured, acts as a check.
(d) Many quantities are observed several times. These repeated measurements and the observation of redundant data serve both as checks and to improve on the precision of the final results.

Checking the survey

Even with all the checking procedures the surveyor employs, errors can still occur in the finished plan and for this reason final independent checks are required. For simple work this would involve inspecting the final plan in the field and comparing some measurements scaled off the plan with their equivalents on the ground.

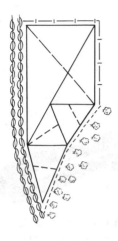

2 Tape and offset surveying

This method of surveying is often referred to as 'chain surveying', deriving its name from the fact that the principal item of equipment traditionally used was a measuring chain. Nowadays, as a result of improvements in manufacturing techniques and a consequential reduction in purchase price, the chain has been replaced by the more accurate steel band.

The technique is fundamental for the collection of detail in all areas of surveying and may be sufficient on its own to cover the requirements of an engineering survey of a small area of land. A good knowledge of tape and offset surveying is therefore essential to a proper understanding of surveying as a whole.

Equipment used in tape and offset surveying

The items of equipment required fall under four broad headings: those used for linear measurement, those used for measuring right angles, that for measuring ground slope, and other items.

1. Equipment for the measurement of lines
The chain

Chains are normally either 20 m or 30 m long. They are made of tempered steel wire, 8 or 10 SWG, and are made up of links which measure 200 mm from centre to centre of each middle connecting ring. Swivelling brass handles are fitted at each end and the total length is measured over the handles. Tally markers, made of plastic, are attached at every whole metre position, and those giving 5 m positions are of a different colour. Older versions might have brass tallies marking every tenth link (Fig. 2.1). The chain is robust, easily read, and easily repaired

Fig. 2.1 Land measuring chain

Fig. 2.2 Surveyor's band (*Courtesy*: Stanley Tools Ltd)

in the field if broken. It is liable to vary somewhat in length, however, owing to wear on the metal-to-metal surfaces, bending of the links, mud between the bearing surfaces, etc.

The surveyor's band (or drag tape)

The steel band is a much more accurate measuring instrument than the chain. It is made of steel strip, some 6 mm in width, and is carried on a four-arm open frame winder. A handle is fitted for returning the band into its frame after use, and this also provides a locking device for retaining the band. Rawhide thongs are supplied for attaching to the small loops at the extremities of the bands to allow them to be pulled or straightened. Alternatively handles, similar to those of the chain, can be fitted: lengths of 30 m or 50 m are normal but 100 m bands may be encountered. BS 4484: Part 1: 1969 requires that metres, tenths and hundredths of metres should be marked, with at least the first and last metres also subdivided into millimetres. The operating tension and temperature for which it was graduated should be indicated on the band (Fig. 2.2).

Tapes

These may be made of synthetic material, glass fibre being typical, or coated steel or plain steel. BS 4484: Part 1 suggests 10 m, 20 m or 30 m as the desirable lengths, and these are generally available.

For the synthetic types the British Standard requires major graduations at whole metre positions and tenths, with minor graduations at hundredths, and 50-mm intervals indicated. Those manufactured of glass fibre have a PVC coating (Fig. 2.3). They are graduated every 10 mm and figured every 100 mm; whole metre figures are shown in red at every metre. These tapes are said to have good length-keeping properties, but it is conventional to use them for relatively short measurements.

Steel tapes may be provided with a vinyl coating or may be plain. The former type has sharp black graduations on a white background. They can be obtained graduated every 5 mm and figured every 100 mm; the first and last metre lengths are also graduated in millimetres. Whole metre figures are again shown in red at every metre.

Fig. 2.3 Typical
measuring tape
(*Courtesy*: Stanley
Tools Ltd)

The latter type have graduations and figures etched on to the steel and they present the same subdivisions as the vinyl-covered types. However, they are generally wider and are contained in a leather case rather than a plastic one.

It is important that tapes and bands be wiped clean and be dry before rewinding into their cases or on to their frames respectively. In addition, plain steel units need occasional treatment with an oily rag to prevent rust formation.

2. Equipment for measuring right angles
The cross staff

A cross staff consists essentially of an octagonal brass box with slits cut in each face so that opposite pairs form sight lines (Fig. 2.4). The instrument may be mounted on a short ranging rod and, to set out a right angle, sights are taken through any two pairs of slits whose axes are perpendicular. The other two pairs then enable angles of 45° and 135° to be set out.

The optical square

There are two types of optical square, one using two mirrors and the other a prism (Fig. 2.5(*a*)). The instrument is compact, rarely measuring more than 75 mm in diameter by about 20 mm thick, and is more accurate than the cross staff.

The mirror type makes use of the fact that a ray of light reflected from two mirrors is turned through twice the angle between the mirrors, which in turn is easily derived using the principles of reflection of light (Fig. 2.5(*b*)). Mirror A is completely silvered, while mirror B is silvered to half its depth, the other half being left plain. Thus, the eye looking through the small eye-hole will be able to see half an object at O_1. An object at O_2 is visible in the upper (silvered) half of mirror B, and when $O_1\hat{X}O_2$ is a right angle (where X is the centre of the instrument), the image of O_2 is in line with the bottom half of O_1 seen direct through the plain glass.

The surveyor stands at X, sights O_1, and directs his assistant to move O_2 until the field of view is as shown in Fig. 2.5(*b*). Then $O_1\hat{X}O_2$ is a right angle for, considering any ray from O_2 incident on mirror A at angle α to the normal, it will emerge at the same angle to the normal.

Fig. 2.4 Cross staff

Fig. 2.5(a) SH4
Optical square

Fig. 2.5(b)

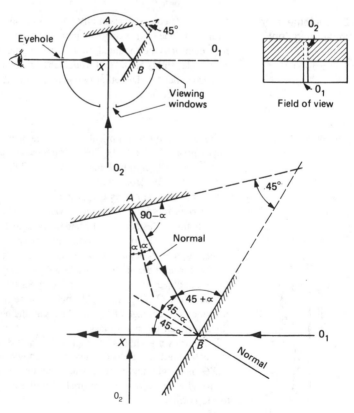

Eyehole

A

45°

0_1

X

B

Viewing
windows

0_2

0_2

0_1

Field of view

A

90−α

Normal

α α

45+α

45−α

45−α

45°

X

B

Normal

0_2

0_1

Therefore $\qquad X\hat{A}B = 2\alpha$

and, from a consideration of the angles,

$$X\hat{B}A = 90° - 2\alpha$$

Therefore $\qquad A\hat{X}B = O_1\hat{X}O_2 = 90°$

i.e. the result is independent of α.

The prismatic type of optical square employs a pentagonal-shaped prism, cut so that two faces contain an angle of exactly $45°$. It is used in the same way as the mirror square, but is rather more accurate.

A variant of this model (Fig. 2.5(a)) has two such prisms, one mounted above the other. For setting out right angles the top one only is used, but the prisms are so positioned that both of them can be used to set out two points O_1 and O_2 such that $O_1\hat{X}O_2$ is 180°.

3. Equipment for measuring ground slope
The Abney Level

This versatile little instrument can be used either as a hand level or for the measurement of vertical angles.

As shown in Fig. 2.6(a), which illustrates a typical instrument, the Abney Level consists essentially of a sighting tube, to which is attached a graduated arc. An index arm, pivoted at the centre of the arc, carries a small bubble tube, whose axis is normal to the axis of the arm, so that as the tube is tilted the index moves over the graduated arc. By means of an inclined mirror mounted in one-half of the sighting tube, the bubble is observed on the right-hand side of the field of view when looking through the eyepiece of the sighting tube. The milled head is used to manipulate the bubble. The arc itself is graduated in degrees

Fig. 2.6 Abney Level

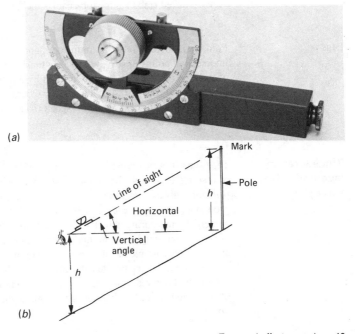

(a)

(b)

from 0 to $+90°$ and $-90°$, and the scale is read by the vernier on the index arm to 10 minutes. Also engraved on the arc is a gradient scale giving gradients from 1:10 to 1:1.

In using the instrument for the measurement of vertical angles, when measuring mean ground slope, it will be seen from Fig. 2.6(*b*) that if the sight is taken on to a point whose height above the ground is the same as the observer's eye, then the line of sight will be parallel to the mean ground surface. A ranging pole with a mark on it at the required height makes a suitable target. To measure the angle, a sight is taken on to the mark, and the bubble brought into the field of view by means of the milled head to be bisected by the sighting wire at the same time as the wire is on the target. The angle is now read on the vernier.

4. Other equipment
Ranging rods

These are poles of circular section 2 m, 2.5 m or 3 m long, painted with characteristic red and white bands which are usually 0.5 m long, and tipped with a pointed steel shoe to enable them to be driven into the ground. They are used in the measurement of lines with the tape, and for marking any points which need to be seen. A sectional tubular type is also made, the short bottom section of which, apart from its robustness, is often useful where headroom is restricted.

In hard, or paved, ground a tripod is used to support the rods.

Arrows

When measuring the length of a long line, the tape has to be laid down a number of times and the positions of the ends are marked with arrows, which are steel skewers about 40 mm long and 3−4 mm diameter. A piece of red ribbon at the top enables them to be seen more clearly.

Pegs

Points which require to be more permanently marked, such as the intersection points of survey lines, are marked by nails set in the tops of wooden pegs driven into the ground by a mallet. A typical size is 40 mm × 40 mm × 0.4 m long. In very hard or frozen ground, steel dowels are used instead, while in asphalt roads, small 5 or 6 mm square brads are used.

Procedures in tape and offset surveying

Tape and offset surveying consists in measuring with the tape the lengths of a series of straight lines, the positions of which are governed by principles given later in this chapter, and then locating points on the ground relative to these lines by the methods given in Chapter 1, namely, by measuring two other lines, known as *ties*, or by measuring *offsets* at right angles to the main survey line. Since ties and offsets are short measurements, they can be made with the glass fibre tape and rarely exceed one tape length. Thus the two basic procedures which need to be known at this stage are (1) the measurement or 'ranging' of lines, and (2) the setting out of the right angles in connection with offsets.

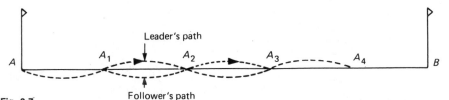

Fig. 2.7

1. Measuring the length of a line

The operation is carried out by two assistants, known as chainmen because the measurement was traditionally made by chain, one acting as *leading chainman* and the other as *follower*. The chainmen take one end each of the steel band and the band is pulled out full length and examined for defects. The leader equips himself with ten arrows and a ranging rod, and the follower also takes a ranging rod. Then (Fig. 2.7), to measure line *AB* having previously positioned ranging rods at both *A* and *B*:

(a) The leader drags his end of the band forward to A_1, and holds his ranging rod about 0.3 m short of the end.

(b) The follower holds his end of the band firmly against station *A* and the surveyor lines in the leader's pole between *A* and *B* by closing one eye, sighting poles *A* and *B*, and signalling the leader till he brings his pole into line *AB*. The system of signalling usually adopted is to swing the left arm out to the left as an instruction to the leader to move his pole in that direction: the right arm is similarly used to indicate movement to the right, while both arms extended above the head, then brought down, indicates that the pole is on line.

(c) The leader straightens the band past the rod by sending gentle 'snakes' down the band.

(d) The follower indicates the band is straight, and the leader puts an arrow at the end, A_1. (At this stage offsets or ties may be taken from known positions to required detail.)

(e) The leader then drags his end to A_2, taking nine arrows and his pole.

(f) The follower moves to A_1, and puts his pole behind the arrow, and the surveyor again lines in from here or from *A*.

The above procedure is repeated, the follower picking up the first arrow before he moves from A_1. The leader moves to A_3, carrying eight arrows. The follower moves to A_2, carrying the arrow from A_1.

If the line measured is longer than ten times the band length, the leader will exhaust his supply of arrows, so that when the 11th band length is stretched out, the follower will have to hand back the ten arrows to the leader. This fact is pointed out to the surveyor who notes it in his field book. The number of arrows held by the follower serves as a check on the number of full band lengths measured in the line.

2. Setting out right angles

Since this operation is often required in connection with the measurement of offsets, this is a convenient point at which it may be

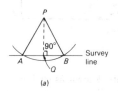

Fig. 2.8

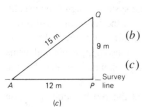

(a)

(b)

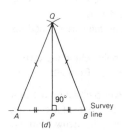

(c)

(d)

discussed. There are two cases to consider: (1) dropping a perpendicular from a point to a line; and (2) setting out a line at right angles to the survey line from a given point on the steel band.

1(a) For short offsets, the end of the tape is held at the point to be located, and the right angle is estimated by eye. Although a usual method in practice, it is not so accurate as the following methods.

(b) Again for offsets, the tape is swung with its zero as centre about the point and the minimum reading at which it crosses the band is noted. This occurs, of course, when the tape is perpendicular to the band, but the method can be used only on smooth ground where a free swing of the tape is possible.

(c) Where the above method is not applicable, with the free end of the tape at centre P (the point), strike an arc to cut the band at A and B (Fig. 2.8(a)). Bisect AB at Q. Then $P\hat{Q}A = 90°$.

(d) Run the tape from P to any point A on the band (Fig. 2.8(b)). Bisect PA at B, and with centre B and radius BA strike an arc to cut the band at Q. Then $A\hat{Q}P = 90°$, being the angle in a semicircle.

2(a) Cross staff: this is mounted on a short ranging rod which is stuck in the ground at the point at which the right angle is to be set out. The cross staff is turned until a sight is obtained along the survey line and the normal is then set out by sighting through the slits at right angles to this.

(b) Optical square: this is used as already described, being either held in the hand or else propped on a short ranging rod.

(c) Pythagoras' theorem (3, 4, 5 rule or any multiple thereof, say 9, 12, 15): with the zero end of the tape at P take the 24 m mark of the tape to A, where $AP = 12$ m on the band. Take the 9 mark on the tape in the hand and, ensuring that the tape is securely held at A and P, pull both parts of the tape taut to Q. Then $A\hat{P}Q = 90°$ (Fig. 2.8(c)).

  Note: A convenient combination is 8, 8.4, 11.6. If the 8 is set off on the band then $8.4 + 11.6 = 20$ m can be set off on a 20 m tape.

(d) Take A and B on the band so that $PA = PB$ (Fig. 2.8(d)). Strike arcs from A and B with equal radii to intersect at Q. Then $A\hat{P}Q = 90°$.

As exercises, the above methods of setting out right angles should be practised. Case 2(c) in particular might well arise in the setting out of right angles on the construction site. For instance, P could be the corner of a building whose sides contain PA and PQ. Steel tapes up to 30 m long are normally used for setting out and it is worth noting that an error of 5 mm in AQ (Fig. 2.8(c)) causes an error of approximately 2 minutes of arc in P.

Errors in linear measurement and their correction

In all surveying operations, as indeed in any operation involving measurement, errors are likely to occur, and so far as is possible they must be guarded against or their effects corrected for. The types of error which can occur have been classified in Chapter 1 under the following three headings: (i) mistakes, (ii) systematic errors, (iii) random errors. The three types will be dealt with briefly, with examples of their occurrence and the remedies for them in tape and offset surveying.

Mistakes

These are due to inexperience or to carelessness on the part of the surveyor or the chainmen and are, of course, quite random in both occurrence and magnitude. If allowed to pass unchecked, mistakes could lead to a faulty plan being produced. By careful work, however, and by taking suitable check measurements, it should be possible to make a survey which is free from mistakes. Typical mistakes in measuring the length of a line are:

(1) Omitting an entire band length in booking. This is prevented by noting down each band length, and by the leader keeping careful count of the arrows, as described earlier.
(2) Mis-reading the steel band. It is best if two people make important readings.
(3) Erroneous booking sometimes occurs; it is prevented by the chainman carefully calling out the result and the surveyor repeating it, paying attention when calling 5 or 9, 7 or 11.

Systematic or cumulative errors

These arise from sources which may be taken to act in a similar manner on successive observations, although their magnitude can vary. Their effects, when known, may be eliminated.

(1) Standard

The most careful measurements will not produce an accurate survey if, for example, the band has been damaged and is therefore of incorrect length, because every time the band is stretched out it will measure not 30 m but 30 m ± (some constant or systematic error). If uncorrected, such an error could have serious effects. By checking the band against a standard, such as two marks measured for the purpose, the exact error per band length is known. If this error cannot be eliminated, a correction can be applied which will enable the effect of the error to be removed.

Correct length of line

$$= \text{measured length of line} \times \frac{\text{length of band used}}{\text{length of standard}}$$

The length of the standard is, of course, usually 30 m or 50 m. It would be good practice to discard any steel band that differed from the standard by more than 1 in 5000, i.e. 6 mm in 30 m.

(2) Sloping ground

As was stated in Chapter 1, all measurements in surveying must either be in the horizontal plane, or be corrected to give the projection on this plane. Lines measured on sloping land must be longer than lines measured on the flat, and if the slope is excessive, then a correction must be applied. There are two methods.

(*a*) ·*Stepping* · On ground which is of variable slope this is the best method, and needs no calculation. The measurement is done in short lengths of 5–10 m, the leader holding the length horizontal. The point on the ground below the free end of the band is best located by plumb bob, as shown in Fig. 2.9. It will be seen that it is easier to work downhill when 'stepping' than to work uphill, the follower then having the difficult job of holding the band taut, horizontal, and with the end vertically over the previous arrow. The leader has therefore to line himself in.

Fig. 2.9

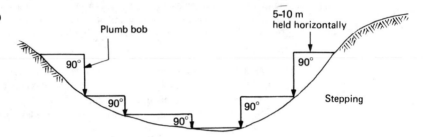

(*b*) *Measuring along the slope* This method is applicable where the ground runs in long regular slopes. The slope is measured either by an instrument like the Abney Level, or by levelling, a procedure which gives the surface height at points along the slope.

(i) Measurement of slope angle, α.

Correct length = measured length $\times$ cos α
where $\qquad \alpha$ = angle of slope (*see* Fig. 2.10(*a*))
Thus the correction = $-L\,(1 - \cos \alpha)$
where $\qquad L$ = the measured length.

This method corrects only the total length of the line, and if intermediate measurements are to be correctly made, adjustments must be made during measurement. These can be readily effected. In Fig 2.10(*b*), *AB* represents one tape length, say 30 m, measured along the slope. What we require is the point *C* beyond *B* such that a plumb bob at *C* will cut the horizontal through *A* at *D*, where *AD* is 30 m on the horizontal; i.e. we require the correction *BC* which is to be added to each tape length measured along the slope.

Fig. 2.10

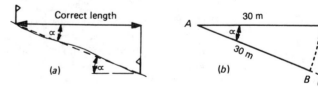

(*a*)   (*b*)

Now

$$AC = AD \sec \alpha$$

$$BC = AC - AB = AC - AD = AD \sec \alpha - AD$$

$$= AD \left(1 + \frac{\alpha^2}{2} + \frac{5\,\alpha^4}{24} + \dots \right) - AD$$

where α is in radians. Therefore,

$$\text{correction } BC = \frac{\alpha^2}{2} \cdot AD \quad \text{say}$$

(ii) Slope can be expressed also as 1 in n, which means a rise of 1 unit vertically for n units *horizontally*: for small angles $\alpha = 1/n$ radians.

Figure 2.11 shows the relationship between corrections and slopes for 20 m lengths.

Fig. 2.11 Relationship between corrections and slopes for 20 m lengths

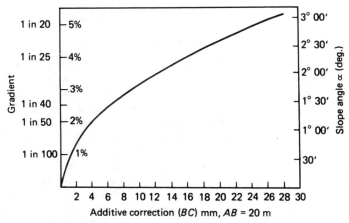

Example 2.1 In measuring the length of a line, what is the maximum slope (*a*) in degrees and (*b*) as 1 in *n* which can be ignored if the error from this source is not to exceed 1 in 1000?

(*a*) Permissible error = 1 in 1000 = $\dfrac{AD}{1000}$

Therefore
$$\frac{AD}{1000} = AD\,(sec\ \alpha - 1)$$

$$\frac{1}{1000} = sec\ \alpha - 1$$

whence
$$\alpha = 2.6°$$

(*b*) Permissible error = $\dfrac{AD}{1000}$

Therefore
$$\frac{AD\alpha^2}{2} = \frac{AD}{2n^2} = \frac{AD}{1000}$$

$$n = \sqrt{500}$$

Therefore, maximum slope is **1 in 22.4**.

Fig. 2.12

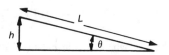

(iii) Slope can also be expressed in terms of the difference in level, h, between two points (Fig. 2.12).

$$\text{Correction} = - [L - (L^2 - h^2)^{1/2}]$$

$$\simeq - \frac{h^2}{2L}$$

an approximation which is acceptable if h/L is less than 1 in 25.

(iv) Finally, Pythogoras' Theorem may be used, (Fig. 2.12).

$$\text{Correct length} = \sqrt{(L^2 - h^2)}$$

(3) Tape standardization

For very accurate work a spring balance should be attached to one end of the steel band. The purpose of this is to ensure that the band is tensioned up to the value at which it was standardized, i.e. if the band is 30 m long at 20°C under a 5 kg pull on the flat, then a tension of 5 kg should be applied to eliminate any correction for pull. The balance is usually attached to a ranging rod or arrow, and thence to one end of the band, by means of a short cord. The far end of the band will then be attached to a second rod, and if these rods be set firmly on or in the ground and levered backwards, the tension applied to the tape can be regulated to any value. All good quality bands should have a standardization certificate, which, for example, might say that the band, nominally 30 m long, is in fact only 29.999 m on the flat at 20°C with a tension of 5 kg applied. This data is used to make corrections to the length as taped to refine the procedure.

In addition to the standardization and slope corrections mentioned above the following factors might have to be considered:

(a) Elasticity and thermal changes in those cases where the field conditions differ from those at which the tape was standardized.
(b) Deviation from the straight.
(c) Height above mean sea level.
(d) Sag, if the tape has been standardized on the flat, not in catenary.

(a) As mentioned, the correct tension can be applied to the band by attaching a spring balance to the handle at one end. If the standard tension is not applied a correction should be made since the length of the tape will have changed:

$$\text{Correction} \qquad = \frac{(P - P_s)}{A \cdot E} \cdot L$$

where P, P_s = field and standard tension respectively

A = cross-sectional area of band
E = Young's modulus of elasticity for the band
L = length measured.

Similarly a correction is required if the tape temperature, t, is not equal to the standard temperature, t_s.

Correction $\qquad = \alpha L(t - t_s)$

where α = coefficient of linear expansion.

If the temperature during measurement is greater than the standard temperature, the tape will expand so that more ground is taped out than is shown by the graduations, i.e. the reading will be too low, indicating a negative error. A positive correction is thus required. If the temperature is lower than standard, a negative correction is required, and it will be noted that the correct sign for the correction is given by the formula. This is also true for the correction for pull since the tape extends when $P > P_s$.

(b) The higher the accuracy required the more critical the alignment of the band becomes. Corrections can be applied for misalignment but since this would require actually measuring the misalignment as in Fig. 2.13, it is generally easier to take more care and line in the band with a theodolite.

Fig. 2.13

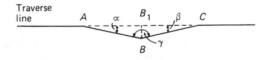

The correction would be

$$- [AB(1 - \cos \alpha) + BC(1 - \cos \beta)]$$

in the case shown in Fig. 2.13.

(c) The length of the line as measured can be reduced to its equivalent length at mean sea level. In Fig. 2.14

$$l = (R + H)\theta \qquad \text{and} \qquad l_1 = R\theta$$

Fig. 2.14

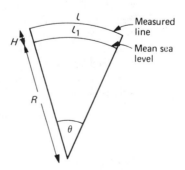

Hence

$$l_1 = l \cdot \frac{R}{R + H}$$

$$\text{Correction} = l - l_1 = l \left(1 - \frac{R}{R + H} \right) = l \cdot \frac{H}{R + H}$$

$$\simeq l \cdot \frac{H}{R} \text{ and is deducted}$$

(*d*) If the highest accuracy is required, rather than lie the band along the ground, it can be suspended between tripod heads, i.e. hung in catenary, and a correction for the sag in the tape applied if the tape has been standardized on the flat. Figure 2.15 shows a simple arrangement that could be used.

Fig. 2.15

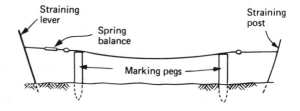

When measuring a line the pegs are aligned, preferably by theodolite (Chapter 4), and after being driven, zinc strips are tacked on. The levels of the tops are found and a transverse scratch mark is made on the first peg to serve as the beginning of the base line, a longitudinal scratch may also be lined in. Lever-type straining arms and a spring balance may conveniently be used for tensioning and supporting the tape, which is adjusted so that the first zero is aligned with the scratched reference mark. A transverse scratch is then made on the second peghead against the second zero, and this serves as reference mark for the second bay, the process being repeated. The tape itself is aligned by theodolite and temperatures are measured as before. Note that in this operation the tape should float just clear of the stakes. Instead of aligning the zero of the tape with the scratch mark made when taping the previous bay, it is also possible to make another scratch to mark the beginning of the new bay, the necessary correction to the bay length then being made by measuring the distance between the two scratches.

Before the introduction of EDM instruments (Chapter 5), base lines of several kilometres length had to be measured in this way in order to scale national triangulation schemes; accuracies better than 1:500 000 were achieved.

$$\text{Correction for sag} = - \frac{w^2 l^3}{24 P^2}$$

where w = weight per unit length of the tape and l is the measured length of span. If the tape has been standardized in catenary no correction is required for sag so long as the field tension P is the same as the standard tension, P_s. Further details of the measurement of base lines in catenary are given in Chapter 7.

It is the surveyor's job to decide which of these modifications to the basic measurement procedure are necessary to achieve the required accuracy. The best way for this to be done is to calculate the expected magnitude of the various corrections and to see if they are significant.

Example 2.2 During the measurement in catenary of a survey line of four bays the following information was obtained:

Bay	Measured length (m)	Temp. (°C)	Difference in level between ends (m)	Tension (N)
1	29.899	18.0	+0.064	178
2	29.901	18.0	+0.374	178
3	29.882	18.1	−0.232	178
4	29.950	17.9	+0.238	178

The tape has a mass of 0.026 kg/m and a cross-sectional area of 3.24 mm². It was standardized on the flat at 20°C under a pull of 89 N. The coefficient of linear expansion for the material of the tape is 0.000011/°C, and Young's modulus is 20.7×10^4 MN/m². The mean level of the line is 26.89 m above mean sea level. Determine the absolute length of the survey line reduced to sea level.

Bay	L	L^3	h	$h^2/2L$
1	29.899	26 728.22	+0.064	0.0001
2	29.901	26 733.58	+0.374	0.0023
3	29.882	26 682.65	−0.232	0.0009
4	29.950	26 865.23	+0.238	0.0009
	119.632	107 009.67		0.004

Tension Correction

$$(P - P_s)\,\frac{\Sigma L}{AE} = (178 - 89) \times \frac{119.632}{3.24 \times 20.7 \times 10^4}$$

$$= +0.016 \text{ m}$$

Sag Correction

$$\frac{w^2 \Sigma (L)^3}{24P^2} = -\frac{(0.026 \times 9.806)^2 \times 107\ 009.67}{24 \times 178^2} = -0.009 \text{ m}$$

Temperature Correction Based on average of 18°C, since there is little variation

$$\alpha \Sigma L(t - t_s) = 0.000011 \times 119.632 \ (18 - 20)$$
$$= -0.003 \ \text{m}$$

Slope Correction

$$\Sigma \frac{h^2}{2L} = -0.004 \ \text{m}$$

Reduction to Mean Sea Level

$$\Sigma L \cdot \frac{H}{R} = \frac{-119.632 \times 26.89}{6 \ 367 \ 000} = -0.001 \ \text{m}$$

Absolute length   = 119.632 + 0.016 − 0.009 − 0.004
                     − 0.003 − 0.001

                = **119.631 m**

Example 2.3 A nominal distance of 30 m was set out with a steel tape from a mark on the top of one peg to a mark on the top of another, the tape being in catenary under a pull of 150 N and at a mean temperature of 25°C. The top of one peg was 0.442 m above the top of the other. Determine the horizontal distance between the marks on the two pegs reduced to mean sea level if the top of the higher peg is 195.57 m above mean sea level.

The tape which was standardized in catenary under a pull of 120 N and at a temperature of 20°C had a mass of 0.026 kg/m and had a cross-sectional area of 3.25 mm². The coefficient of linear expansion for the material of the tape may be taken as 0.000011 per °C, and E as 20.7 × 10⁴ MN/m². The radius of the earth may be taken as 6367 km.

Let L be the true length of the tape on the flat under a pull of 120 N at a temperature of 20°C. Since the tape has been standardized in catenary, if the sag correction be added to the catenary length then the length on the flat will be found.

Thus $L = 30 + \dfrac{30^3 \times (0.026 \times 9.806)^2}{24 \times 120^2} = 30.005 \ \text{m}$

To revert to the conditions of measurement with the tape now in catenary under a pull of 150 N

Tension Correction $= + \dfrac{(150 - 120) \times 30.005}{20.7 \times 10^4 \times 3.25} = +0.001 \ \text{m}$

Sag Correction $= - \dfrac{30.005^3 \times (0.026 \times 9.806)^2}{24 \times 150^2} = -0.003 \ \text{m}$

$$\text{Slope Correction} = -\frac{0.442^2}{2 \times 30.005} = -0.003 \text{ m}$$

$$\text{Temperature Correction} = +(25-20) \times 30.005 \times 0.000011$$
$$= +0.002 \text{ m}$$

$$\text{Reduction to mean sea level} = \frac{-195.35 \times 30.005}{6367 \times 10^3} = -0.001 \text{ m}$$

$$\text{Required Length} = 30.005 + 0.001 + 0.002 - 0.003$$
$$- 0.003 - 0.001$$
$$= \textbf{30.001 m} \text{ (to three places of decimals)}$$

The corrections could have been based on a nominal length of 30.00 m as there would have been no differences in any of the values.

It will be seen that the procedure adopted reduced the standardization in catenary to standardization on the flat, then the normal corrections were applied to the field measurements. Had the tape been found to be, say, 3 mm longer than nominal, the length on the flat under a pull of 120 N at 20°C would have been 30.005 + 0.003 = 30.008 m and this value would have then been subject to correction.

Random or accidental errors

This third group of errors, acting independently on observations, arises from lack of perfection in the human eye and in the method of using equipment. They are not mistakes and, as there is as much chance of their being positive as being negative, the errors from these sources tend to cancel out, i.e. tend to be compensatory. They do not entirely disappear, however, and it can be shown that they are proportional to $\sqrt{L}$, where L is the length of the line. They are, therefore, second-order errors compared with cumulative ones, which are proportional to L. Usually, no attempt is made to correct for them in tape and offset surveys.

Fieldwork

The basic principle is that if A and B are fixed (Fig. 1.1(a)), C can be located by measuring AC and BC. The length of the three sides of $\triangle ABC$ being known, the triangle can be plotted. Any area of land can be divided into a series of triangles which form a framework, which may be plotted, and which covers the greater part of the area to be surveyed. To locate topographical and man-made features relative to this framework, measurements are made with the tape from the lines during the course of the survey. The two methods for locating detail are shown in Fig. 2.16.

Such measurements should be as short as possible, and in any case not greater than one tape-length, so the survey lines will normally run as close to the site boundary as possible. The accuracy of the work will be increased if the framework of triangles is founded on a backbone line run through the site to be surveyed (Fig. 2.17), bearing in mind that the fewest number of well-shaped (or 'well-conditioned') triangles necessary for the work normally gives the best results. By well-conditioned we mean that all the intersections are clear for plotting purposes, giving nearly equilateral triangles if possible, for where points are plotted by striking arcs, representing the measured tie lengths, the determination of intersections at angles of much less than 30° is difficult. Additional survey lines (check lines), crossing existing triangles, should be incorporated where necessary to check on the measurements taken, hence ensuring that errors in measurement will not go undetected.

The intersection points of the lines are called *stations* and these are established first by placing ranging rods (and, later, pegs if permanency is required), after a preliminary reconnaissance survey of the site has been made by the surveyor. The principles governing the location of the stations and the lie of the survey lines may be summarized:

(1) Although general topography will dictate to a large extent the actual layout of the triangles, as few survey lines as necessary should be used, and obstacles and steep, uneven slopes avoided, as far as possible.

(2) There should be at least one long 'backbone' line in the survey upon which the surveyor may found the triangles.

Fig. 2.16

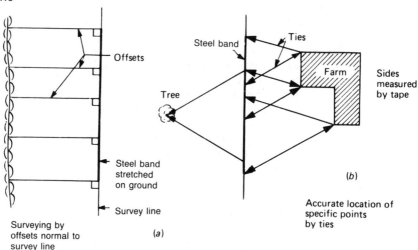

Surveying by offsets normal to survey line

(a)

Accurate location of specific points by ties

(b)

(3) The triangles should have the angles lying between 30° and 120° so as to give clean intersections, and check lines must be provided for all independent figures.

(4) Where possible, avoid having survey lines without offsets unless they are check lines; and keep offsets short, especially to important features.

Fig. 2.17

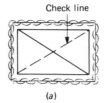

Check line

(a)

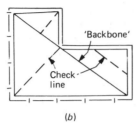

'Backbone'

Check
line

(b)

Figure 2.17 and the illustration at the beginning of this chapter show some layouts of survey lines and possible check lines.

The maximum length of line which can be ranged is normally governed by visibility, in the UK the limit being about 250 m. Also, too long a line may have excessive errors (systematic) due to faulty alignment during measurement. Broadly speaking, however, if systematic errors are compensated for as already outlined, an overall accuracy of the order of 1:1000 can be achieved by careful work.

Booking the survey

Booking is carried out by the surveyor in a field book, which consists of good-quality paper, the pages each approximately 150 mm long by 100 mm wide, say. The field book is bound in the same way as a reporter's notebook. Each page is ruled up the centre with either a single coloured line or with two such lines about 15 mm apart to represent the survey line, and booking starts at the bottom of the page. The salient points to note when starting a survey are:

(1) After a preliminary reconnaissance of the site, bearing in mind the principles enumerated, make a sketch showing the location of the chosen stations and survey lines.

(2) Take enough measurements, generally ties from nearby easily recognizable features, and note enough information to enable each station to be relocated if necessary.

(3) Take the bearing from true or magnetic north of at least one of the lines.

The principal features involved in booking the lines, offsets, etc., may be summarized as:

(1) Begin each line at the bottom of a fresh page.

(2) Take plenty of room and make no attempt to scale the bookings.

(3) Exaggerate any small irregularities which are capable of being plotted.

(4) Make clear sketches of all detail, inserting explanatory matter in writing where necessary: *do not rely on memory*.

(5) Book systematically, proceeding up one side of the survey line and then the other, starting with the side having more detail (and hence more offsets).

These and other features involved in booking are best dealt with by considering the following survey in Peel Park, Salford (Figs 2.18 and 2.19). It will be noted that, to prevent confusion or the risk of mistakes, the decimal point has not been used, i.e. 5.52 m is written as 5/52.

Accuracy of measurements

To a certain extent, this depends on the scale to which the survey is to be plotted, and bearing in mind that the scale might be increased, it is better to be more accurate than may appear to be strictly necessary. Normally we measure survey lines to the nearest 20 mm and offsets to the nearest 50 mm.

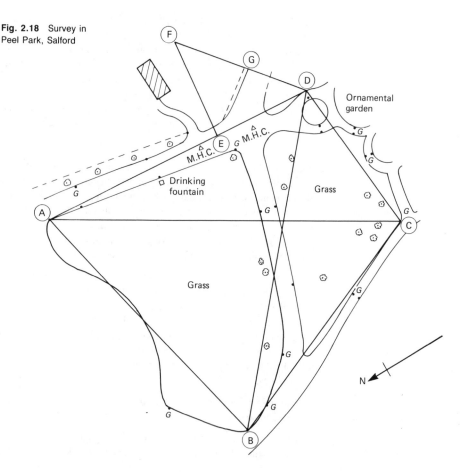

Fig. 2.18 Survey in Peel Park, Salford

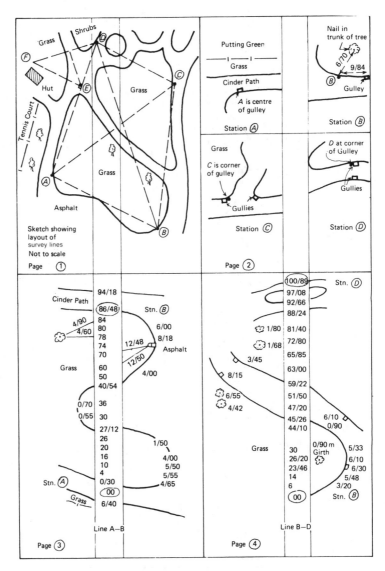

Fig. 2.19 Pages from Field Book showing survey of Peel Park (continued next page)

A good draughtsman can plot a length to within 0.25 mm. Hence, if the scale used is 1/500, i.e. 1 mm represents 500 mm on the ground, then 0.25 mm represents 0.125 m on the ground. Thus the order of accuracy given above is sufficient, but since parts of the survey may be required to be plotted to a larger scale later, it is suggested that measurements be made to 10 mm, i.e. 0.01 m. This is easily effected by means of any of the tapes previously described.

The perpendicularity of short offsets is normally judged by eye or by swinging the tape, but with longer offsets one of the methods given earlier (e.g. optical square), or ties, should be used to prevent error

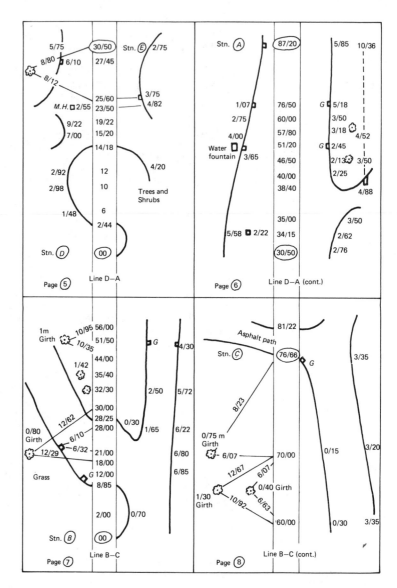

Fig. 2.19 (cont'd)

Note: Whole metre values can be written with or without decimal values, e.g. 21/00 or 21

in location. Where the offset would be longer than one tape-length, a subsidiary triangle should be employed (Fig. 2.20(a)). The use of subsidiary lines and triangles to improve accuracy should also not be forgotten when picking up detail where the offsets to the survey line would be at a very acute angle to the detail (Fig. 2.20(b)).

Office work

Preliminary calculations and checking must first be carried out before plotting can take place, e.g. if any survey lines have been measured along regular slopes, the projection of these lengths on the horizontal

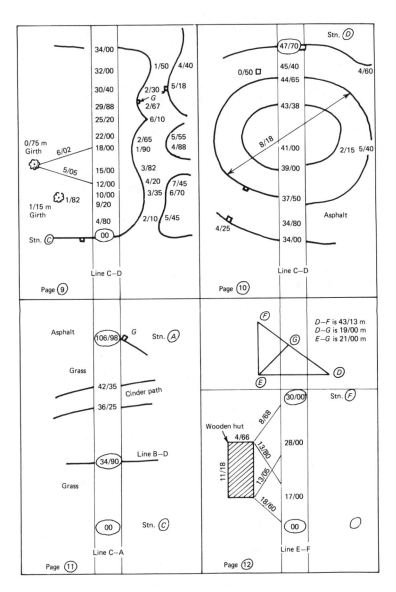

Fig. 2.19 (cont'd)

must be calculated. Check line lengths can be calculated using the appropriate formulae and compared with the measured lengths, or they can be scaled off after the triangles have been plotted. Checking by calculation is preferable, and any errors revealed at this stage will require rectification by repeat measurements in the field, after which plotting may proceed.

Nowadays most surveys are plotted by computer-based plotters as described in Chapter 5. Hand produced surveys should be plotted on good quality paper or, better, on paper mounted on holland cloth to reduce shrinkage. 4H pencils are used for drawing the framework,

Fig. 2.20

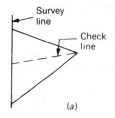

Survey
line

Check
line

(a)

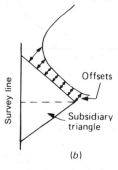

Survey line

Offsets

Subsidiary
triangle

(b)

Fig. 2.21 Use of the
offset scale

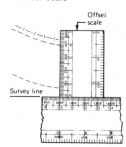

Offset
scale

Survey line

2H pencils for plotting detail. Waterproof Indian ink is used for inking in the completed survey.

Additional requirements to the normal drawing instruments are:

Beam compasses for striking large-radius arcs.
A steel straight edge, 1 or 2 m long.
A set of railway curves.
Set squares, protractor, and French curves.
Colour brushes and a set of stick water colours.

Scales for plotting may be wood-based with celluloid facings, 300 mm long and supplied with small offset scales. The common plotting scales were dealt with in Chapter 1.

It is often good practice to rough out the survey lines on tracing paper, so that by overlaying this on the paper to be used the survey may be properly centred on that sheet. It is preferable that the north be towards the top of the sheet, but by no means essential, since the North point must always be shown.

A line to represent the longest survey line is drawn and the length scaled off. By striking arcs the other stations are established and the network of triangles can be drawn. Check lines are scaled off and compared with the measured distances.

Offsets and ties are plotted systematically in the same order in which they were booked, i.e. working from beginning to end of each line up one side and then the other. The right angles for offsets may be set out separately with a set square, or more conveniently an offset may be used as shown in Fig. 2.21 in conjunction with a scale. The scale is set parallel to the line. Both chainage and offset are scaled off simply in one operation.

When the points are plotted the detail is drawn in, using the symbols which have by now become more or less standardized. Some of the symbols used vary according to the scale of the plan: e.g. a feature of width 400 mm on a 1/500 scale plan can be shown by two fine parallel lines about 0.8 mm apart; on any smaller scale than this, however, a single line will be used. A list of some typical conventional symbols is given in Fig. 2.22, and to supplement this reference may be made to a list supplied in Technical Memorandum H5/78 (Department of the Environment).

After the detail has been drawn in, the plan should be taken to the site and checked. If nothing has been omitted, or there is no error, the detail is inked in, the North point is drawn, and any necessary lettering and titling carried out. It is usual nowadays for stencils to be used, giving uniform and neat printing even in the largest letters, although it undoubtedly lacks the character of good hand lettering. Transfer lettering systems are also now commonly used.

A scale line should be drawn on the plan, relating plan length to ground length; this is not only a convenience in scaling from the plan, but also gives some indication of any shrinkage which may have taken place since production.

Fig. 2.22 Some conventional signs

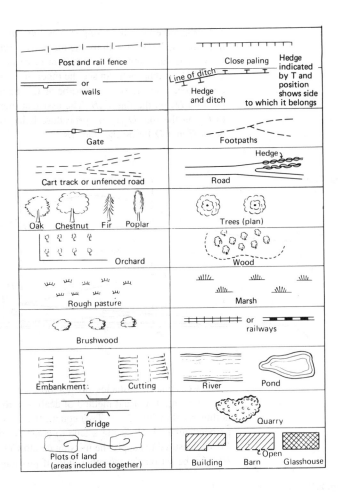

Post and rail fence	Close paling / Hedge indicated by T and position shows side to which it belongs / Line of ditch / Hedge and ditch
or walls	
Gate	Footpaths
Cart track or unfenced road	Road / Hedge
Oak Chestnut Fir Poplar	Trees (plan)
Orchard	Wood
Rough pasture	Marsh
Brushwood	or railways
Embankment: Cutting	River Pond
Bridge	Quarry
Plots of land (areas included together)	Building Barn Open Glasshouse

Miscellaneous problems

1. It occasionally happens that a survey has to be made of a field with a pond, standing crops, or a small wood in the middle, and the normal method of tape and offset surveying as described is not directly applicable. A typical method of solving this problem is shown in Fig. 2.23. Each corner must be tied as shown, and it is better if the corner triangles are checked by suitable check lines, shown dotted.

Fig. 2.23

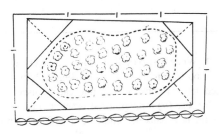

2. When the visibility along survey line *AB* is obscured by a small hill, this can be dealt with by the method of *repeated alignment* (Fig. 2.24). The surveyor and his assistant place themselves with poles at C_1 and D_1 so that each can see the other three poles in addition to his own. Assuming the surveyor to be at C_1, he directs his assistant to position D_2 on the line C_1B. The assistant then ranges the surveyor to C_2 on the line D_2A, the procedure being repeated until the two poles C and D lie on the line *AB*.

Fig. 2.24

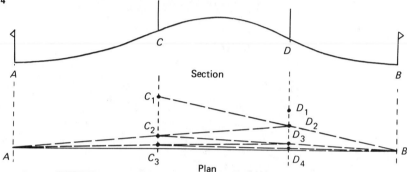

3. Occasionally obstacles do not obscure vision but do prevent measurement. The usual method of dealing with this type of obstacle is illustrated in Fig. 2.25. Two equal offsets *EC* and *FD* are set out perpendicular to *AB*, using a tape to construct the right angles, and *EF* is measured to supply the missing length *CD*. As a check, *GK* and *HL* may be set out on the other side, if possible, and *KL* measured.

Fig. 2.25

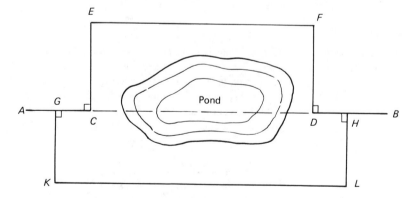

4. When a survey line crosses a river or similar long obstacle, it is impossible to measure round. Figure 2.26 shows three of the possible geometrical constructions that can be used.

In Case (*a*) a ranging rod is set at *H* on the far bank. *CE* is set off on the near bank perpendicular to *AB*, and a pole is ranged in to point

Fig. 2.26

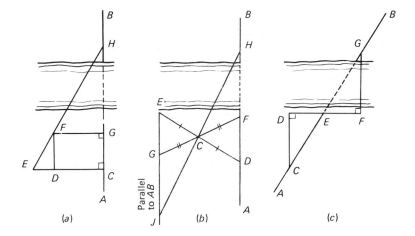

(a)　　　(b)　　　(c)

F between *E* and *H*. The perpendicular is dropped from *F* on to *AB*, meeting at *G*.

Distances *CE*, *FG*, and *CG* are measured and by similar triangles, the distance *GH* across the river can be determined, i.e.:

$$GH = \frac{CG \times FG}{EC - FG}$$

Case (*b*) gives the answer directly, and does not involve setting out any angles. A line *DE* is set out on the near bank and bisected at *C*. The line *FCG* is now set out such that *FC* = *CG*. With a pole *H* on the far bank and on the line *AB*, a pole can be set at *J* on the intersection of lines *EG* and *HC* produced backwards. Then *JG* = *FH*.

In Case (*c*) where the line crosses the river on the skew, poles are placed on line *AB* at *E* and *G* on the near and far bank respectively. A line *DF* is set out along the bank, so that *GF* is perpendicular to *DF*. A perpendicular from *D* is constructed to meet *AB* at *C*. Then

$$EG = \frac{CE \times EF}{ED}$$

Exercises 2

1 Explain with the aid of sketches how you would carry out the following operations:

   (i) Pick up a boundary fence which curves rapidly towards a survey line and then away again.

   (ii) Measure a line where a rise in the ground prevents you from seeing from one station to another.

   (iii) Measure a line which crosses a river of about 40 m width.

   (iv) Survey a field with a copse in the centre.

   (v) Set out a right angle from a survey line.

Fig. 2.27

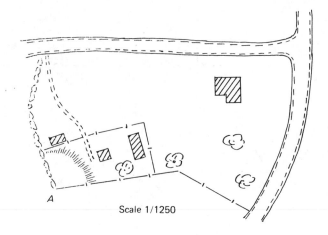

Scale 1/1250

2 Figure 2.27 is a sketch of a small site. Explain the points you would have in mind in deciding the layout of the survey lines for a tape and offset survey and draw suitable lines.

Draw up a field book approximately to size using dimensions roughly estimated from the accompanying plan and showing the booking along the last 60 m of one of the lines, near *A*, which is to include a building. (*London*)

3 Derive expressions for the correction per tape length to be applied when measuring on a regular slope in terms of (*a*) the slope angle θ, and (*b*) the gradient expressed as 1 in *n*.

What is the greatest slope you could ignore if the error from this source is not to exceed 1 in 1500? Give the answer (*a*) as an angle, (*b*) as a gradient.

Answer: (*a*) 2.11°; (*b*) 1 in 27.4.

4 Five bays of a line *AB* were measured under a tension of 53 N and the following data was recorded.

Length of span (m)	Rise between ends of span (m)
29.149	0.027
29.944	0.196
29.474	0.126
29.514	0.055
29.690	0.336

Field temperature = 10°C

The tape was standardized on the flat under a pull of 89 N and at temperature 20°C

Tape details:
Coefficient of thermal expansion 0.000011/°C
Young's Modulus 207 kN/mm^2
Density 7700 kg/m^3
Cross sectional area 6 mm^2

How long is line *AB*?   (*Salford*)
Answer: 147.355 m

5   A steel tape, 50 m long on the flat at 20°C under a pull of 89 N, has a cross-sectional area of 6 mm and a mass of 2.32 kg. It is to be used to measure lengths to an accuracy of $\pm 1/10\ 000$ when supported at mid-span.

   Assuming that (i) the ends of the tape and the mid-span support are at the same level, and (ii) the field temperature is 20°C, determine the change in tension to be applied to ensure that an error greater than +5 mm will not occur when measuring a length of 50 m.

   Take Young's modulus to be 207 000 N/mm^2 and *g* to be 9.81 m/s^2.
   Answer: +88.5 N.

6   A nominal distance of 30 m was set out with a steel tape from a mark on the top of one peg to a mark on the top of another, the tape being in catenary under a pull of 220 N and at a mean temperature of 17°C. The top of one peg was 0.68 m below the top of the other, which was 250.00 m above mean sea level. Determine the horizontal distance between the marks on the two pegs, reduced to mean sea level. The tape, which was standardized in catenary under a pull of 178 N and at a temperature of 20°C, had a mass of 0.026 kg/m and a cross-sectional area of 3.25 mm^2. Take the coefficient of linear expansion as 9×10^{-7}/deg C, Young's modulus as 155 kN/mm^2, and the radius of the earth as 6367 km.
   Answer: 29.994 m.

7   A tunnel is to be constructed beneath a wide river and in order to determine the co-ordinates of the two ends of the tunnel centre-line a minor triangulation scheme has been set out. The base line of the scheme has been measured using a 30 m steel tape stretched flat on the ground. The ground itself has a fairly gentle and uniform gradient. Tension was applied to the tape by means of spring balances attached to each end and held by hand. The tape was held over pegs driven flush with the ground so that a fine line on the top of each peg enabled a reading on the tape to be made. The pegs divided the base line into four bays.

   After the measurement the tape was standardized by measuring the distance between the first two pegs initially with the standard steel tape, and then with the tape that had been used in the field.

   From the following data, determine the corrected horizontal length of the base line to the nearest 0.001 m.

Bay	Tape readings (m)		Temperature (°C)	Difference of level between ends of bay (m)	Field tension (kgf)
1	0.013	29.975	10	0.41	10
2	0.036	29.957	10	0.34	10
3	0.078	29.941	11	0.53	10
4	0.050	17.938	13	0.19	10
		Standardization Test			
1	0.007	29.981	14	Standard tape	10
1	0.004	29.964	17	Field tape	10

Both tapes standardized originally on the flat under a tension of 5 kgf at 20°C.

Cross-sectional area of tape = 0.0355 cm^2.

Coefficient of linear expansion for steel = 1.12×10^{-5} per deg C.

Young's Modulus of Elasticity for steel = 2.81×10^6 kgf/cm^2.

(*I.C.E.*)

Answer: 107.665 m.

3 Levelling

Levelling is the operation required in the determination or, more strictly, the comparison, of heights of points on the surface of the earth. The qualification is necessary, since the height of one point can be given only relative to another point or place. If a whole series of heights is given relative to a plane, this plane is called a *datum*, and in topographical work the datum used is the mean level of the sea, since it makes international comparison of heights possible. In England, mean sea level was determined at Newlyn, Cornwall, from hourly observations of the sea level over a six-year period from 1 May 1915. This level is termed Ordnance Datum and is the one which will normally be used, though on small works an arbitrary datum may be chosen.

The basic equipment required in levelling is:

(*a*) a device which gives a truly horizontal line (the Level),
(*b*) a suitably graduated staff for reading vertical heights (the Levelling Staff).

In addition, equipment is necessary to enable the points levelled to be located relative to each other on a map, plan or section.

Before proceeding with the detailed description of the equipment and its use, however, some definitions are required.

A *level line* is one which is at a constant height relative to mean sea level, and since it follows the mean surface of the earth it must be a curved line.

A *horizontal line*, however, is tangential to the level line at any particular point, since it is perpendicular to the direction of gravity at that point.

Over short distances the two lines are taken to coincide; but over long distances a correction for their divergence becomes necessary. Figure 3.1 illustrates this point. In the figure, h represents the height

Fig. 3.1

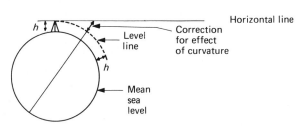

of the instrument above mean sea level. For a distance of 100 m the correction is less than 1 mm in level.

The levelling device must be set up so that its longitudinal axis is at right angles to the direction of gravity (i.e. the line taken by a plumb bob), and the line of sight will then be horizontal, assuming the instrument to be in correct adjustment. Early levelling devices utilized the plumb bob. If a semicircular protractor, with a plumb bob attached to its centre, is held vertically, flat edge uppermost, then when the string of the plumb bob cuts the 90° graduation, the flat edge is horizontal and a sight can be taken along it. This, broadly, was the principle of the earliest practical levels, dating to pre-Roman times and, as shown later, some modern self-levelling instruments employ a form of pendulum as part of the self-levelling mechanism. (It is interesting to note in this context that another of the early levelling devices, the water level, was self-levelling. It consisted of a U-tube partly filled with water, and it was only necessary to sight along the two free water surfaces to obtain a horizontal sight.) Many levelling instruments do not use a plumb bob, this being replaced by a spirit level, a glass tube curved internally in longitudinal profile and partly filled with fluid, as described in the next section. The spirit level acts in effect as a very long plumb bob.

The differences in the readings on the vertically-held graduated staff where intersected by the horizontal line of sight is a direct measure of the difference in height between the two staff stations (Fig. 3.2). The reader can refer to page 64 for amplification.

Fig. 3.2

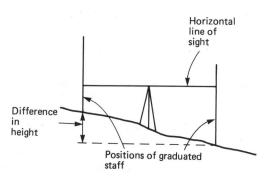

Elements of the surveyor's level

The general features of the conventional level are:

(a) a telescope to give extended lines of sight in the horizontal plane, and

(b) a bubble tube to enable those lines to be brought horizontal.

The bubble tube

The tubes vary in length between 50 mm and 125 mm, and are ground to a circular profile in longitudinal section. By increasing the radius to which the tube is ground, the 'sensitivity' of the bubble is increased, since the distance through which the bubble is displaced by any specific

Fig. 3.3

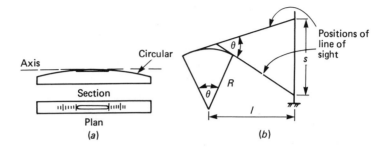

Axis

Circular

Section

Plan

(a)

Positions of line of sight

(b)

tilt of the vertical axis becomes greater. With too flat a curvature, however, the time taken for the bubble to come to rest becomes excessive.

The top surface of the tube is graduated symmetrically about its centre, as shown in Fig. 3.3.(a), the two larger lines representing the positions of the ends of the bubble at normal temperature. The other graduations are necessary because the length of the bubble can vary with temperature, though for many years manufacturers made bubble tubes which had negligible temperature variation. For example, the bubble of the Watts self-adjusting level had an approximately elliptical cross-section, and the volumes of spirit and vapour (i.e. bubble) were so designed that the expansion of the spirit due to temperature rise was counter-balanced by the decrease of surface tension; the bubble thus maintained a constant length over a temperature range of 55°C, and it was necessary to read only one end of the bubble. However, the use in modern levels of prismatic reading devices (which, as described later, enable both ends of the bubble to be read side by side) removed the need for constant-length bubbles. A plane tangential to the zero or centre of the graduations contains the axis of the bubble tube (Fig. 3.3(a)). The bubble surface is always horizontal: thus a line tangential to a point on the internal surface of the tube, and mid-way between the ends of the bubble, will always be horizontal since it is parallel to that plane in which the ends lie. If the bubble tube is rotated in a vertical plane so that the bubble is in the centre of its run, then the axis of the bubble tube must be horizontal.

The individual graduations are often at 2 mm spacing and the angular value of one division is generally indicated on the surface of the bubble tube. Thus, the movement of the bubble centre through one division could imply rotation of the axis by, say, 20 seconds of arc.

Determination of sensitivity

To determine the sensitivity of the bubble on an instrument, sets of readings are taken on a staff held a convenient distance (l), say 30 m, from the instrument with the bubble centre displaced from the centre of its run as far as possible to the object-glass end of the tube, and then to the eyepiece end using the instrument footscrews (Fig. 3.3.(b)). Thus the difference in staff readings (s) is obtained and when it is divided by distance (l), the angle (θ) through which the line of sight has been rotated is determined. The bubble centre must have undergone the same movement which is deduced by noting the positions of both ends of the bubble with respect to the graduations on the tube.

If f_1 and f_2 be the readings of the object glass end of the bubble and r_1 and r_2 the readings of the rear end of the bubble respectively before and after rotation of the line of sight, the distance of the bubble centre from the centre of the graduations is

$$\frac{f_1 - r_1}{2}$$

and

$$\frac{r_2 - f_2}{2}$$

in each case.

Then the rotation of the line of sight

$$= \theta = \frac{s}{l} \text{ radians} = 206\,265\,\frac{s}{l} \text{ seconds}$$

The bubble centre has moved a total distance of

$$\left(\frac{f_1 - r_1}{2} + \frac{r_2 - f_2}{2} \right),$$

Therefore $\quad 206\,265\,\dfrac{s}{l} = \left(\dfrac{f_1 - r_1}{2} + \dfrac{r_2 - f_2}{2} \right) = q$, say.

Thus the angular value of one division $= 206\,265\,\dfrac{s}{lq}$ seconds.

If the angular value of one graduation, length z, of the tube be ϕ seconds, and R be the radius of the internal curved surface, then

$$\phi = 206\,265\,\frac{z}{R} = 206\,265\,\frac{s}{lq} \text{ seconds}$$

Example 3.1 During levelling it was noticed that the bubble had been displaced two divisions off centre when the length of sight was 100 m. If the angular value of one division of the bubble tube is 20 seconds, find the consequent error in the staff reading. What is the radius of the bubble tube if one graduation is 2 mm long?

Rotation of line of sight $= 2 \times 20 \text{ sec} = 206\,265\,\dfrac{s}{l}$

where s is the error in staff reading.

Therefore $\quad s = \dfrac{100 \times 40}{206\,265} = 0.019 \text{ m} = 19 \text{ mm}$

Also $\qquad R = 206\ 265\ \dfrac{z}{\phi}$ (ϕ is expressed in seconds)

$$= \dfrac{0.002 \times 206\ 265}{20} = \mathbf{20.627\ m}$$

The surveyor's telescope

The Kepler type of telescope (Fig. 3.4) is the one used in surveying and this consists essentially of two convex lenses mounted so that their principal axes lie on the same line to form the optical axis of the instrument.

Fig. 3.4

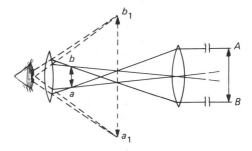

The converging object lens, i.e. the one nearest to the object, AB, forms a real image, ab, the rays from which pass on to the eyepiece, where they are refracted again and form a virtual image at some convenient distance in front of the eye. Note that this image, a_1b_1, is inverted and magnified. Magnification is an important property of surveying telescopes, the resolving power depending directly upon magnification. However, the field of view diminishes with increase in magnification and, accordingly, in order to obtain a bright image of the staff, the clear aperture of the objective needs to be increased as magnification increases.

The diaphragm

To provide positive and visible horizontal and vertical reference lines in the telescope, a diaphragm is inserted in front of the eyepiece in a plane at right angles to the optical axis. There are many forms of diaphragm (alternatively termed cross-hairs, graticule or reticule), but nowadays it is usually a thin glass plate on which the lines are engraved. The imaginary line passing through the intersection normal to the cross-hairs, and through the optical centre of the object glass, is called the line of collimation of the instrument, and all level readings are taken to this line. A typical design of the lines on a diaphragm and the view of a levelling staff through a telescope is shown at the chapter heading.

The diaphragm is held inside the telescope tube by four adjusting screws which enable (*a*) the cross-hairs to be adjusted so that the horizontal cross-hair is truly horizontal, (*b*) the line of collimation to be moved vertically and laterally.

In focusing this simple telescope, the real image formed by the objective lens is made to lie in the same plane as the diaphragm. If this is not done some serious errors in reading will ensue due to the phenomenon known as parallax. It is a matter of common observation and can be readily confirmed by the student that if, when viewing two distant objects which lie approximately along a straight line with the eye, the eye is moved to one side, then the more the distant object moves relative to the nearer one in the same direction as the eye, and this is known as *parallax*. If the image is not formed in the plane of the diaphragm and parallax is observed when the eye is moved slightly when viewing through the telescope different readings will be given depending upon the position of the eye (Fig. 3.5).

Fig. 3.5

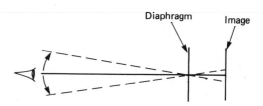

Diaphragm   Image

Internal focusing

Early instruments were of the external focusing type, i.e. they had the same basic construction as the simple one shown in Fig. 3.4. The eyepiece and the object lens were mounted in two tubes arranged so that one could slide inside the other, and focusing was achieved by moving one of the systems relative to the other. This type is now superseded by the internal focusing telescope (Fig. 3.6) in which the eyepiece and object lens are mounted in a tube of fixed length. A movable concave lens is usually inserted between them, Fig. 3.22 shows one variation of this.

Fig. 3.6 Internal focusing telescope

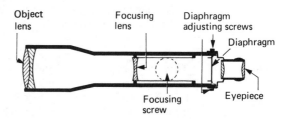

Object lens   Focusing lens   Diaphragm adjusting screws

Diaphragm

Focusing screw   Eyepiece

The concave lens is moved by means of a rack and pinion gearing, the pinion being connected by a spindle to the focusing screw, and the image focused on the cross-hairs without any movement of the object lens (Fig. 3.7). The image from the objective would form at P' if the concave lens was absent and it is treated as the virtual object for that lens, the actual image forming at P.

Although this extra lens absorbs some light, the disadvantage of this is more than offset by having a closed tube into which dust and moisture

Fig. 3.7

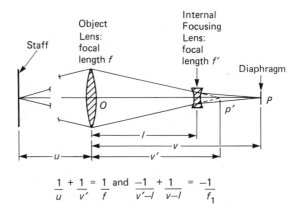

$$\frac{1}{u} + \frac{1}{v'} = \frac{1}{f} \quad \text{and} \quad \frac{-1}{v'-l} + \frac{1}{v-l} = \frac{-1}{f_1}$$

have no access. In addition, the internal focusing telescope is much more compact, while wear on the sliding surfaces is much less serious than in the external focusing type where it causes 'droop' with consequent loss of alignment in the principal axes of the eyepiece and objective.

Some manufacturers introduce prism systems into the optical path inside the telescope which re-inverts the image so that the eye will be presented with an upright picture. In automatic levels, discussed later in this chapter, devices to ensure that the line of collimation is horizontal are introduced in the optical path. There is in fact, in modern surveying instrumentation, a great number of different designs of telescopes, some looking more like periscopes than telescopes. In all instruments, however, the purpose of the telescope remains the same; to define precisely a line of sight and to magnify a target.

Defects in lenses

In modern surveying telescopes, single lenses are rarely used since they suffer from defects which limit the quality of the image.

The two most important defects are (*a*) chromatic aberration, (*b*) spherical aberration.

So far as (*a*) is concerned, it will be recalled that when white light is refracted through a glass prism it is split into its component colours, the red end of the spectrum being refracted less than the violet end. This phenomenon, known as *dispersion*, makes accurate focusing difficult, the image being surrounded by a rainbow-like boundary. The defect is remedied by using two lenses which are cemented together, one being a concave lens of flint glass and the other a convex lens of crown glass.

The larger lens in a surveying telescope, i.e. the object lens, is invariably constructed so that it forms its image in the same way as a single convex lens.

Spherical aberration, as its name implies, arises from the use of spherical surfaces for the lenses, and again prevents accurate focusing due to rays incident on the edge of the lens being refracted more than rays incident on the centre. Such aberration may be reduced by 'stopping down' the lens so that only the centre portion is used, but this also cuts down the amount of light entering the eye, so that the usual remedy is to use two lenses so arranged that the aberration of one eliminates that of the other. The pair may be a concave lens and a convex lens cemented together as objectives, or two identical plano-convex lenses a fixed distance apart as in eyepieces.

It is possible to obtain combinations which will eliminate both spherical and chromatic aberrations though, in practice, a compromise has to be made. For example, the type of eyepiece used in English surveying telescopes is based on the Ramsden eyepiece, made up of two identical plano-convex lenses with their curved faces facing, and separated by a distance equal to two-thirds the focal length of either. Such an eyepiece is free from spherical aberration but not from chromatism.

The levelling staff

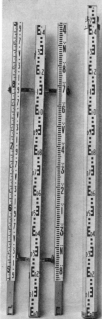

Staffs used for ordinary levelling work are sectional and are assembled either telescopically or by slotting onto one another vertically (Fig. 3.8). Most modern designs are manufactured in an aluminium alloy although staffs made of mahogany are still available. BS 4484: Part 1: 1969 requires lengths of either 3 m, 4 m or 5 m on extension, upon which the closed lengths naturally depend. It is possible that an extended length of 4.267 m will also be encountered since this was a typical equivalent Imperial dimension for the Sopwith staff shown in the left of Fig. 3.8. BS 4484: Part 1 requires upright figuring with graduations 10 mm deep spaced at 10 mm intervals, the lower three graduations in each 100 mm interval being connected by a vertical band to form an E shape, natural or reversed. The 50 mm or 100 mm intervals are therefore located by these shapes. The graduations of the first metre length are coloured black on a white background, with the next metre length showing red graduations and so on alternately.

Two such staffs are shown in Fig. 3.8 together with another form in which the number of dots above the decimetre values indicates the number of whole metres. Readings can be estimated to 1 mm over certain sighting distances, or recourse made to the parallel plate micrometer. To assist in holding the staff truly vertical, a small circular spirit level and a pair of handles are sometimes incorporated.

Fig. 3.8 Surveying staffs. Note telescopic construction of fourth staff

Precise levelling staff

This staff is used for more accurate work and Fig. 3.9 shows a typical model. A graduated invar strip is fitted in a wooden frame, being fastened at the bottom and spring-mounted at the top so that the calibrated length is unaffected by temperature effects on the frame. Two folding handles and a circular level are also provided. The staff

Fig. 3.9

 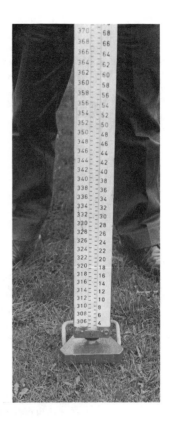

is always set up on a steel base plate and may be supported by a pair of struts, hinged to the top and of adjustable length, so that the staff can be maintained in a vertical position. Moreover, it can be rotated under the struts to face the level.

The invar strip has two sets of graduation lines as 10 mm intervals, but these are displaced and numbered differently so that two different readings can be obtained for each sighting. A field check is possible because the two readings must differ by 3.0155 m (see Fig. 3.21).

Matched pairs of staves are obtainable for use in geodetic levelling.

The surveyor's level

Three basic types of level can be distinguished.

(1) Dumpy level

In a Dumpy level the telescope and vertical spindle are cast as one piece. The levelling head shown consists essentially of two plates, the telescope being mounted on the upper plate while the lower plate screws directly on to a tripod. The two plates are held apart by three levelling screws or foot screws, and adjustments to these enable accurate levelling of the instrument to be carried out. When this has been effected, using the bubble attached to the telescope, the instrument should remain level no matter in which direction it is pointed (Fig. 3.10(*a*)).

Fig. 3.10(*a*) Dumpy level

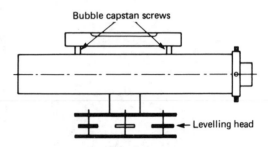

Bubble capstan screws

Levelling head

(2) Tilting level

In this level the telescope is not rigidly fixed to the vertical spindle, but is capable of a slight tilt in the vertical plane about an axis placed immediately below the telescope. This movement is governed by a fine-setting screw at the eyepiece end, and the bubble is brought to the centre of its run for each reading of the level. Thus the line of collimation need not be perpendicular to the vertical axis, as it must in the Dumpy level.

Tilting levels are robust and capable of the highest accuracy, and many modern levels, whether for precise work or for ordinary levelling, incorporate the principle (Fig. 3.10(*b*)).

Fig. 3.10(*b*) Tilting level

Bubble tube Diaphragm screw

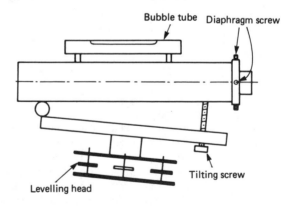

Tilting screw

Levelling head

(3) Automatic level

There is now a great range of levelling instruments available with no precise bubble attached. The telescopes of these instruments need only be approximately levelled and then a compensating device, usually based on a pendulum system inside the telescope, corrects for the residual mislevelment. These levels are very popular because of their ease of use, but they suffer from instability as mentioned later.

Fig. 3.11

Fig. 3.12 Watts SL432 Level

The tilting level

Figure 3.11 shows a tilting level mounted on a tripod, another type is shown in detail in Fig. 3.12. The lower part of the level assembly consists of a three-screw base which must be set-up so that a small circular spirit level bubble is in adjustment. Other versions of the level design which use a quick-release 'ball and socket' base instead of the three screws are often referred to as 'quick-set' levels.

Above the base the instrument consists of a telescope and sensitive bubble tube. The telescope may be directed along any particular line of sight by means of a clamping screw, slight lateral movement to the left or right of this line being made with the 'slow motion' or 'tangent' screw. Once the telescope has been sighted onto the levelling staff the upper assembly must be accurately levelled using the 'fine-setting' or 'tilt' screw mounted below the eyepiece of the telescope. As this is turned the telescope and bubble tilt vertically together and small movements are made until the bubble is centred, thus defining a horizontal line of sight on an instrument in correct adjustment.

On some older instruments the bubble tube is simply viewed through an inclined mirror but those shown in Figs 3.11 and 3.12 have a prismatic coincidence reader. This refinement is now fitted to most modern instruments. Figure 3.13(a) shows the principle, the prisms reflect an image of both ends of the bubble into the eyepiece, the image is split down the centre as shown in Fig. 3.13(b) so that one half of each end is viewed next to the other half. As the telescope is tilted, the two halves appear to move in opposite directions. This magnifies the actual movement twofold and, in the Watts model for example, this is further magnified by the optical system, which makes very exact setting possible. Consequently, bubbles of lower sensitivity than would otherwise be necessary can be employed, and this greatly improves stability and ease of setting.

Fig. 3.13

(a)

(b)

Wild

Setting up

Fig. 3.14

The following sequence of operations is required to bring a tilting level ready for use.

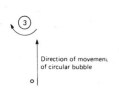

(3) Footscrew

Direction of
movement of
circular bubble

(1)　　(2)

Plan (a)

(3)

Direction of movement
of circular bubble

(1)　　(2)

Plan (b)

(a) Screw the lower plate of the instrument on the head of the tripod, whose legs have been opened and firmly fixed on the ground.

(b) The circular bubble should be brought to its central position, using the foot screws or 'ball and socket' assembly. If using an instrument fitted with foot scews, the best procedure is as follows. Referring to Fig. 3.14, by rotating foot screws 1 and 2 in opposite directions at the same time the bubble can be set on the line 12. Now, by rotating foot screw 3 only, the bubble can be centred in the target ring.

(c) It is essential that parallax between the cross-hairs and the image of the levelling staff be eliminated, for reasons already explained. There is no doubt that failure to do so is responsible for much error in levelling. To eliminate parallax: (i) turn the telescope to the sky, or hold a piece of white paper in front of the objective, and focus the eyepiece so that the cross-hairs appear clear and distinct. This is usually achieved by turning the eyepiece, which is threaded into the telescope barrel. It must be realized that the eyepiece setting depends on the characteristics of the surveyor's eye, so that it will vary from one person to another; for one given operator, the setting will not vary; (ii) now sight the levelling staff and focus its image with the focusing screw so that when the eye is moved slightly there is no relative movement between the image and the cross-hairs.

(d) Centre the sensitive bubble using the tilting screw before every reading. Ensure that the tripod itself is untouched when taking readings.

(b), (c) and (d) can be referred to as temporary adjustment.

Permanent adjustment of the tilting level

There is only one adjustment necessary to this type of instrument to ensure that it is in good adjustment: the bubble tube axis must be made parallel to the telescope axis so that when the bubble is central the line of sight is horizontal.

Two-peg test

There are many variations of this test. Just one will be described here, but the reader's attention is drawn to Exercise 3.1 at the end of this chapter, which gives one alternative.

Fig. 3.15

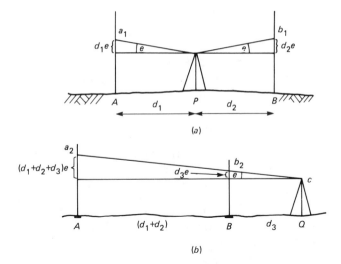

(a)

(b)

On a relatively flat site establish two pegs A and B about 50 metres apart and set up the instrument at P, a point half-way between them. After careful levelling and focusing, sight on the staff held vertically at A and take reading a_1. Repeat with the staff held at B and record reading b_1. Assuming the line of collimation is not horizontal but inclined at an angle e, the collimation error, Fig. 3.15(a), then the true difference in height between A and B is given by

$$\Delta h_{AB} = (a_1 - d_1 \cdot e) - (b_1 - d_2 \cdot e)$$

Since the instrument is midway between A and B, d_1 and d_2 are equal and so

$$\Delta h_{AB} = a_1 - b_1$$

It is worth noting here the importance of this result. Even when the levelling instrument is not in correct adjustment, the difference in height measured between two points by a level, equidistant from each, is the true difference in height.

Now move the instrument to Q, a point which extends the line AB by d_3 (about 25 m). Repeat the observations on to a staff at A and B recording the readings a_2 and b_2 (Fig. 3.15(b)). The line of collimation will again be inclined to the horizontal by the angle e. In this case the true difference in height between A and B is given by

$$\Delta h_{AB} = [a_2 - (d_1 + d_2 + d_3) \cdot e] - [b_2 - d_3 \cdot e]$$
$$= (a_2 - b_2) - (d_1 + d_2) \cdot e$$

By equating the two measures of the height difference

$$(a_1 - b_1) = (a_2 - b_2) - (d_1 + d_2) \cdot e$$

Therefore $\quad e = \dfrac{(a_2 - b_2) - (a_1 - b_1)}{(d_1 + d_2)}$

For a tilting level of average precision, e, the collimation error, should be less than $\pm$ 0.00005 rad (equivalent to a height error of $\pm$ 0.5 mm per 10 m).

Adjustment

If the error is greater than this the level should be adjusted. With the instrument still set at Q a horizontal line of collimation would give a reading on the staff at A of

$$a_2 - (d_1 + d_2 + d_3) \cdot e$$

Using the tilting screw, the line of collimation is lowered (or raised if e is negative) to the correct staff reading and then the bubble is brought to its central position using the capstan screws, which alter the alignment of the bubble with respect to the telescope.

The same principle is used in all other variations of the two-peg test but the distances d_1, d_2 and d_3 are varied.

Fig. 3.16

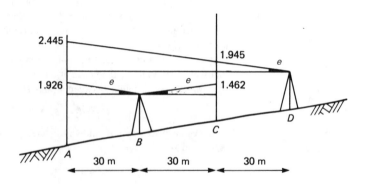

Example 3.2 In order to check the adjustment of a tilting level the following procedure was followed. Pegs A, B, C and D were set out to lie on a straight line such that $AB = BC = CD = 30$ m (Fig. 3.16). The level was set up at B and readings taken to a staff at A and then at C. The level was then moved to D and readings again taken to a staff held first at A and then at C. The readings are given below. Determine whether or not the instrument is in adjustment and, if not, explain how the instrument can be corrected.

Instrument position	Staff reading (m)	
	A	C
B	1.926	1.462
D	2.445	1.945

Difference in height determined from B

$$\Delta h_{AC} = 1.926 - 1.462 = 0.464 \text{ m}$$

difference in height determined from D

$$\Delta h_{AC} = 2.445 - 1.945 = 0.500 \text{ m}$$

If the instrument had been in adjustment and the line of collimation horizontal, both determinations of the height difference would have given the same answer.

To adjust, first calculate the collimation error

$$e = \frac{0.500 - 0.464}{60} = 0.0006 \text{ rad}$$

Note that the difference in height between A and C is greater when deduced from the staff readings with the level at D than at B. This implies that the staff reading of 2.445 is too high and so the collimation error is upwards.

With the instrument set at D a horizontal line of collimation would bisect staff at A at

$$2.445 - 0.0006 \times 90 = 2.391 \text{ m}$$

This staff reading is set using the tilting screw and the bubble brought to its central (level) position using the bubble adjusting screws.

Example 3.3 A tilting level is set up with the eyepiece vertically over a peg A (Fig. 3.17). The height from the top of A to the centre of the eyepiece is measured and found to be 1.516 m. A level staff is then held on a distant peg B and read.

This reading is 0.696 m. The level is then set over B. The height of the eyepiece above B is 1.466 m and a reading on A is 2.162 m.

(1) What is the difference in level between A and B?
(2) Is the collimation of the telescope in adjustment?
(3) If out of adjustment, can the collimation be corrected without moving the level from its position at B?

Fig. 3.17

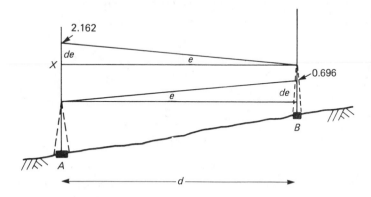

(1) Level at A. Apparent difference in level

$$= 1.516 - 0.696$$
$$= 0.820 \text{ m } (Note: B \text{ is higher than } A)$$

$$\text{True difference} = 0.820 + de$$

Level at B. Apparent difference in level

$$= 2.162 - 1.466$$
$$= 0.696 \text{ m}$$

$$\text{True difference} = 0.696 - de$$

Therefore the true difference in level

$$= \frac{0.820 + 0.696}{2}$$

$$= \mathbf{0.758 \text{ m}}$$

Note that in each set of readings error de occurs; comparison of the true difference in level with the apparent difference in level reveals that de is opposite in sense to that assumed in Fig. 3.17 and in the calculations.

(2) The collimation of the telescope is not in adjustment, the line of sight being inclined downwards.

(3) With the level at B, the height of the eyepiece (1.466) is correct.

Therefore reading X of staff at A should be given by

$$X - 1.466 = 0.758$$
$$X = 2.224 \text{ m}$$

Make the reading equal 2.224, using the tilting screw and then bring the bubble to its central position using the bubble adjusting screws.

Finally, to illustrate the effect of maladjustment of the line of collimation of a level, imagine that the unadjusted level of Example 3.3 had been used to establish the reduced level of a point 540 m up the slope from A (Fig. 3.17), the backsights being twice the length of the foresights.

The level reads $2.224 - 2.162 = 0.062$ m downwards in a length of d. Thus if each foresight is x metres long and each backsight $2x$ metres long, the corresponding staff readings will be $0.062 \ x/d$ and $0.062 \ 2x/d$ lower than they should be. Consequently, the measured rise between two staff positions $3x$ apart is reduced by $0.062 \ x/d$, and accordingly the total error over a length of 540 m is:

$$0.062 \ \frac{x}{d} \ \frac{540}{3x}$$

The reduced level of the point, as deduced from the levelling, would need to be increased by this amount to compensate for the error in collimation.

Reversible levels

Fig. 3.18

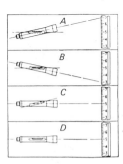

In some instruments the telescope and main spirit level can be rotated through 180° about the optical axis, thus enabling the mean of two such readings to give the true level. This method of using a reversion bubble appears to have been first suggested in 1859 by Amsler, and introduced in 1910 by Heinrich Wild in the famous Zeiss Level. It is essential that the two bubble axes (corresponding to the graduations on the two sides) are strictly parallel, and this parallelism is guaranteed by the makers to be correct to within two seconds of arc.

If a reversion instrument be set up and levelled, a reading taken on a staff, and the telescope then roated through 180°, re-levelled by the tilting screw and a second reading taken, any error due to lack of parallelism between the collimation axis and the bubble axis will cause one reading to be too high and the other to be too low by exactly equal amounts (A and B in Fig. 3.18). The mean of the two readings will thus enable the true level to be ascertained for a particular staff position.

Though the above results are free from instrumental errors, the instrument is readily adjusted to remove errors in the collimation adjustment by tilting the telescope until the reading corresponds to the mean of two readings taken as above (C in Fig. 3.18), and then bringing the bubble central by means of the capstan screws (D in Fig. 3.18).

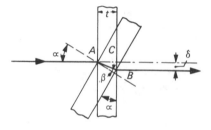

Fig. 3.19

Precise levels

The Watts SL40 and SL432 are stated to have closing errors of 8 mm and 2 mm respectively when 'double levelling' over a distance of 1 km. They can be referred to as 'construction levels' and 'engineer's levels' respectively, the latter being eminently suitable for engineering surveys and for construction work requiring the higher precision available.

There is a further group of levels known as precise levels and these, possessing a high magnification, measure to the highest accuracy. Not only are they used in geodetic levelling (see later) and for very precise engineering surveys, but they can be used (i) to measure small vertical displacements of structures, and (ii) in the checking and alignment of machinery, including small changes in inclination.

The parallel plate micrometer

The parallel plate micrometer is an essential feature of these instruments and may be either an integral part of the device, or an attachable unit. In this latter case the level itself could well be referred to as a universal level since it can be used at different accuracies. This unit enables the interval between the horizontal cross-hair and the nearest staff division to be read directly to 0.1 mm rather than be estimated. The device consists essentially of a parallel glass plate fitted in front of the object lens and given a tilting motion by the rotation of a micrometer head at the eye end of the telescope. Due to refraction, a ray of light parallel to the telescope axis is displaced upwards or downwards, according to the direction of tilt, and by an amount varying with the angle of tilt; when the plate is vertical no displacement occurs. Figure 3.19 illustrates the theory.

Let the displacement of the line of sight to the nearest graduation be δ when the plate is tilted through α from the vertical. Then, in triangle ABC,

$$AB = \frac{t}{\cos \beta} \quad \text{and} \quad BC = \delta$$

Therefore $\delta = \dfrac{t}{\cos \beta} \sin (\alpha - \beta)$

$$= t \left(\sin \alpha - \cos \alpha \, \frac{\sin \beta}{\cos \beta} \right)$$

But $\mu \sin \beta = \sin \alpha$, μ being the refractive index for the glass used in the plate.

Therefore $\delta = t \sin \alpha \left[1 - \dfrac{\cos \alpha}{\sin \alpha} \dfrac{\dfrac{\sin \alpha}{\mu}}{\sqrt{1 - \dfrac{\sin^2 \alpha}{\mu^2}}} \right]$

$$= t \sin \alpha \left(1 - \frac{\cos \alpha}{\sqrt{\mu^2 - \sin^2 \alpha}} \right)$$

$$= t \sin \alpha \left(1 - \frac{\sqrt{1 - \sin^2 \alpha}}{\sqrt{\mu^2 - \sin^2 \alpha}} \right)$$

if α be small (in radians)

$$\delta = t\alpha \left(\frac{\mu - 1}{\mu} \right)$$

Figure 3.20 shows the Wild N3 Precision Level. Its three footscrews rest on a circular baseplate which is attached to the tripod, and a circular bubble is available for initial setting up. After pointing to the staff the tubular level is observed by viewing from the eyepiece end and the two half-images are brought into coincidence using the small graduated tilting screw. This is mounted concentric with the (larger) micrometer knob which activates the parallel plate. A clamp and tangent screw allows motion about the vertical axis of the instrument and the vertical graticule line can be set on the centre of the staff using these devices, after approximate alignment by an optical sight on the other side of the telescope. The parallel plate forming the optical micrometer gives a maximum deflection of 10 mm (one graduation interval on the Wild invar staves). Its movement is registered by a scale which again can be observed in the same eyepiece as the 'split bubble'. The scale readings range from 0 to 100 with 50 as the vertical position for the plate, and estimation is possible to 0.01 mm.

The graticule contains two short stadia lines apart from the vertical line, and the line of sight is defined partly by a pair of wedge lines, which can be set symmetrically across a staff graduation by rotating the parallel plate (Fig. 3.21). The Wild staff contains two sets of graduations at 10 mm intervals and as mentioned previously the higher-numbered row acts as a check against errors of observation during geodetic levelling since it is displaced with respect to the other.

A typical sighting drill when two staves, with erect numbering, are available is:

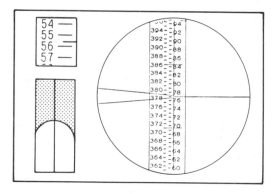

Fig. 3.21 Staff reading (graduation centred within wedge) 77 cm; Micrometer reading 0.556 cm; Book reading 77.556 cm, i.e. 0.77556 m

(a) read right graduation of rear staff,
(b) read right graduation of front staff,
(c) read left graduation of front staff,
(d) read left graduation of rear staff.

We adopt the mean of the two values of height difference given by those readings.

In all cases the micrometer drum reading could be added to obtain the relevant booking. This can be illustrated by reference to the following cases, using for simplicity an ordinary staff graduated to 10 mm. Let the line of sight fall between 1.03 and 1.04 when the parallel plate is vertical and the relevant scale reading is therefore 50. If the horizontal line of the graticule be set on the 1.03 reading, the micrometer reading would be say 82.5: this indicates a movement of 3.25 mm upwards and the corresponding staff reading could be booked as 1.03325 m. Let the forward staff reading then lie between 1.47 and 1.48 when the plate is vertical, and assume the micrometer gives a reading of 28.0 when the horizontal line is positioned on the 1.48 graduation. The line of sight has now been deflected downwards through 2.20 mm and the staff reading could be deduced as 1.47780 m, with a staff station level difference of 0.44455 m. If the readings were not established by considering the actual deflections, but the micrometer readings were simply added, we should book 1.03825 m and 1.48280 m respectively, resulting in the same level difference.

The horizontality of the line of sight is checked by the two-peg test. If an adjustment is required, the micrometer knob is actuated to set the required reading on the micrometer scale, the split bubble placed in coincidence using the tilting screw and then the protecting glass cover at the objective end of the telescope is rotated until the relevant main staff graduation is centred within the wedge-shaped graticule lines.

Automatic levels

Instruments which include some device to correct for any sight residual deviation of the line of collimation from the horizontal are termed automatic levels. There is a wide range of designs but in the interests of conciseness it is proposed to describe just one in detail.

Fig. 3.22

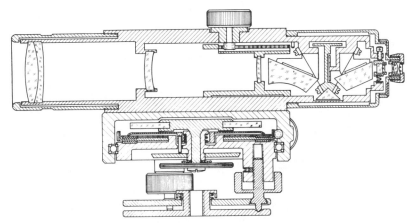

The Zeiss Ni2 self-levelling instrument

'Automatic' levels are a post-Second World War production and the Zeiss Ni2 self-levelling instrument, Fig. 3.22, was the forerunner of its type. It has the appearance of a conventional level but, possessing a number of novel features, it is somewhat more complex in construction. These features can be summarized as follows.

(1) Absence of bubble tube, preliminary levelling being carried out using the conventional three-screw levelling head and a small target bubble mounted on the tribrach; this brings the line of collimation to within 10′ of the horizontal.

(2) Further levelling up is unnecessary, correction for any slight tilt of the collimation axis being automatically made by a prismatic compensator fitted between the diaphragm and the focusing lens. The optical components of the compensator consist of three prisms, of which the centre one (with its air-damped cushioning device) is suspended by two wires from the top of the telescope barrel and is thus free to swing. The two outer prisms are fixed to the barrel, and the one nearest the eyepiece is also a roof prism.

The swinging prism weighs only 5 g so that inertia effects are absent, and the motion is frictionless and free from wear. The damping device prevents prolonged oscillation and enables readings to be made within one-half a second after disturbances.

(3) Because of the three reflections in height and the lateral inversion in the roof prism, a correct, *upright image* of the staff is obtained.

(4) Focusing is by a moving convex lens, the telescope containing also a fixed concave lens. With this arrangement of lenses, sufficient space is obtained between the focusing lens and the reticle to accommodate the compensator; also the instrument is almost completely anallatic (*see* Chapter 4).

(5) The focusing screw enables quick preliminary focusing to be made, one complete turn of the screw covering the complete focusing range of the telescope. A friction brake-operated gearing then operates automatically allowing a quarter turn of the focusing screw at a much slower (1:5) gearing, so that very fine parallax-free focusing can be made.

Fig. 3.23

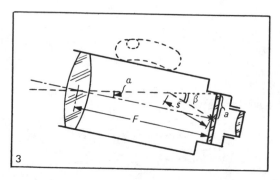

(6) A rotatable flat cog-wheel mounted on the tribrach bears the upper part of the instrument, and the tangent screw worm gear, mounted to the upper part, mates into this cog. The friction between the cog bearing and the tribrach is made greater than that between the upper-part bearing and the cog, so that when the tangent screw is turned, the upper part will rotate relative to the cog-wheel, thus giving an endless tangential slow-motion device without any clamping. For rough hand setting, of course, the telescope is turned in the usual manner and the worm gear pulls the cog-wheel round with it.

Operation of the compensator   Figure 3.23 indicates the idealized operation of the instrument. When the vertical axis is inclined at a small angle to the vertical, if the horizontal rays of light passing through the optical centre of the objective can be deflected so as to strike the horizontal cross-hair of the diaphragm, as shown, then the compensated staff reading will be given despite tilt. From the figure:

$$a = F\alpha = s\beta$$

Therefore
$$\frac{F}{s} = \frac{\beta}{\alpha}$$

When the telescope barrel tilts through α, the centre prism of the compensator takes up a new position in which the vertical line through its centre of gravity passes through the intersection point of the two supporting wires. The geometry of the quadrilateral formed by the prism base, the two wires and the top of the telescope is such that the prism base is now inclined at 3α to the horizontal. By the laws of reflection (the effects of other reflections and refractions being compensating)

$$\beta = 6\alpha \quad \text{and} \quad s = \frac{F}{6}$$

Note that in practice it will be difficult to position the compensator exactly, so that some element of under-compensation or over-compensation can occur.

Vibrational effects in automatic levels

Despite the action of the dampers fitted to the compensating devices in the instruments just described, it is possible for periodic vibrations induced by wind, traffic or plant to affect the reading accuracy. The advantages of automatic levelling are then to some extent offset by the disadvantages of sensitivity.

It is essential with such instruments that the tripod be set up with the legs pushed firmly into the ground. Vibrations can then be stopped or damped by lightly restraining the tripod with the hands. Such a practice could not be tolerated in conventional spirit levelling but is permissible with automatic levels. With the levels firmly bedded, the collimation height will be unaffected and any slight tilt induced is automatically corrected for by the compensator. In some locations, however, where for example continuous vibrations due to mechnical engineering equipment are liable to be encountered, it may be preferable to use a spirit level. It is not possible to generalize, and in any given case the decision must be made having regard for the circumstances. The damping devices fitted to automatic levels are extremely efficient, and it is only occasionally that higher frequency vibrations cause resonant vibrations in the compensator of sufficient magnitude to affect accuracy.

Wild NA 2000

In Chapter 4 we discuss some aspects of the automatic reading of theodolites. Over the years similar interest has been devoted to the automatic registration of levelling observations and recent advances in digital electronics have facilitated the design and production of the Wild NA 2000 digital level. By virtue of electronic scanning this instrument evaluates staff images and like the Zeiss Ni2 is a forerunner of its type.

Fig. 3.24 Wild NA2000 digital level connected to a Husky Hunter data logger running Optimal software (*Courtesy*: Optimal Solutions Ltd)

Fig. 3.25 Wild
NA2000 bar coded staff
(*Courtesy*: Leica UK
Ltd)

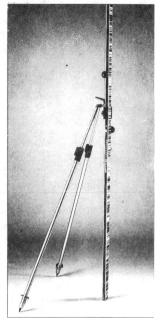

The NA 2000 (Fig. 3.24) is a compensated universal level which allows measurements to be made electronically, between distances of 1.8 m and 100 m, onto a levelling staff provided with a bar code scale (Fig. 3.25). The standard deviation in this mode of operation is stated by the manufacturer to be ± 1.5 mm for a double run of levels over 1 km. The instrument can also be used in a normal optical mode.

Having set up the level, using its conventional three screw levelling head, the next step after focusing is to press a key on a control panel to the right of the eyepiece, and the staff reading is displayed digitally after about 4 seconds. The level is provided with a diaphragm of the type shown at the chapter heading, thereby allowing the distance to the staff station to also be displayed (*see* Chapter 4) and the manufacturer quotes the distance accuracy as 3−5 mm per 10 m. A plug-in module allows the data to be stored electronically for automatic booking and checking and offloading into other computer programs.

The instrument has a beam splitter in the optical path which arranges for infra-red radiation to reach a photodiode array whilst visible light passes through a reticule and eyepiece. When being used in scanning mode the photodiodes effectively replace the eye of the observer when the bar-coded staff is scanned. The photodiode array, consisting of 256 diodes within a length of some 6.5 mm, converts the bar-code image into a video signal which is then amplified, digitized electronically and passed to a microprocessor. At the same time the position of the internal focusing lens is established by the encoder, whilst the compensator action is vetted by an electronic position detector. A two-stage correlation procedure is carried out in the evaluation of the signal, the measured data being compared to an internal reference, so that ultimately the staff reading and distance are derived.

The Dumpy Level

As shown by Fig. 3.26 the Dumpy Level is of a relatively simple design, consisting essentially of a telescope and bubble assembly mounted directly on a levelling head. Unlike a tilting level it does not have the facility to finely adjust the bubble centring with a tilting screw to level the telescope but instead, before any readings are taken, the telescope has to be set horizontal in all directions using the foot screws. The instrument is again provided with a clamp and tangent screw to set the telescope in any direction.

Fig. 3.26 Dumpy Level (*Courtesy*: Leica UK Ltd)

The procedure for levelling the instrument is exactly the same as for a theodolite and is described in Chapter 4. Figure 3.27 indicates

Fig. 3.27

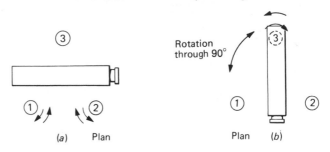

the general principle of this temporary adjustment, the telescope, and hence the bubble-tube, is placed parallel and at 90° to a pair of footscrews, the bubble being centred in each case in a manner analogous to Fig. 3.14. It is usual to repeat the operations and the bubble should remain central in whatever direction the telescope is pointed if the instrument is in adjustment. The bubble is now said to traverse. The length of time taken for this levelling compared to that for tilting and automatic levels, and the loss in accuracy due to dislevelment occurring between the initial setting and subsequent readings are two reasons why the dumpy level is not particularly popular with the engineer.

Permanent adjustment of the Dumpy Level

There are two adjustments required:

(1) The bubble-tube axis must be set perpendicular to the vertical axis, i.e. the bubble must traverse. The test and adjustment for this are the same as for the plate bubble on a theodolite and are described in Chapter 4.

(2) The line of collimation must be parallel to the bubble-axis. The test for this is the same as for the tilting level, i.e. the two-peg test. However, the adjustment procedure is slightly different. Any change to the line of collimation must be made by a vertical movment of the horizontal cross-hair using the diaphragm adjusting screws since the bubble-tube axis has been positioned in (1).

Procedure in levelling

The basic operation is the determination of the difference in level between two points. Consider two points A and B as shown in Fig. 3.28. Set up the level, assumed to be in perfect adjustment, so that readings may be made on a staff held vertically on A or B in turn.

If the readings on A and B are 3.222 m and 1.414 m respectively (Fig. 3.28(a)), then the difference in level between A and B is equal to AC, i.e. $3.222 - 1.414 = 1.808$ m, and this represents a *rise* in the height of the land at B relative to A. If the reading at B is greater than that at A (Fig. 3.28(b)), say 3.484 m, then the difference in level would be $3.222 - 3.484 = -0.262$ m, and this would represent a *fall* in the height of the land at B relative to A. Thus, we have that in any two successive staff readings:

2nd reading less than 1st represents a *rise*,
2nd reading greater than 1st represents a *fall*.

Fig. 3.28

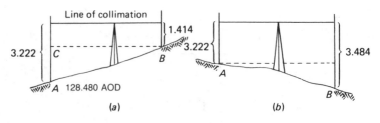

(a)   (b)

If the actual level of one of the two points is known, the level of the other may be found by either adding the rise or subtracting the fall, e.g. if the level at A is 128.480 m above Ordnance datum (AOD), then

(a) Level at B = Level at A + Rise
         = 128.480 + 1.808
         = 130.288 m above datum.

(b) Level at B = Level at A − Fall
         = 128.480 − 0.262
         = 128.218 m above datum.

The levels at A and B are known as *reduced levels* (RL) as they give the level of the land at these points 'reduced' or referred to a datum level (in this case, Ordnance datum, which is the mean sea height at Newlyn), and this method of reducing the staff readings gives a system of booking known as the *Rise and Fall Method.*

A second method, known as the *Height of Collimation Method*, also exists, and since the two methods are in common use they must both be known. In the second method, the height of the line of collimation above the datum is found by adding the staff reading obtained with the staff on a point of known level to the RL of that point. Thus in Fig. 3.28 the height of collimation is 128.480 + 3.222 = 131.702 m AOD, and this will remain constant until the level is moved to another position. The levels of points such as B are determined by deducting the staff reading at these points from the height of collimation:

(a) Level at B = Height of Coll. − Reading at B
         = 131.702 − 1.414
         = 130.288 m AOD.

(b) Level at B = Height of Coll. − Reading at B
         = 131.702 − 3.484
         = 128.218 AOD.

General procedure   This is best dealt with by means of an example, and we will consider the line of levels down the centre line of the road as shown in the plan of Fig. 3.29. The object of such a line of levels is the production of a *longitudinal section* and this is shown schematically in Fig. 3.29 with the level readings marked thereon.

The instrument is set up at a convenient position P such that a *bench mark* (BM) may be observed. Bench marks are points of known elevation above Ordnance datum which have been established by surveyors of the Ordnance Survey. The commonest types are chiselled in the form of a broad arrow on permanent features such as bridge parapets, church plinths, etc., and the centre of the bar across the arrow gives the level (to 0.01 m on the 1/1250 and 1/2500 scale maps) at which the toe of the staff should be held, as shown in Fig. 3.30.

This first reading, made with the staff on a point of known reduced level (which need not, of course, be a bench mark), is known as a

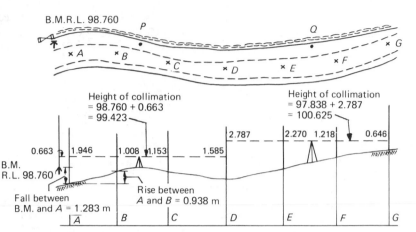

Fig. 3.29

B.M.R.L. 98.760

P

Q

× G

× A   × B

× C

× D

× E

× F

Height of collimation
= 98.760 + 0.663
= 99.423

Height of collimation
= 97.838 + 2.787
= 100.625

2.787   2.270   1.218   0.646

0.663   1.946   1.008   1.153   1.585

B.M.
R.L. 98.760

Fall between
B.M. and A = 1.283 m

Rise between
A and B = 0.938 m

A   B   C   D   E   F   G

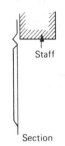

Bench mark

Staff

Section

Fig. 3.30

backsight (BS), and this term will now be used to denote that reading taken immediately after setting up the instrument, with the staff on a point of known level. The staff is now held at points *A*, *B* and *C* (*see* Fig. 3.29) in turn, and readings, which are known as *intermediate sights*, are taken. It is found that no readings after *D* are possible, due either to change in level of the ground surface or some obstruction to the line of sight, and it is necessary therefore to change the position of the instrument. The last reading on *D* is then known as a *foresight* (FS), and is the final reading taken before moving the instrument. The point *D* itself is known as a *change point*, because it is the staff position during which the position of the level is being changed.

The instrument is moved to *Q*, set up and levelled, and the reading, a backsight, taken on the staff at the change point *D*, followed by intermediate sights with the staff on points at which levels are required, until a further change becomes necessary, resulting in a foresight on point *G*. This procedure is repeated until all the required levels have been obtained.

Bench marks are provided at various densities throughout the country, 5 per km^2 and 40 per km^2 being typical maxima in rural and urban areas respectively.

In addition to the type shown above, flush-bracket bench marks can be encountered. These, which are uniquely numbered, take the form of metal plates, approximately 90 mm × 175 mm in size, cemented into the walls of buildings. They are positioned at intervals of about 1.5 km, wherever possible, along lines of geodetic levelling. A broad arrow is marked on the face of the plate, and the bench-mark level is given with respect to a small horizontal platform at its head.

The United Kingdom is divided into areas of about 400 km^2 by geodetic and secondary levelling lines, these areas being relevelled at intervals of twenty to forty years depending upon their character. Fundamental bench marks act as control in such operations and they are established at selected sites, being founded on stable rock and soil deposits to minimize any movement. Their altitudes are referred to brass bolts in the tops of pillars set alongside underground chambers which contain two reference points.

Booking
(1) Rise and fall method

The readings are booked in a *level book* which is specially printed for the purpose, as shown in Table 3.1. The reduction of these readings is carried out in the same book, so that two types of ruling are available, corresponding to the two methods of booking. The booking of the readings given in Fig. 3.29 on the Rise and Fall system would be as shown in Table 3.1.

It will be noted that estimations have been made to 1 mm when making the staff readings. The levels referred to in Figs 3.11 and 3.12 would allow this; alternatively, the parallel-plate micrometer could be used in certain cases. In view of the precision quoted for the bench-mark levels, it is doubtful if there is any merit in working to 1 mm in some of the more usual civil engineering projects; the reduced levels could, of course, be rounded off to 0.01 m later to fit in with the precision of the bench-mark levels. A compromise of reading to 0.005 m might be considered sometimes, as will be seen in later examples, but the engineer will have to decide in certain cases, such as say sewer construction at flat gradients, to what accuracy to work. However, it may often be possible to base the levellings on one bench mark only, whose assigned value can be expressed to the third place of decimals as in the levelling book examples in Table 3.1, rather than to two places (in this case 98.76 m) as on the Ordnance Sheets. Staff readings taken to 1 mm will then be in accordance with the assigned value.

Note that each reading is entered on a different line in the applicable column, except at change points, where a foresight and a backsight occupy the same line. The reason for this is made quite clear by

Table 3.1

Back-sight	Inter-sight	Fore-sight	Rise	Fall	Reduced level	Distance (m)	Remarks
0.663					98.760		BM on gate, 98.76m AOD Staff
	1.946			1.283	97.477	0	station A
	1.008		0.938		98.415	20	B
	1.153			0.145	98.270	40	C
2.787		1.585		0.432	97.838	60	D (change point)
	2.270		0.517		98.355	80	E
	1.218		1.052		99.407	100	F
		0.646	0.572		99.979	120	G
3.450	2.231	3.079	1.860		99.979		
2.231		1.860			98.760		
1.219		1.219			1.219		

referring to the remarks column — the change point occurs at staff station D, and since the staff is not moved only one reduced level is involved, requiring the one line. The RL is obtained by applying the rise or fall shown by the foresight compared with the previous intersight. The backsight is taken with the staff still on this point of known level, and the next rise or fall is obtained by comparing this backsight with the next intersight.

As a check on the arithmetic involved in reducing the levels, the backsights and foresights and the rises and falls must be summed up. The checks are then:

$$\Sigma \text{ (Backsights)} - \Sigma \text{ (Foresights)}$$
$$= \Sigma \text{ (Rises)} - \Sigma \text{ (Falls)}$$
$$= \text{Last RL} - \text{First RL.}$$

The application of the checks is shown in the example, and the student is advised to prove the rules for the general case, i.e. by using letters instead of numbers. It must be stressed that these checks concern only the accuracy of the reductions, and have no effect on the accuracy of the readings themselves. Note also that in this example Σ (Backsights) exceeds Σ (Foresights) which implies that overall a rise has occurred between A and G.

(2) Height of collimation method

The height of collimation is obtained by adding the staff reading, which must be a backsight, to the known RL of the point on which the staff stands. All other readings are deducted from the height of collimation, until the instrument setting is changed, whereupon the new height of collimation is determined by adding the backsight to the RL at the change point (Table 3.2).

Table 3.2

Back-sight	Inter-sight	Fore-sight	Height of collimation	Reduced level	Distance (m)	Remarks
0.663			99.423	98.760		BM on gate, 98.76m AOD
	1.946			97.477	0	Staff station A
	1.008			98.415	20	B
	1.153			98.270	40	C
2.787		1.585	100.625	97.838	60	D (change point)
	2.270			98.355	80	E
	1.218			99.407	100	F
		0.646		99.979	120	G
3.45~~0~~	2.~~2~~31			99.97~~9~~		
2.~~2~~31				98.~~7~~60		
~~1~~.219				~~1~~.219		

The arithmetical checks to be applied to this system of booking are

$$\Sigma \text{ (BS)} - \Sigma \text{ (FS)} = \text{Last RL} - \text{First RL}$$

Σ (all RLs except the first) $=$

$$\Sigma \text{ (each instrument height)} \times \text{(no. of Inter. Sights and}$$
$$\text{FSs deduced from it)} - \Sigma \text{ (FS + IS)}$$

This second check is cumbersome and is often ignored so that, as a consequence, the intermediate RLs are unchecked. In this case, errors could go unchecked (compared with the Rise and Fall method where errors in all RLs are detected).

Reduction is easier with the height of collimation method (or height of instrument method, as it is sometimes called) when levelling for earthworks, and large numbers of intersights are taken from each position of the instrument.

In lengthy levelling operations where pages of readings are involved, each page should be checked separately. Book the last reading on each page as a foresight, and repeat it as a backsight at the top of the next page.

Misclosure

Suggested values of allowable misclosures are given later in the chapter, (page 83), and when the specified value has been exceeded, relevelling should be undertaken. However, if the misclosure is acceptable then it can be eliminated by adjusting the backsights and foresights such that half the total misclosure is applied, in an equal and opposite sense, to each of their totals. The sense of the correction ($+$ or $-$) will depend upon the sense of the misclosure; a reduction in ΣBS with a corresponding increase in ΣFS will cause an overall negative correction in the quantity (First RL $-$ Last RL).

Example 3.4 The following figures were extracted from a level field book, some of the entries being illegible owing to exposure to rain. Insert the missing figures and check your results. Re-book all the figures by the Rise and Fall method, and state the advantage of this method of booking.

BS	IS	FS	H of I	RL	Remarks
?			279.08	277.65	OBM
	2.01			?	
	?			278.07	
3.37		0.40	?	278.68	
	2.98			?	
	1.41			280.64	
		?		281.37	TBM

(London)

Each line in a level book can be regarded as an equation which can be drawn up by applying the principles just outlined.

Thus, line 1, H of I = RL + BS
Hence,   BS = H of I − RL
                    = 279.08 − 277.65
                    = 1.43

line 2,   RL = H of I − IS
                    = 279.08 − 2.01
                    = 277.07.

In this way, the field book is completed as follows; it is suggested that the reader carries out the re-booking by applying the principles outlined. The total rise and total fall are 4.30 m and 0.58 m respectively.

BS	IS	FS	HI	RL	Remarks
1.43			279.08	277.65	OBM
	2.01			277.07	
	1.01			278.07	
3.37		0.40	282.05	278.68	
	2.98			279.07	
	1.41			280.64	
		0.68		281.37	TBM
4.80		1.08		281.37	
1.08				277.65	
3.72				3.72	

The reader must realize that this is a purely academic exercise and, in practice, under such circumstances the whole of the work would be re-levelled.

The term TBM implies 'temporary' or 'transferred' Bench Mark. Such bench marks are usually established in the proximity of construction work when a conventional bench mark is not conveniently available. Further discussion on their construction and location is given in Chapter 10.

Headroom of bridges

Reduced levels of bridge soffits and similar measurements are made by using the staff in an inverted position (Fig. 3.31), taking care when using a telescopic staff that the catches are properly engaged.

The inverted staff reading is booked in the relevant column of the level book with a negative sign, so that when reducing this reading

Fig. 3.31

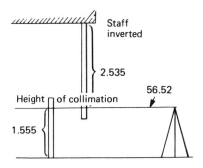

from the height of collimation of the level we get

$$\text{RL of soffit} = 56.52 - (-2.535)$$
$$= 59.055 \text{ m AOD}$$

This value could be rounded off to 0.01 m, upwards or downwards, without much loss of accuracy here. The headroom of the bridge in Fig. 3.31 is

$$1.555 + 2.535 = 4.09 \text{ m}$$

Uses of levelling

Apart from the general problem of determining the difference in level between two points, which has already been fully dealt with, the main uses of levelling are:

(1) the taking of longitudinal sections,
(2) cross-sections,
(3) contouring,
(4) setting out levels. (*See* Chapter 10)

Longitudinal sections

An example of such a section has been given in Fig. 3.29, from which it will be seen that the object is to reproduce on paper the existing ground profile along a particular line — often, though not invariably, the centre line of existing or proposed work, e.g. the centre line of a railway, road or canal, or along the proposed line of a sewer, water main, etc. Staff readings to 0.01 m should be generally adequate for this purpose.

The accuracy with which the ground profile is represented on the section is dependent on the distance between staff stations, and this in turn depends on the scale of the section. As a general basis, however, levels should be taken at:

(1) every 20 m,
(2) points at which the gradient changes, e.g. top and bottom of banks,
(3) edges of natural features such as ditches, ponds, etc.,
(4) in sections which cross roads, at the back of the footpath, on the kerb, in the gutter, and on the centre of the road.

The sections are usually plotted to a distorted scale, a common one for road work being 1/500 scale horizontal and 1/100 vertical.

The following points should be borne in mind during the actual levelling, particularly when levelling long sections, to avoid the build up of error:

(*a*) Start the work from a bench mark if possible, and make use of any nearby bench marks which lie within the length being levelled. Bench marks may be disturbed by subsidence and rebuilding, so any mark used as a reference height should be checked by levelling to another mark nearby and comparing the measured height difference with the quoted height difference. Because of discrepancies that do occur, all reduced levels for any one project

Fig. 3.32(*a*)

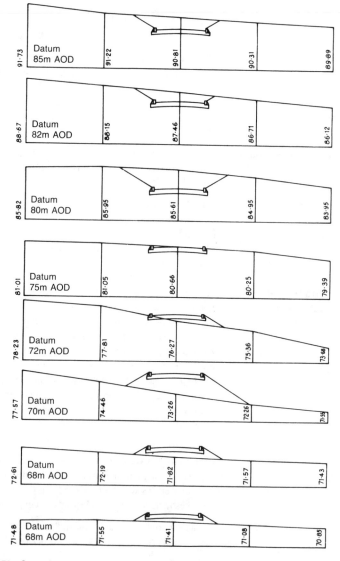

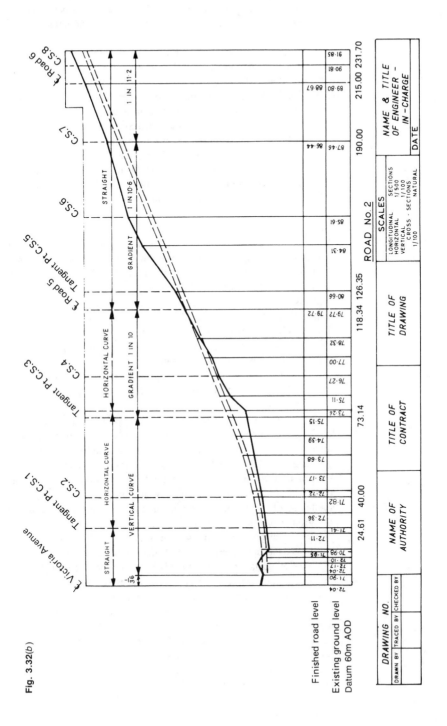

Fig. 3.32(b)

Finished road level

Existing ground level
Datum 60m AOD

should be based on the quoted value of just one bench mark and the reference height so used then given on any plan, cross-section, etc.

(b) Try to keep backsights and foresights equal in length to minimize errors which will occur if the line of collimation is not parallel to bubble-tube axis.

(c) Make all changes on firm ground, preferably on identifiable features on which check levels can be taken if required.

(d) Take the final foresight on a bench mark or, better, close back on the starting point by a series of 'flying levels', i.e. a series of equal foresights and backsights, each about 40–60 m long. Note that 20–30 m is often adopted in precise levelling.

(e) Do not work with the staff extended in high wind.

(f) Exercise great care when setting up the level.

Cross-sections

Works of narrow width such as sewers and pipelines require only one line of levels along the centre line of the proposed trench, since there will generally be little change of the ground surface level over the proposed width. Wider works, however, such as roads, railways, embankments, large tanks, etc., will necessitate the use of ground on either side of the centre line, and information regarding relative ground levels is obtained by taking *cross-sections* at right angles to the centre line. The width of these must be sufficient to cover the proposed works, e.g. 15 m either side of the centre line for a normal road. The longitudinal spacing of the sections depends on the nature of the ground, but should be constant if earthworks are to be computed. A spacing of 20 m is common.

The centre line is first set out, pegs being placed at points where cross-sections are required, and the cross-sections themselves may then also be set out — using an optical square or similar instrument where precision is required in setting out the right angle — with white arrows or ranging rods marking the points where levels are required. The choice of points will be governed by the same principles as govern the taking of longitudinal sections, and the aim is again to reproduce the ground profile accurately.

In the actual levelling, cross-sections may be completed one at a time, setting up the instrument as many times as is necessary.

This method facilitates booking but may be tedious in that where the ground has steep cross-gradients an excessive number of change points will be required. In this case it is often quicker to take staff readings on other cross-sections as the ground allows, though it is obvious that great care must be exercised in booking such readings, that none are forgotten, and that they are identified with the correct longitudinal and lateral measurements.

It is common to plot cross-sections to a natural, i.e. undistorted, scale and, since only the ground profile and a limited depth are required, the plots can be kept compact by judicious choice of datum or base height. Figure 3.32 gives examples of plotting.

Contouring

A *contour* is a line joining points of equal altitude. Contour lines are shown on plans as dotted lines, often in distinctive colour, overlaying the detail. The vertical distance between successive contours is known as the *vertical interval*, and the value of this depends on (*a*) the scale of the plan, (*b*) the use to which the plan is to be put. For example, a 1/5000 plan prepared by photogrammetric methods (Chapter 13) for the planning of a highway project may have contours at 5-m intervals.

As regards the interpretation of contours, when they are close together, steep gradients exist, and as they open, the gradients flatten. Two contour lines of different value cannot intersect, and a single contour cannot split into two lines having the same value as itself. A contour line must make a closed circuit even though not within the area covered by the plan. The main value of a contour plan, therefore, is that it enables an assessment to be made of the topography; such plans will be commonly prepared when layouts of large projects such as housing estates are under consideration. Accurate contour plans are invariably prepared when reservoir projects are being designed, but for general civil engineering work, a vertical interval of 1 m, say, may be required, and the preparation of such a contour plan is a lengthy and tedious job. The following methods are probably the most commonly used ones for contouring.

'Gridding'

This method is ideal on relatively flat land, especially on comparatively small sites. Squares of 10 to 20 m side are set out (according to the accuracy required) in the form of a grid, and levels are taken at the corners. To save setting out all the squares, two sets of lines may be established using ranging rods as shown in Fig. 3.33 and to locate any particular square corner the staff man lines himself in, using pairs of ranging rods. For booking the reading, the staff man is at $E4$ in the example shown, and this would be noted in 'Remarks'. The reduced levels are then plotted on the plan, which has been gridded in the same manner, and by any suitable means of interpolation (e.g. radial dividers) the required contours are plotted.

Fig. 3.33

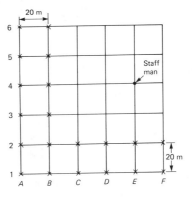

Radiating lines

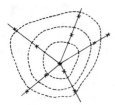

Fig. 3.34

Rays are set out on the ground from a central point, the directions being known. Levels are taken along these lines at measured distances from the centre (*see* Fig. 3.34). Again, interpolation is used to give the contour lines; this method is particularly suitable for contouring small hills or knolls.

Direct contouring

This method may be applied in any case, but is used to best advantage in hilly terrain. The actual contour is located on the ground and marked by coloured laths or other convenient means. The levelling technique is shown in Fig. 3.35, the level is set up and levelled at some convenient position, and the height of collimation established by a backsight on to some point of known level. In Fig. 3.35 this gives a height of collimation of 33.99 AOD. Thus, any intermediate sight of 0.99 means that the staff is on a point 33.99 − 0.99 = 33 m AOD. The surveyor directs the staff man up or down the hill until the staff reads 0.99 and, on being given a signal, the staff man sticks a lath into the ground (one colour being used for one particular contour line), and moves to another point on the contour. A series of staff readings of 33.99 − 32 = 1.99 would enable the 32 m contour to be set out, and so on. Speed of setting out depends largely on the skill of the staff man; staff readings to 0.01 m are quite adequate for this purpose.

The positions of the laths are later surveyed by some suitable method, e.g. tape and offset survey, traverse survey, plane tabling, tacheometry, etc.

Fig. 3.35

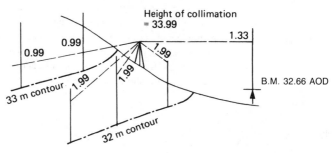

Curvature and refraction

Referring again to the definitions given at the beginning of this chapter, it will be remembered that the line of collimation is not a level line but is tangential to the level line. As a consequence, corrections must be applied when the sights are long, i.e. when the deviation of the tangent from the circle becomes appreciable.

Taking a level line as being the circumference of a circle whose centre is the earth's centre (an assumption which is accurate enough for this purpose), the required correction can be deduced as follows:

Let L be the position of the instrument, which is directed towards a staff held vertically at X — i.e. held along the extension of radius $OX(=R)$ (*see* Fig. 3.36).

Fig. 3.36

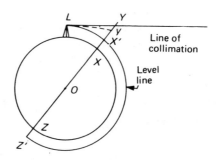

Now
$$(LY)^2 = X'Y \cdot Z'Y$$
$$= X'Y(X'Y + X'Z')$$
and $\quad X'Z' = XZ + 2XX'$

where $\quad XZ \simeq 12\ 734$ km

and $\quad XX' \simeq 1.5$ m so that $X'Y \not> 2.5$ m if a 4 m staff be used

Therefore $\quad (LY)^2 = X'Y \cdot XZ$, neglecting $(X'Y)^2$ as a second-order term

i.e. $\quad (LY)^2 = 2R \cdot X'Y$

and $\quad X'Y = $ correction for *curvature* $= \dfrac{(LY)^2}{2R}$

$$= \dfrac{(LY)^2}{12\ 734}\ \text{km}$$

$$= 0.078\ (LY)^2\ \text{m}$$

where LY is expressed in kilometres.

Owing to the gradual reduction in air density with altitude, light rays from a higher altitude are generally considered to be refracted downwards as they pass through progressively denser air so that the line of collimation actually cuts the staff at y. If R_1 be taken as the radius of curvature of that line then, as before,

$$Yy = \frac{LY^2}{2R_1}$$

Therefore the total correction for refraction and curvature is

$$X'y = LY^2 \left(\frac{1}{2R} - \frac{1}{2R_1} \right)$$

If we take the coefficient of refraction to be

$$k = \frac{\text{radius of curvature of earth}}{\text{radius of curvature of line of collimation}} = \frac{R}{R_1}$$

then $\qquad X'y = \dfrac{LY^2}{2R} (1 - k)$

Conventionally, $\quad k = 1/7$, over land whence

$$X'y = 0.078(1 - \tfrac{1}{7})(LY)^2$$

$$= 0.067\ D^2\ \text{m} \quad \text{where} \quad D = LY\ \text{km}$$

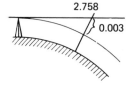

2.758

0.003

Correction $\qquad$ = 0.067 (0.2)^2
$\qquad\qquad\qquad\qquad$ = 2.8 mm

Corrected reading $\qquad$ = 2.758 − 0.003
$\qquad\qquad\qquad\qquad$ = **2.755 m**

Figure 3.37 indicates why the correction is applied in this manner.

Fig. 3.37

Reciprocal levelling

By means of reciprocal levelling, the need for applying these corrections may be avoided. If taking levels across a wide ravine, for example, an instrument would be set up near a staff held at X and sights taken on this staff and on another staff held on the far bank, Y, as in Fig. 3.38(a). Immediately afterwards, similar readings are made with a second instrument set up on the opposite bank, as in Fig. 3.38(b).

Fig. 3.38

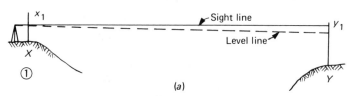

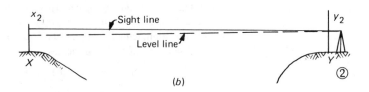

Note that with this method *two* similar instruments in correct adjustment are required.

Level (1). Apparent difference in level between X and Y

$$= Xx_1 - Yy_1$$

True difference in level will be the apparent difference less the total correction for curvature and refraction ($= \alpha$, say)

Therefore the true difference

$$= Xx_1 - [Yy_1 - \alpha]$$
$$= Xx_1 - Yy_1 + \alpha$$
$$= \text{Apparent difference}_{(1)} + \alpha$$

Level (2). Apparent difference in level

$$= Xx_2 - Yy_2$$

Therefore the true difference

$$= [Xx_2 - \alpha] - Yy_2$$
$$= [Xx_2 - Yy_2] - \alpha$$
$$= \text{Apparent difference}_{(2)} - \alpha$$

Add equations (1) and (2),

True difference in level $= \frac{1}{2}$ [sum of apparent differences].

Example 3.6 In levelling across a river, reciprocal levelling observations gave the following results for staffs held vertically at X and Y from level stations A and B on each bank respectively:

Staff reading of X from $A = 1.753$ m
Staff reading of X from $B = 2.080$ m
Staff reading of Y from $A = 2.550$ m
Staff reading of Y from $B = 2.895$ m

If the RL of X is 90.37 AOD, obtain that of Y.

It will be noted from the staff readings that Y is lower than X.

Instrument at A. Apparent difference

$$= 2.550 - 1.753 = 0.797 \text{ m}$$

Instrument at B. Apparent difference

$$= 2.895 - 2.080 = 0.815 \text{ m}$$

Therefore the true difference $= \dfrac{0.797 + 0.815}{2}$

$$= 0.806 \text{ m}$$

Therefore   RL of $Y = 90.37 - 0.81$

$$= \textbf{89.56 AOD}$$

Note that where only *one* instrument is used, this method corrects not only for curvature and refraction but also for any lack of adjustment in the collimation axis. In these circumstances, however, it must be appreciated that temperature change, and hence change in refraction effects, may take place while the level is being moved. Where two levels not in perfect adjustment are available, reciprocal levelling is carried out, after which the instruments are interchanged and another set of readings taken. In general terms we can say that, if the height of collimation is L, the reduced level of a station at which the staff reading is h_1, is $(L - h_1) + D^2K + Dc$: K is the combined correction for refraction and curvature, and c is the collimation error in radians. The reader should be aware that temperature gradients and hence refraction problems can be generated by heat from plant and machinery. In Chapter 10 refraction problems are described during the control survey for the Channel Tunnel that joins England to France.

It has been assumed in this section that the line of sight is concave towards the earth, as in Fig. 3.37. Geisler, in a paper, *Refraction near the Ground and its Influence on Precise Levelling*, given to the 12th Congress of FIG in London, discussed the influence of topography and ground cover upon the air layers near to the ground. Light rays from staff to level pass through such layers, and climatic conditions in this zone differ from those in the overlying atmosphere. The refraction correction of one-seventh the curvature correction is in fact based upon the latter consideration.

Geisler showed that the line of sight can be convex towards the earth, depending upon the temperature gradient. If the ground is warmer than the adjacent air layers, the temperature gradient (dT/dh) is negative and then the line of sight curves upwards (Fig 3.39) rather than downwards as shown in Fig. 3.37 which is in fact the case when dT/dh is positive and which can occur mainly after sunset.

Fig. 3.39

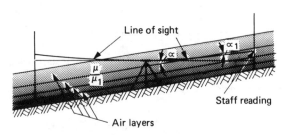

In Fig. 3.39 we have $\mu \cos \alpha = \mu_1 \cos \alpha_1 =$ a constant, and on differentiation we obtain $d\mu = \mu \tan \alpha \, d\alpha$. The relationship between μ and μ_0 (the refractive index at 0°C and 760 mm of mercury) is

$$\mu - 1 = \frac{\mu_0 - 1}{1 + kt} \cdot \frac{p}{760}$$

where $k = 1/273$.

Also we have

$$d\mu = \frac{\mu - 1}{1 + kt}(-k)\, dt$$

and

$$d\mu = (\mu - 1)\frac{dp}{p}$$

It is known that change in pressure is of much less influence than change in temperature and accordingly one can write

$$d\mu = -(\mu - 1)\frac{dT}{T}$$

where $T = 273 + t$ (i.e. absolute temperature).

Thus

$$\mu \tan \alpha \cdot d\alpha = -(\mu - 1)\frac{dT}{T}$$

Therefore

$$d\alpha = -\frac{\mu - 1}{\mu T} \cdot \frac{dT}{dh} \cdot dL$$

where $\tan \alpha$ is written as dh/dL, dh being the rise over a length dL before refraction through to α_1. The expression can also be written as

$$d\alpha = \frac{1}{R_1} \cdot dL$$

where
$$\frac{1}{R_1} = \left(\frac{\mu - 1}{\mu T}\right)\left(\frac{-dT}{dh}\right)$$

At 20°C, $T = 293°$ and $\mu = 1.00027$

$$\frac{1}{R_1} = -\frac{0.00027}{1.00027 \times 293} \cdot \frac{dT}{dh} = -10^6 \frac{dT}{dh}$$

It will be seen that if $dT/dh = 0.16°$ per metre then R_1 is virtually equal to the earth's radius.

The line of sight to staff at the higher station in Fig. 3.39 deviates more than that to the lower station since it passes through layers in which dT/dh is of greater magnitude and, accordingly, the apparent difference in level is reduced; this would be true even if the two sighting lengths were the same, nor would the error be eliminated by also levelling in the other direction. If a sighting length of 50 m were used and dT/dh was $-0.16°$ per metre 'upwards' and $-0.04°$ per metre 'downwards', then using the expression $L^2/2R_1$ as the effect of curvature on distance, deduced as previously, the error in levelling is

$$\frac{1}{2} \times 10^{-6} \times 50^2 \times (0.16 - 0.04)$$
$$= \frac{1}{2} \times 10^{-6} \times 2500 \times 0.12$$
$$= 0.15 \text{ mm per } 100 \text{ m}$$

Only if the temperature gradients were the same for each direction would refraction have had no influence.

Geisler carried out experimental work using two points 100 m apart on an asphalt surface and having a height difference of some 2.3 m, the levels being of the Zeiss Ni2 type, and similar, with parallel plate micrometers attached. Amongst his conclusions were (i) for 50 m sighting distances with upper staff reading of about 0.4 m, refraction may reduce height differences by, say, 2.5 mm per kilometre, and (ii) for best results, short sights with upper staff readings greater than 0.5 m should be observed during morning hours.

Three-peg test

Angus-Leppan, in a paper in *Survey Review*, No. 159, in fact proposes the use of a three-peg test to eliminate or minimize influences of curvature and refraction when adjusting levels. His data show that the line of sight can become concave upwards (i.e. K is negative) as in Fig. 3.39. The staff and level positions he suggests are shown in Fig. 3.40. If the staff readings, when the instrument is established at P, are h_A, h_B and h_C then, in accordance with the expression given on page 79,

$$H_A - H_B = (h_B - h_A) + (a - 2b)c + K[(a - b)^2 - b^2]$$

Fig. 3.40

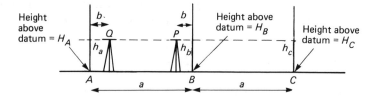

where c is the collimation error, in radians, and

$$H_A - H_C = (h_C - h_A) - 2bc + K[(a - b)^2 - (a + b)^2].$$

Similar expressions can be derived when the instrument is established at Q, and this allows the two unknowns c and K to be determined from the sets of staff readings and distances a and b.

Accuracy in levelling

The main factors affecting accuracy in levelling, apart from the above, are:

(1) reading of staff
(2) bubble not being central
(3) instrument not being in adjustment
(4) differential settlement of the tripod
(5) tilting and settlement of the staff
(6) sensitivity of bubble or compensator.

Assuming a tripod on firm ground with its legs well dug in, we can ignore the effects of tripod settlement. Similarly, the effect of staff tilt and settlement can be kept to the minimum by use of a staff with a target bubble to ensure verticality standing on firm ground on a change plate.

The effects of maladjustment are eliminated in long runs of levelling by equal backsights and foresights with the level set up such that its line of sight is as high as possible.

Reading errors

These depend on (1) the magnification and image clarity afforded by the telescope, (2) on the manner in which the staff is marked, and (3) the length of sight.

Bubble displacement

The accuracy of bubble centring depends on the method used to view the bubble. The less accurate instruments are fitted with viewing mirrors while those of higher accuracy are fitted with prismatic viewers.

The uncertainty caused by bubble mislevelment in tilting levels is comparable to the uncertainty due to mislevelment of the compensator in automatic instruments. Manufacturers usually quote the setting accuracy of the bubble or compensator for their instruments. The accuracy of setting for a standard engineer's level would lie somewhere in the range of $\pm 1.5''$ to $\pm 0.5''$, with the more precise instruments capable of being set horizontal to $\pm 0.2''$. The precision of staff reading over any range can be determined from this value, e.g. an angle of 1″ subtends a distance of 0.3 mm at 60 m.

Misclosure

The normally accepted maximum misclosure for a line of ordinary levelling closing back on the starting point is $\pm 12\sqrt{K}$ mm where K is the length of the circuit in km. This value has been derived empirically but can be justified theoretically. Assuming the standard error of a single staff reading due to all factors to be ± 1 mm then the standard error in the difference in height between two points determined as the difference in two staff readings from the same instrument position is $\pm\sqrt{2}$ mm (*see* Chapter 8). Assuming an average sight distance of 60 m then there will be about eight change points in each kilometre of levelling resulting in a standard error per km of $\pm\sqrt{8}\sqrt{2} = \pm 4$ mm. Two-thirds of all misclosures should fall within these limits but, since it is normally accepted that the maximum discrepancy should not exceed three times the standard error, then the limits of acceptability should be $\pm 12\sqrt{K}$ mm.

In rough terrain where a greater number of change points are required, the maximum acceptable misclosure may be increased to $\pm 24\sqrt{K}$ mm.

There is no doubt, however, that by carefully following the rules given earlier in this chapter (Longitudinal sections) and using a modern tilting level with coincidence bubble reader, greater accuracy than this is possible. The errors arise mainly from field technique as can be noted, for example, when comparing those values with that of ± 2.5 mm per km of single levelling quoted for the Watts SL432. In a round of levels, taken in North Lancashire in connection with an engineering project, a total length of some 16 km closed within 40 mm. Great care was taken to keep backsights and foresights equal and approximately 75 m in length, while only firm features were used to change points.

The error was thus $\pm 10\sqrt{K}$ mm and this value in fact is quoted for the accuracy of engineers' levels in BS 5606: 1990. *Code of Practice for Accuracy in Building*. The value of ± 10 mm is mentioned when establishing temporary bench marks from Ordnance bench marks.

That error of $\pm 10\sqrt{K}$ mm lies within the limits of second- and third-order accuracy suggested by Prior in a paper *Accuracy of Highway Surveys* delivered to the 11th Congress of the International Society for Photogrammetry, i.e. $8.4\sqrt{K}$ mm and $12\sqrt{K}$ mm respectively. First-order accuracy was quoted as $4\sqrt{K}$ mm and fourth-order accuracy as $120\sqrt{K}$ mm.

Precise levelling

By further refinement of field technique and by using precise staffs and levels of the types described earlier, much greater accuracy is attainable and the term 'precise levelling' is then applied. In 1948 the International Association of Geodesy adopted a number of resolutions and defined accidental errors and systematic errors in the context of levelling. The former are those due to causes acting independently on all successive levelling observations and obeying Gaussian laws (*see*

Chapter 8). The latter are due to causes acting in a similar manner on successive or adjacent levelling observations. They do not obey Gaussian laws but the Association considered that beyond a certain distance, some tens of kilometres, their efects became accidental so that the two types of error could be combined in terms of $\sqrt{K}$.

Typical rules governing field technique in precise levelling are given below and these can be usefully applied to levelling where medium accuracy is required:

(1) Backsights and foresights to be of equal length, two rods being used;
(2) All lines of levels to be run twice in opposite directions, each run to be made on different days with different change points;
(3) Levelling to be carried out only during favourable conditions of light and temperature, and not during high wind or in the rain;
(4) All change points to be made on a special footplate;
(5) The operations to be considered in such a manner as to reduce as much as possible the influence of causes which might introduce systematic errors. Atmospheric refraction and the gradients of the levelling lines are of importance here as mentioned previously: staff readings should be above the 0.5 m level. A typical sighting drill has been suggested in an earlier section of this chapter. It is good practice to alternate the order of reading backsights and foresights at successive instrument stations in order to eliminate staff settlement between the readings as well as tripod displacement. Staff displacement between instrument changes cannot be eliminated by such observation techniques, although the net effects tend to compensate when double levelling is carried out.

Both spirit levels and automatic levels are affected by radiant heat. During precise levelling in particular, the instrument should be protected by an umbrella. Systematic errors in automatic levelling can arise due to (i) compensator-drag (friction or hysteresis in the suspension or damping system) and (ii) centring the spherical level. A systematic difference occurs between settings made in opposite directions, partly due to the effects of surface tension on the deformed bubble and partly due to not allowing sufficient time for the bubble to attain its equilibrium position. An accidental error can arise when centring the bubble and the observer needs to be able to view squarely downwards, either directly or by some form of mirror aid.
(6) If the standards laid down regarding allowable error (in the form $\pm C\sqrt{K}$ mm) are not complied with, the work shall be repeated.

For some engineering purposes, precise levelling may be required. If followed carefully, the procedures described should be sufficient to ensure that the maximum misclosure should be less than $\pm 6\sqrt{K}$ mm in a circuit K km in length, i.e. tend towards the value for first-order accuracy quoted previously. For work of the highest standard required by organizations such as the Ordnance Survey to establish a countrywide network of benchmarks, difference between forward and backward levelling is not allowed to exceed $\pm 2\sqrt{K}$ mm.

During the Second Geodetic Levelling all bench marks were connected to the Observatory Bench Mark at Newlyn, which is 4.751 m above Ordnance Datum. The heights of the fundamental bench marks were then held unchanged during the Third Geodetic Levelling carried out between 1951 and 1958 which produced values of about 1.2 mm for C. Values of the intermediate bench marks were adjusted during this operation in England and Wales: further levelling was carried out in Scotland to give complete coverage, all the heights derived during the Second Geodetic Levelling then being accepted.

The Association advocates that geodetic levellings should be carried out at intervals of 25 years and reference may be made to the paper *Background and Experiments in Preparation for a Fourth Geodetic Levelling of the UK* given by Lt.-Col. J. W. Williams to the 1979 Conference of Commonwealth Surveyors.

Exercises 3

1 A level, set up at C in a position 30 m from a staff held at A and 60 m from a staff held at B, gave readings of 1.914 m and 2.237 m respectively, the bubble having been brought to the centre of its run before each reading. The level was then taken to D, 30 m from B and 60 m from A and staff readings of 1.874 m and 2.141 m respectively obtained, at A and B. The level stations and staff stations lay in a straight line.

Determine the collimation error of the level, the corrected difference in level between A and B, and the staff readings obtained from D when the instrument had been placed in adjustment.

Answer: 0.00093 radians upwards, 0.295 m, 1.818 m, 2.113 m.

2 Book and reduce the following levels, the first number in a pair of levels at a change point being a backsight.

1.632		Point A, RL 54.173 m
3.467	1.124	Change point
	0.568	Point X
1.835	0.381	Change point
	−2.473	Point Y
1.732	3.941	Change point
	2.484	Point B, RL 54.893 m

Apply suitable corrections so that the specified level of B is attained and hence determine the difference in level between X and Y.

Answer: Error +0.016 m, 4.503 m

3 In order to find the rail levels of an existing railway, a point A was marked on the rail, then points at distances in multiples of 20 m from A, and the following readings were taken:

Backsight 3.39 on OBM 23.10.

Intermediate sights on A, $A + 20$ and $A + 40$, 2.81, 2.51 and 2.22 respectively.

$A + 60$: change point: foresight 1.88, backsight 2.61.

Intermediate sights on $A + 80$ and $A + 100$, 2.32 and 1.92 respectively; and finally a foresight of 1.54 on $A + 120$, all being in metres.

Tabulate the above readings on the collimation system; then assuming the levels at A and $A + 120$ were correct, calculate the amounts by which the rail would have to be lifted at the intermediate points to give a uniform gradient throughout.

Repeat the tabulation on the Rise and Fall system, and apply what checks are possible in each case. (*London*)

Answer: 0; 0.03; 0.08; 0.07; 0.11; 0.05; 0 m.

4   Q and R are two staff stations on opposite banks of a river. A level is set up at P on RQ produced and then at S on QR produced. The following readings were taken onto the staves at Q and R:

Reading of staff at Q from P = 1.859 m
Reading of staff at Q from S = 2.217 m
Reading of staff at R from P = 0.824 m
Reading of staff at R from S = 1.168 m

Determine the collimation error of the level and estimate the anticipated staff readings had that error not been present, given that $PQ = 12$ m, $QR = 85$ m and $RS = 16$ m.

Answer: 17″ downwards, 1.858 m, 2.209 m, 0.816 m, 1.167 m.

5   A parallel-plate micrometer attached to a certain level shows a displacement of 0.005 m when rotated through 20° on each side of the vertical. Calculate the thickness of the glass when the relevant refractive index is 1.6. Observations were taken with this level on to a conventional staff held vertically in turn at two stations, and the following data obtained:

Station	Staff reading	Micrometer reading	Remarks
1	1.720 m	3.2	Next lower graduation
2	1.250 m	8.4	Next upper graduation

The micrometer divisions read from 0 to 10 with 5 representing the vertical position of the parallel plate. Sketch typical fields of view (i) for the above readings, and (ii) for micrometer readings of 5.0 in each case. Also determine the difference in level between the two stations.

Assume that inverted images are formed by the telescope.

Answer: 0.4648 m.

6   A level was set up mid-way between two staves held vertical at A and B. With the bubble tube central in each case respective staff readings of 2.085 m and 1.633 m were obtained. The level was then set up at C, as close as possible to B, such that A, B and C were collinear. Again with the bubble centred, readings of 1.957 m and 1.484 m were noted on the staves at A and B respectively.

Calculate the magnitude and direction of the collimation error given that A and B were 90 m apart.

Answer: 48.1″ upwards.

7 A level sight on a staff held 80 m from the instrument reads 2.378 m and the bubble is found to be two divisions off the centre of the run towards the staff. If the level tube is in adjustment and has a sensitivity of 40 seconds, what is the true reading on the staff? Take sin 1″ = 1/206 265.

Answer: 2.347 m.

8 Reciprocal observations whilst levelling across a wide river gave the following readings onto staves held vertical at R and S from instruments stationed at P and Q.

Reading of staff at R from P = 1.332 m
Reading of staff at S from P = 1.016 m
Reading of staff at R from Q = 1.614 m
Reading of staff at S from Q = 1.278 m

R and P were close to each other on one bank with S and Q similarly situated on the other bank. If the reduced level of S is 10.376 m AOD, what is the level of R?

Answer: 10.055 m AOD.

9 A line of levels at 20 m centres, running from a point at the top of a quarry face, around the edge of the excavation, and then across it to a point immediately below the starting point, yielded the bookings given in the table below. Complete the level book by the Rise and Fall Method, and apply the usual checks.

Direct measurement down the face indicated the level difference to be 0.43 m greater than that indicated by the levelling. Subsequent investigation revealed that the level had been tampered with, and was out of adjustment, and that the levelling party had consistently read the staff near the top and bottom at change points, and had set up the instrument on the line of travel with unequal backsight and foresight distances. If the gradient is regular enough at any instrument setting for one to assume reading distance proportional to the difference between the staff reading and the average instrument height of 1.37 m, estimate the instrument error, and the true levels at A and L.

Station	Dist. (m)	BS	IS	FS	Rise	Fall	Red. lev.	Remarks
A	0	0.44						Top of face
B	20		1.64					
C	40		2.94					
D	60	0.17		3.98				
E	80		2.12					
F	100	0.32		4.11				
G	120	0.54		3.88				

Station	Dist. (m)	BS	IS	FS	Rise	Fall	Red. lev.	Remarks
H	140	0.11		4.08				
I	160	4.00		3.96			149.73	BM at commencement of cut
J	180		2.44					
K	200	3.86		0.11				
L	220			0.59				Bottom of face below *A*
								(*London*)

Answer: 0.00949 radians downwards, 168.82 m, 157.12 m.

4 The theodolite and its use

The theodolite is used to measure horizontal and vertical angles. It is without doubt the most important instrument for exact survey work, and many types are available to meet varying requirements of accuracy and precision, ranging from say direct readings of the circles from 5 min to 0.1 sec. There is thus a wide selection from which to choose to satisfy the surveyor's needs and in this chapter representative types which are likely to be met in modern engineering practice in this country and elsewhere will be discussed.

The inventor of the name 'theodolite' was Leonard Digges, and his description of the instrument under the title *The Construction of an Instrument Topographicall Serving most Commodiously for all Manner of Mensurations* was published in the sixteenth century by his son, Thomas. From this time, English surveyors concentrated their attention on the development of the instrument and, about 1785, Ramsden produced his famous telescopic theodolite which was used by Roy in 1787 for the first tie-up between the English and French triangulation systems. The horizontal circle was three feet in diameter, and by means of micrometers, readings could be made to single seconds. Vertical angles could not be measured as the instrument had no vertical circle, though less accurate instruments, equipped with these, were made early in the nineteenth century, culminating in the so-called 'transit' instruments which are described below.

The transit vernier theodolite

Although this instrument is unlikely to be encountered in surveying practice, many of its features are incorporated in glass-circle instruments and its 'open' construction is still used by teaching establishments to assist students to conceptualize the operation of theodolites in general.

As will be seen in Figs 4.1 and 4.2, a *principal* component of the instrument is a telescope which may be revolved through 360° about its transverse horizontal axis (an operation known as transitting). The name *transit theodolite* is then applied to the instrument.

The telescope is provided with an object glass, diaphragm and eyepiece, as described in the previous chapter, and in modern instruments is focused internally. When elevated or depressed it rotates about its transverse horizontal axis (*trunnion axis*), which is placed at right angles to the line of collimation, and the vertical circle, which is connected to the telescope, rotates with it. The trunnion axis is supported at its

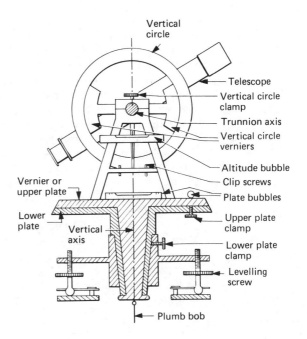

Fig. 4.2

Vertical circle

Telescope

Vertical circle clamp

Trunnion axis

Vertical circle verniers

Altitude bubble

Clip screws

Plate bubbles

Vernier or upper plate

Lower plate

Vertical axis

Upper plate clamp

Lower plate clamp

Levelling screw

Plumb bob

ends on the standards which are carried by the horizontal vernier plate (*upper plate*). Vertical angles are measured on the graduated vertical circle by means of a pair of diametrically-opposite stationary verniers which are carried independently of the circle and telescope, but centred on the trunnion axis, on a vertical frame shaped roughly in the form of a letter T. A bubble tube is often found on this frame, which can be moved by the clip screws. A clamp is provided for fixing the telescope in any position in the vertical plane, and a slow-motion or tangent screw then allows some small angular movement for final coincidence of the image of the point observed with the cross-hairs.

The vernier plate carries one or two small bubble tubes (*plate bubbles*) and also the verniers which are used in the measurement of horizontal angles. This plate is carried on an inner bearing from which a plumb bob may be suspended so as to allow the placing (*centring*) of the instrument over a survey station. The plumb bob should lie on the vertical axis of the instrument, and this in turn should be at right angles to the trunnion axis. The vertical plane in which the telescope revolves should coincide with the vertical axis of the instrument as should the centre line of the inner axis.

The inner bearing rotates within an outer sleeve bearing which carries the horizontal circle (*lower plate*) on which are marked the graduations by which horizontal angles are read. By means of the lower clamp the outer bearings can be prevented from rotating, and similarly the upper clamp fixes the upper and lower plates together. Provision for small angular movements is made by tangent screws for both clamps. If the lower clamp be tightened, the inner bearing may rotate inside

◁ **Fig. 4.1** Vernier
◁ theodolite

the outer bearing so long as the upper clamp is free: if the upper be tightened and the lower clamp is freed, both bearings move together as one: with both clamps tightened, no movement will occur. The standards, and hence the telescope, will rotate about the vertical axis of the instrument so long as the upper plate is free to turn. This rotatable portion of the theodolite is known as the *alidade*. A thin film of oil is required between the bearing surfaces to prevent stiff movement.

The instrument must be set up with the vertical axis truly vertical when angular measurements are made, and this is effected by the levelling head which has a footplate and three footscrews (in old patterns four footscrews may be found). The ball ends of the footscrews fit into recesses in the footplate and the footscrews operate on bushes fitted on the levelling head. For ease in centring the instrument over a station, a centring device is usually provided whereby the instrument and plumb bob can be moved independently of the levelling head when the footplate has been screwed on to the tripod; this device is seen just above the footscrews in Fig. 4.1 and in the illustration at the start of this chapter. It is important that there should be no movement of the footscrew feet on the footplate when this has been screwed to the tripod and the instrument levelled.

The vernier

A typical position of the vernier and main scale is illustrated in Fig. 4.3. It will be noted that there are 60 subdivisions between the index (zero) and 20-minute graduation of the vernier. These subdivisions take up the same length as 59 subdivisions of the main scale, since in a direct reading vernier n subdivisions have the same length as $(n - 1)$ divisions of the main scale. The particular vernier shown allows circle readings to be taken to 20 seconds of arc (the least count of the theodolite).

Fig. 4.3

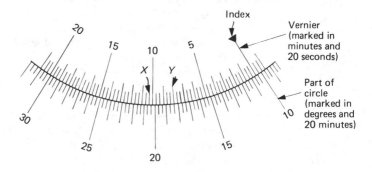

To read the vernier the observer first notes the position of the index: in Fig. 4.3 it lies between 10° 20′ and 10° 40′. The observer now looks along the vernier scale until he finds a graduation on that scale which coincides with a graduation on the main circle. Inspection of the figure will show that this occurs at the eighth division Y (or the twenty-fourth subdivision) on the vernier. This reading, which is 8′0″

(or 24 × 20 sec) is added on to the reading 10° 20′, which is the *lower* of the two graduations straddling the index mark.

It will be realized that were the vernier not present, the observer would be forced to estimate a reading of the index to a fractional part of 20 minutes, and even though small eyepieces are fitted to assist the reading it would be impossible to read to 20 seconds.

To obtain a reading of, say, 10° 30′ 20″, the index mark would be moved into the region of just over 10° 20′ and the upper clamp then fixed (the lower clamp must be tight throughout). Then, using the slow-motion screw, the index mark is made to travel alongside the main scale so that the graduation *X* in Fig. 4.3 coincides with a graduation on the main circle. Note that in both reading and setting out angles, after noting the next lowest main-circle reading nearest the index mark, the remainder of the reading is obtained on the vernier, the value of the coincident main-circle graduation being of no account whatsoever.

Modern instruments

Fig. 4.4 Watts Microptic ST400 (*Courtesy*: Hall & Watts Ltd)

The vernier instrument has been replaced in modern times by the glass-circle instrument. These instruments differ in the first instance from the vernier instrument previously described in that the metal scale plates read by the vernier are replaced by glass circles read by internal optical systems, the standards being enclosed. However, depending upon the manufacturer and model, there are features still common to both the glass-circle and vernier theodolites, i.e. relevant plate clamps and tangent screws are provided by most manufacturers, although a friction system similar to that mentioned for the Zeiss level in Chapter 3 can be met. The glass circles are photographic copies of glass master circles, which in turn have been graduated by means of an automatic dividing engine.

Another difference is that although the plumb bob might be provided for centring purposes, more accurate facilities are also likely to be available. One of these, the built-in optical plummet, is a downward sighting system, and it can be positioned either in the alidade as in Fig. 4.4 or in the tribrach. The instrument is first set up, fairly closely over the station, either by eye or by plumb bob. The tripod legs must be firmly pressed into the ground so that no movement can occur in the instrument as the surveyor moves round or when traffic moves nearby, and the wing nuts clamping the legs must be tight.

When the instrument has been roughly centred, it must be levelled. Assuming three footscrews only, and using the bubble on the horizontal circle, the procedure is as follows:

(*a*) Rotate the inner axis so that the bubble tube is parallel to two of the footscrews. Turning those footscrews, the bubble is brought to the centre of its run. The footscrews are turned simultaneously with the thumbs moving towards each other or away from each other. The left-thumb movement gives the direction of the consequent movement of the bubble.

(*b*) Rotate the inner axis so that the bubble tube is at right angles to its former position, when it should be parallel to a line joining the third footscrew to the mid-point of the line joining the other two. Bring the bubble to the centre of its run using the third screw only.

A correctly-adjusted instrument will now be levelled, and as the horizontal circle rotates and takes the bubble tube round, then the bubble should remain at the centre of its run. In practice, the above procedure is carried out at least twice, the telescope being wheeled successively through 90° back to position (*a*) and then, after checking with the two screws, to position (*b*).

The optical plummet is then focused as sharply as possible, with no parallax, on the station. By moving the instrument in parallel shifts its index mark is centred over the station: a check on the plate bubble should now be made. When the plummet is within the alidade, observations are taken from two diametrically opposite positions and the mean adopted (if there be any discrepancy due to relative inclination of the vertical axis of the theodolite and line of sight of the plummet) to give the correct centring.

In addition to the optical plummet, other features which can be encountered are the provision of (i) a centring device above the footscrews, e.g. in the Watts ST456 20-second Microptic Theodolite, shown at the start of the chapter, and (ii) a detachable tribrach base for use in three-tripod surveying discussed in Chapter 6 (the Watts ST402 one-second instrument has this facility). Figure 4.4 shows the ST400 model, also reading directly to one second, and it illustrates points made

Fig. 4.5

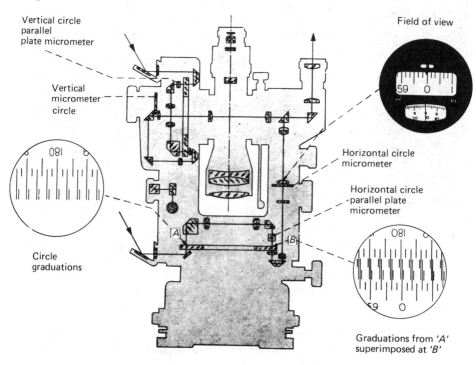

Vertical circle parallel plate micrometer

Vertical micrometer circle

Field of view

Horizontal circle micrometer

Horizontal circle parallel plate micrometer

Circle graduations

Graduations from 'A' superimposed at 'B'

previously in this section. The altitude bubble, which can be viewed by a prism-reader, is located in the 'far' standard carrying the vertical circle, its slow-motion, or clip screw, being placed below the vertical circle. Figure 4.5 shows this bubble etc. in cross-section.

Accuracy is obviously a most important consideration when selecting instruments for particular tasks. It is suggested that instruments reading directly (or estimating) to, say, 5 or 6 seconds be used in the main for topographical surveys and general setting out. Those reading directly to one second have more universal applications, e.g. precise traversing, precise setting out and second-order triangulation surveys. Lateral displacement of the line of sight by, say, 5 mm occurs at distances of 200 m and 1 km for the two values. Other instruments reading to finer limits or to coarser limits of least count are available, but the two mentioned in this paragraph are very commonly encountered by the civil engineer in the course of his work.

Methods of circle reading

Generally speaking, we can say that the glass-arc instruments will have their circle readings, or registrations, described as either micrometer readings or scale readings. In either case, the image of the vertical circle and/or the horizontal circle is brought into the field of view of the reading eyepiece which may be on an upright (Fig. 4.5) or alongside the telescope. Either the letters V and H (or Hz) or some other distinguishing characteristics, such as colour, will appear in the field of view to minimize the risk of possible reading mistakes.

Light is admitted by suitably placed circle illumination mirrors and in the one-second theodolites, by means of systems of lenses and prisms, it passes through diametrically opposite points in the two circles, through a micrometer or micrometers to the reading eyepiece, as shown in Fig. 4.5. As mentioned previously, this figure refers to the Watts one-second instrument which has two separate lens systems with one micrometer for the vertical circle and one micrometer for the horizontal circle, each under different control.

In this instrument, the circles are graduated at 10-minute intervals, alternate lines are lengthened inwards and short parallel lines are marked at their ends. In the field of view three apertures are seen: the large central aperture gives the reading of one side of the selected circle, the upper aperture shows one or two sets of three lines each, and the bottom aperture gives the relevant micrometer reading. The two outer lines of the sets in the upper aperture are the images of pairs of the short parallel lines, previously mentioned, which are at 20-minute intervals. When a circle rotates these double lines move across the field of view in one direction, whilst the images of single lines from the diametrically opposite position move across in the other direction. When a reading is being taken, the circle will be stationary so that the single lines will not move, but the images of the double lines can be shifted by means of the parallel plate micrometer and placed symmetrically astride the single line images: this is known as the *bisection method* of circle reading. The image of the relevant small micrometer circle then appears in the bottom aperture and is added to the reading in the

Horizontal circle

Main scale	183° 20′
Micrometer	7′ 26″
	183° 27′ 26″

Fig. 4.6

central aperture to give the required reading for that pointing of the telescope, this being the mean of the diametrically opposed points (see Fig. 4.6).

When a specific zero setting is wanted, say 45° 07′ 26″, the value 07′ 26″ is set on the micrometer scale using the micrometer knob after the initial pointing has been made on A. The circle repetition knob is then carefully actuated so that the light gap is bisected when the combined reading of main scale and micrometer is 45° 07′ 26″. A somewhat similar operation is effected when setting out $A\hat{B}C$, since after establishing our zero pointing to A the micrometer reading is set to its corresponding value, then the alidade is rotated until the angle is nearly correct. The light gap bisection is now finalized by means of the azimuth slow motion device, and the telescope will be pointing to C.

An alternative approach to the bisection method of circle reading is the *coincidence method* in which the images of graduations from opposite points of the circle are made to coincide in the field of view of the reading microscope when a micrometer knob is actuated. Figure 4.7(*a*) shows the Wild T2-A theodolite and Fig. 4.7(*b*) readings obtained with it: the coincident graduations are shown in the upper window, degrees and tens of minutes in the middle window, with the relevant micrometer reading (up to 10 minutes) in the lower window. This particular instrument uses a selector knob, like the Watts No. 2 microptic, to give readings of either the horizontal circle or the vertical circle, but the same micrometer (shown in Fig. 4.7(*c*)) serves for each. When this micrometer is actuated, the parallel plates are inclined through equal and opposite amounts until the images of adjacent graduations of the diametrically opposite points of the circle are exactly coincident.

Many former T2 instruments had a slightly different registration in that tens of minutes were not directly registered but had to be determined by the observer. Figure 4.8 illustrates a typical coincidence of diametrically opposed graduations. The index line reads 265° 40′, but in fact this reading should be deduced from the numbers of divisions between main graduations 180° apart, and it will be observed that there are four such between 265° and 85°. This indicates a reading of 265° plus 4 × 10′ and the micrometer reading is added afterwards. It will be noted that the interval between graduations is 20 minutes on the circles: the two sets move in opposite directions in the field of view when the telescope is rotated, thereby giving coincidence every ten minutes but the parallel plates ensure that coincidence is possible within that interval. In Fig. 4.8, it will be seen that the parallel-plate micrometers move the images through a distance of α for coincidence, i.e. each scale image moves through $\alpha/2$; the drum reading, not shown in the diagram, appears below the scale reading, as in Fig. 4.7(*b*).

The universal one-second instruments invariably have circle readings from diametrically opposed points, but for most of the instruments reading to a lower precision, e.g. 20 seconds, only one position of a circle is observed and registered. Consequently, the optical systems are relatively simpler. Some of these theodolites use scales rather than

Fig. 4.7 (a) The Wild
T2-A theodolite
(*Courtesy*: Leica UK
Ltd)

(a)

Fig. 4.7 (cont'd) (*b*)
(*c*) Micrometer

(*b*) Reading example
360° model on vertical
circle:

Degree number	094
Tens of minutes	1
Micrometer scale	2′ 44″
Booking	94° 12′ 44″

(*c*)

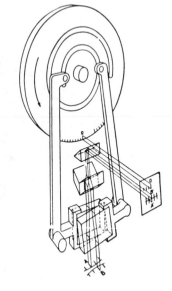

(*b*)

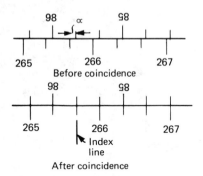

Fig. 4.8 Former
method of reading the
Wild T2 Universal
theodolite. Upright
numbers indicate the
required circle readings

micrometers to achieve direct readings with the aid of an index line.
Figure 4.9 shows the circle readings of the Kern K1-S theodolite whose
horizontal circle has both clockwise and counterclockwise graduations
reading directly to 30 seconds. This system is useful for the setting
out of circular curves (Chapter 11).

**Automatic index
for vertical circle
readings**

The vertical circle on older theodolites has to be set to horizontal before
every reading using an altitude bubble, and if carried out accurately
then the mean of face-left and face-right readings of the vertical circle
gave the required value for the vertical angle being observed. To obviate

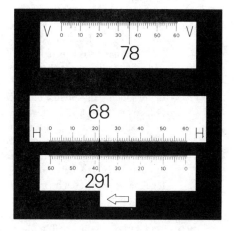

Fig. 4.9 Vertical circle reading 78° 35.7′. Horizontal circle reading — clockwise 68° 21.8′; anticlockwise 291° 38.2′

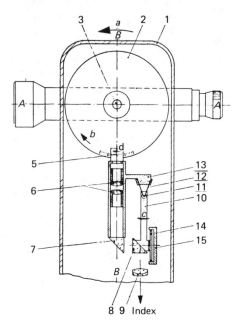

Fig. 4.10
AA, Line of Collimation;
BB, Vertical Axis.
1 Dust casing
2 Vertical circle
3 Trunnion axis
5–9 Path of reading rays
5 Index
6 Reading objective
8 Pendulum prism
10 Pendulum
11 Pivot
12–13 Pivot mounting
14 Damping piston
15 Damping cylinder

the need for this centring operation, which reflects inclination of the vertical axis, Messrs Askania of West Berlin introduced one of the more radical innovations of the post-war years insofar as theodolite design was concerned, namely the provision of a pendulum compensator to replace the altitude bubble. This proved so popular that nowadays one would expect this form of compensator, or its liquid equivalent mentioned later, to be fitted to all theodolites.

The principle of the Askania compensator is shown in Fig. 4.10. The vertical circle is read through an optical arrangement which includes a prism attached to a pendulum, and the path of the reading rays (5—9 in the diagram) is so adjusted that, with the vertical axis *BB* truly vertical, vertical angles are correctly obtained. If the theodolite tilts through *a* from the vertical, however, then, assuming for convenience that *AA* is to be horizontal for a given sight, *AA* must be rotated in direction *b* in order to remain horizontal. The vertical circle is carried round with the telescope in direction *d*, but owing to the influence of the pendulum, the reading obtained at the index is not thereby changed. Due to the tilt of *BB*, the pendulum (10) carrying the prism (8) swings in direction *c* and the path of the reading rays is so influenced that the correct circle reading is still obtained.

With a sufficiently large range of action of the pendulum, large tilts of the vertical axis *BB* could be accommodated, but since preliminary levelling is carried out by means of a target bubble which, when central, ensures that *BB* is plumb to within ±3 min, then it is only essential for the pendulum to have this range. Vertical-circle readings are then automatically converted by the pendulum prism arrangement. An air damping device is incorporated to damp out oscillations in the pendulum.

As an alternative, liquid type compensators have been developed, some manufacturers including both types within their ranges. For example, Wild incorporate a pendulum compensator and a liquid type of compensator in their T2 and T1 theodolites respectively. The T1 theodolite, which has a detachable tribrach, is illustrated in Fig. 4.11(*a*). Both circle readings are displayed through an eyepiece alongside the telescope, the horizontal circle being yellow and the vertical circle white. The same micrometer serves both circles and in the sexagesimal system the readings are given directly at intervals of 6 seconds, whilst an index mark allows closer estimation.

The action of its compensator is indicated in Fig. 4.11(*b*). When the vertical axis is truly vertical, the surface of the liquid, which is silicon oil, is parallel to the base of its transparent container and rays pass through without deviation. If the vertical axis be displaced through δ the oil surface makes this angle with the base but the ray is deviated through $\gamma = (\mu - 1)\delta$, μ being the refractive index. Thus, if the line of sight of the telescope be horizontal, say, a reading of 90° can still be registered (providing μ remains constant).

As mentioned earlier, automatic indexing ensures that the correct vertical-circle reading is given for a particular telescope pointing when the vertical axis is not truly vertical. If the telescope clamp remains locked the registered circle readings now alter as the alidade is rotated; in fact, the component of the angle of inclination of the vertical axis at right angles to the direction of the initial pointing is half the difference of the two circle readings obtained on orienting the line of sight to be 90° right and 90° left respectively of that direction.

Measurement of horizontal angles	To measure angle *ABC*, the instrument is set up over station *B* in the manner described, and carefully levelled by means of the footscrews. The face of the instrument must be checked at this stage. Most telescopes have sights similar to those on a gun, fitted on top of the barrel, to assist in sighting the target. With these sights on top and the telescope

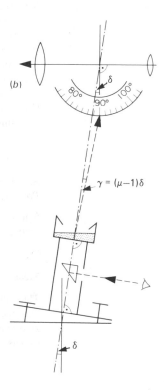

Fig. 4.11 Wild T1 theodolite (*Courtesy*: Leica UK Ltd)

pointing to the target, the vertical circle, which is known as the *face* of the instrument, will be left or right of the telescope. Suppose it is to the left; the theodolite is said to be in the *face left* position. By rotating the telescope through 180° in the vertical plane (i.e. about the trunnion axis), and then through 180° in the horizontal plane, the telescope will again be pointing at the signal, but with the gunsights on the underside of the barrel, and the vertical circle to the right — i.e. the theodolite is in the *face right* position. Starting with all clamps tightened, then:

(a) The plates are unclamped and the horizontal circle set to zero or any arbitrary value near zero. The upper clamp is locked, holding the two plates together.

(b) The telescope is directed to station *A* using the gunsight. When closely pointing on *A* the lower clamp is also locked and the vertical hair of the diaphragm is accurately sighted onto the station using the lower tangent screw. The horizontal circle reading is now taken.

(c) With the lower clamp fixed, the upper clamp is released and the telescope swung in a clockwise direction until directed towards station *C* using the gunsight.

(d) The upper clamp is then fixed and the upper tangent screw used to accurately align the telescope onto station *C*. The horizontal circle reading at *C* can then be obtained.

(e) The upper clamp is released and the theodolite turned through 180°, the telescope is then also turned through 180° in the vertical plane and the gunsight used to roughly sight onto station *C*.

(f) The upper clamp is locked and the upper tangent screw used to align the telescope onto station *C* and the horizontal circle reread.

(g) The upper clamp is unlocked and the telescope directed towards station *A* with the gunsight.

(h) The upper clamp is locked and the upper tangent screw used to align the telescope onto station *A*. The horizontal circle can then be read for this pointing on *A*.

Angle *ABC* is obtained as shown in the following example:

At station B

Pointing	*Face left*	*Face right*
Station *C*	93° 34′ 40″	273° 34′ 40″
Station *A*	01° 15′ 20″	181° 15′ 40″
	92° 19′ 20″	92° 19′ 00″

Mean value 92° 19′ 10″

Thus, two measurements of the angle are obtained during this set and their mean can be found. Further sets can be taken after changing the zero setting (*a*) by about 180°/*n* each time, *n* being the required number of sets.

Measurement of vertical angles	The angles of elevation (+) or depression (−) are measured with respect to the horizontal plane containing the trunnion axis of the instrument. When the angle of elevation becomes large, a diagonal eyepiece may be needed to facilitate the pointing of the telescope. This accessory is shown fitted on the instrument illustrated at the beginning of the chapter.

After setting up over the station, the telescope is directed to one of the signals and exact coincidence on the mark obtained using both horizontal and vertical tangent screws. If a horizontal angle is being observed at the same time as a vertical angle then the procedure discussed previously is adopted. Assuming that the instrument possesses an automatic index to the vertical circle, the reading will give the angle relative to the instrument zero. This may be a + or − angle related to the horizontal plane or, on most modern instruments, the angle related to zero vertically upwards.

After the telescope has been directed to the other signal at *C*, obtaining coincidence as before, the vertical circle reading gives a further angle

(+ or −), and applying this reading to the other gives the vertical angle subtended at the instrument.

If the instrument is not provided with an automatic index, the altitude bubble should always be in the centre of its run when reading the vertical circle. It will be so, or nearly so, when the instrument is in adjustment and it can be so positioned using the tangent screws to the vertical circle.

Observations should be made on both faces of the instrument, irrespective of whether or not an automatic index is fitted, the mean value giving the required vertical angle. This eliminates any index error (see page 98) due to lack of adjustment. It must be remembered that the mean of face left and face right readings does not eliminate errors in horizontal circle readings caused by inclination of the vertical axis: this is very important as altitude increases. When the vertical axis is not truly vertical, the inclination of the trunnion axis varies with the pointing direction and this affects the reading of the horizontal circle. An example of this can be found in Bannister and Baker, *Solving Problems in Surveying* (Longman), in which the correction i tan (altitude) is used, i being the inclination of the trunnion axis. The same expression applies for horizontal circle readings when, even though the vertical axis is truly vertical, the horizontal axis is inclined to it. However, in this situation the mean of face left and face right readings is the correct value of the horizontal angle. It is worth mentioning that a trunnion axis micrometer could be built into the Kern DKM-2A theodolite so that angle i could be evaluated and the correction of each reading applied directly.

Permanent adjustments of glass-circle theodolites

The following adjustments may be required:

(1) the plate bubble
(2) horizontal collimation (to place the line of sight at right angles to the trunnion axis)
(3) vertical collimation (to ensure that the vertical circle reads the correct value when the line of sight is horizontal: also to check the altitude bubble when no compensator is fitted).

(1) Adjustment of plate bubble — to set vertical axis truly vertical

This bubble, set on the upper plate, must be central when the plate is horizontal; and the vertical axis must then be vertical, since the plate and vertical axis are set at right angles by the maker. If this adjustment is correct, then the plate bubble will remain in the centre of its run throughout a complete rotation of the telescope in azimuth.

The instrument is set up and levelled as accurately as possible with the aid of the tripod legs. The plate-bubble tube is placed over a pair of footscrews and by means of these the bubble is brought to the centre of its run (Fig. 4.12(*a*)). The instrument is rotated through 180° in azimuth (Fig. 4.12(*c*)) and any deviation of the bubble noted. Half this deviation is corrected by the same pair of footscrews and the remainder by means of the plate-bubble capstan nuts. The bubble tube is now arranged over the other footscrew, and half any deviation is

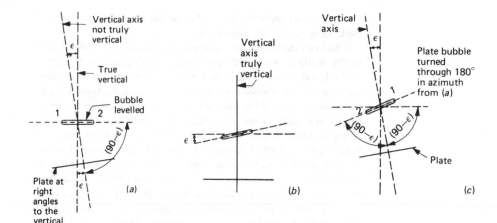

Fig. 4.12

taken out by that screw. This procedure is repeated until the bubble remains central for any position and it is then said to traverse.

In Fig. 4.12(c), the plate has been turned through 180° in azimuth; the bubble tube is shown with ends 1 and 2 reversed in position, and the angle 90° − ε between the vertical axis and the bubble is as indicated. This means that an angle of 2ε will be given by the intersection of the horizontal plane and the plane through the bubble axis. Using the footscrews, the movement of the bubble through an angle ε will evidently move the vertical axis into its true position and the capstan nuts will take out the remainder and place the line 1.2 parallel to the plate (since inspection of Fig. 4.12(b) shows the angle between the bubble axis and the plate to be ε after use of the footscrews).

Permanent adjustment 1 for the Dumpy level (Chapter 3) is carried out in this way.

(2) Setting the telescope sighting line at right angles to the horizontal axis

The sighting line joining the vertical cross-hair to the optical centre of the object glass must be at right angles to the horizontal axis. If the vertical hair is displaced to one side or the other of its true position, then the line of sight will no longer be along the telescope and at right angles to the horizontal axis (Fig. 4.13). The line will trace a flat cone and not a plane if this adjustment is not carried out.

The instrument is set up and levelled on reasonably level ground in such a position that a field of view up to 100 m is available on both sides of the instrument.

(a) A small sharp object (e.g. a chaining arrow) is placed at A about 100 m from the instrument and the telescope, say face left, is directed towards it, obtaining exact coincidence on the cross-hairs using the tangent screws.

(b) The telescope is transitted and a second arrow B set at about the same distance away from the instrument as A on the line of sight. A and B should be at the same level if possible so that any error due to the next adjustment is avoided. The instrument is now face right (Fig. 4.14(a)).

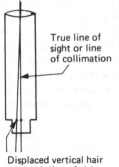

True line of
sight or line
of collimation

Displaced vertical hair
gives this line of sight

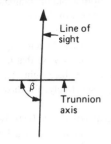

Line of
sight

β

Trunnion
axis

Fig. 4.13

(c) The clamping arrangement is freed and the telescope rotated through 180° in azimuth and exact coincidence obtained at *A* with the tangent screw after clamping. The instrumnet is still face right.

(d) The telescope is again transitted towards the position of *B*. The vertical hair position is again noted. If the line of sight bisects *B*, then the adjustment is correct; if not, a second arrow B_1 is placed at the same distance from the instrument as *B* and on the line of sight. The instrument is now face left. The line of sight is directed to some point *C* positioned such that $B_1 C = \frac{1}{4} BB_1$ (i.e. the line of sight is turned through an angle δ relative to the horizontal axis, Fig. 4.14(*b*)). This re-direction is effected by means of the two side diaphragm screws. Repeat the adjustment as necessary until correct. It is as well to check that the vertical hair is truly vertical at this stage. If the telescope is directed on to a very small sharply-defined nearby point, this point should appear to move along the vertical hair as the telescope is moved in a vertical plane. The diaphragm adjusting screws enable the cross-hairs to be turned until the vertical hair is truly vertical.

Alternatively, the telescope can be directed on to a sharp sighting mark at a minimum distance of 100 m at about instrument height, and the horizontal circle reading noted on one face. After transitting and resighting, it is then noted on the other face. The two readings should differ by 180° and any discrepancy is twice the error. The circle reading which should have been noted can therefore be deduced and set by the micrometer in conjunction with the azimuth tangent screw: the graticule is then adjusted to ensure that the vertical line lies on the sighting mark. A repeat check can then be undertaken.

Fig. 4.14

(a)

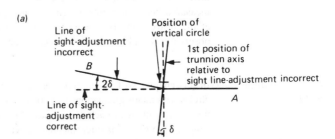

Line of
sight-adjustment
incorrect

Position of
vertical circle

B

1st position of
trunnion axis
relative to
sight line-adjustment incorrect

2δ

A

Line of sight-
adjustment
correct

δ

(b)

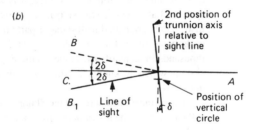

2nd position of
trunnion axis
relative to
sight line

B

2δ

2δ

C.

A

B_1

Line of
sight

δ

Position of
vertical
circle

(3) Vertical collimation

The adjustment to remove the index error can be effected by first sighting a distant target and reading the vertical circle on both faces. (The altitude bubble, if fitted, should be at the centre of its run each time.) Half the discrepancy in those readings can be allotted to the second vertical circle reading to obtain the corrected reading. If a compensator is fitted the micrometer can be actuated to register its part of that reading, and then an adjustment screw allows the remainder of the corrected vertical circle reading to be presented. When the theodolite is not compensated the corrected circle reading is deduced as described above, and the mean of the discrepancy can be allotted to the face left reading to obtain a mean vertical reading. The telescope is then directed back to the sighting mark and the corrected vertical angle registered using the micrometer and the altitude bubble tangent screw. The altitude bubble is now centred by means of the relevant adjusting screw.

Most theodolites read zero on the vertical circle when the telescope is pointing vertically upwards in the face left position, say, so that the horizontal sighting will register 90° face left and 270° face right. The Watts ST400, however, reads zero and 180°, respectively, for those horizontal sightings.

Adjustments (1) and (2) ensure that the vertical axis should be truly vertical and that the line of sight will trace out a plane and will be at right angles to the horizontal axis (trunnion axis). It is also necessary for the trunnion axis to be at right angles to the vertical axis so that the telescope will transit in a true vertical plane. On the older vernier theodolites there was an adjustment for this, but nowadays the adjustment for horizontality of the trunnion axis is discounted by most manufacturers — Watts, for instance, stating that precise machining ensures that this axis will remain at right angles to the vertical axis with sufficient accuracy. If there is a lack of perpendicularity between the trunnion axis and the vertical axis, errors arise in horizontal angles measured between points which subtend appreciable vertical angles at the instrument. However, the mean of face left and face right observations will be free from such errors.

The centesimal system

In this chapter so far, angular measurements have been referred to circles with major graduations from 0° to 360° with secondary graduations which subdivide each degree into 10-minute or 20-minute intervals (Fig. 4.3). Vernier or micrometer subdivision then gives the reading down to seconds and, since there are sixty minutes in a degree and sixty seconds in a minute, the system is known as the *sexagesimal* system (Latin: *sexaginta*, sixty). It is possible, however, to obtain instruments graduated in 400 major parts from 0 to 400, each of these parts being termed a *gon*. The gon can be subdivided down to ten-thousands in steps of ten, angles being expressed in decimals in this system which is known as the *centesimal* system (Latin: *centum*, hundred).

It can be mentioned here that the SI unit of plane angular measurement is the radian. Some instruments are available which carry graduations in radian measure, each being related to the mil (0.001 rad) rather than the degree or the gon.

The automatic recording of circle readings has received much attention over the years. One of the earliest developments was the introduction by Askania of their Precision Theodolite Tpr, after Gigas, whose circle readings could be obtained directly to 0.2 second either visually or photographically. However, in the latter case, the readings were obtained following a second step in which an accessory known as a film reader was used. Thus, although booking errors tended to be eliminated, the whole operation was not automatic in so far as the provision of the field data was concerned.

Code theodolites were then developed by certain manufacturers, i.e. Fennel's FLT instrument, and these were such that the field measurements could be processed directly via the computer. The circle readings were obtained in a coded form on film which, after processing, was passed to an electronic converter in which the recorded data was transferred into a five-channel punched-tape code. This, in turn, was fed into a computer for the evaluation of the required information, including co-ordinates.

Developments in electronics have allowed the photography stage to be eliminated, encoders now being used in the readout system. In so far as theodolite circle readings are concerned, angular or rotary encoders are incorporated to measure angular displacement. They comprise circular plates suitably coded so that, with the aid of a reading head, a digital readout for an individual distinct increment of displacement can be obtained.

Two types, absolute encoders and incremental encoders, may be encountered in the various electronic theodolites and total station instruments now marketed. The code pattern of the former type ensures a unique digital response for each indvidual increment of displacement, whilst that of the latter type gives rise to an output based upon the number of periods, or counts, between the start and finish of the whole displacement. Thus, in the case of incremental encoders, the start position has to be known for the determination of an overall measurement.

Based upon the type of reading head, the instruments can be further classified as photo-electric, magnetic or brush.

Figure 4.15 shows the principles of a system, devised by Kern, which utilizes an incremental rotary encoder for the measurement of horizontal or vertical angles. Its design is based upon the Moiré phenomenon for which two gratings have to be formed so that they can create interference fringes. Each main circle is graduated (but not coded or numbered) so as to form one grating, the spacing between the lines being equal to the width of the lines. A section of 200 lines, enlarged 1.01 times, is projected onto the diametrically opposite section thereby producing the two gratings. As the telescope position changes, the interference fringes move and the circle reading can be established (i) coarsely by counting the number of Moiré periods, and (ii) finely by an interpolation procedure. Four light-sensitive diodes cover a

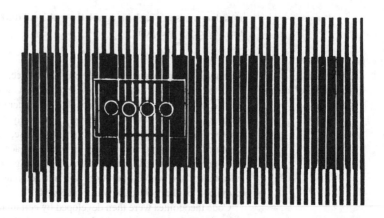

Fig. 4.15 Moire periods (*Courtesy*: Leica UK Ltd)

distance equal to three-quarters of the Moiré period which is assumed to have a sine-wave form. The position of the bright fields or the dark fields relative to a readout window is a measure for the relative positioning of a pair of diametrically opposed graduation lines when deciding the fine readings, whilst the coarse readings are derived by counting the number of periods passing a specific spot at that window.

Absolute encoders are incorporated within the earlier versions of Zeiss Elta equipment, such as the Elta 2 Electronic Tachometer. Diametrically opposite graduations are projected together and by means of a plane parallel plate these are placed in coincidence with a photo-electric micrometer (Fig. 4.16). During the measurement, a micro-processor controls the switching on and off of the illumination of the circle, the action of the micrometer, the reading and storage of the circle and micrometer code after coincidence, and finally the evaluation of the circle reading from the coarse reading given by the circle code and the fine reading given by the micrometer code.

Fig. 4.16 Zeiss Elta Codes (*Courtesy*: Carl Zeiss)

Fig. 4.17 Wild T1600
electronic theodolite
(*Courtesy*: Leica UK
Ltd)

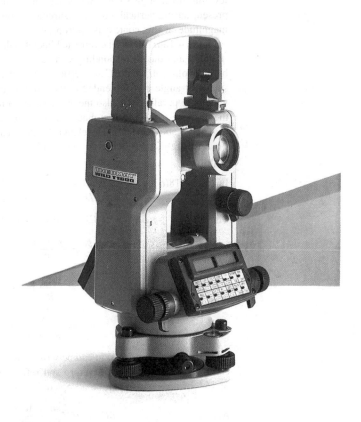

Most manufacturers produce electronic theodolites and in many cases
they are now cheaper to purchase than the optical equivalent. They
are typified by Fig. 4.17 which illustrates the Wild T1600 electronic
theodolite. This measures horizontal and vertical angles by means of
absolute encoders with a one second display, although the accuracy
of the mean of two face observations is quoted as 1.5 seconds. Corrections for errors in horizontal collimation and vertical index are automatically performed. Angle measurements are displayed continuously
as the telescope pointing is changed, horizontal angles being measured
clockwise or anticlockwise at the push of a button. The instrument is
controlled through a keyboard where the horizontal angle can be reset
to zero, or any other value, for reference pointings.

**Trigonometrical
levelling**

Trigonometrical levelling is a technique for the determination of
differences in height by measuring vertical angles. In Chapter 3 the
influences of curvature and refraction were discussed in relation to

levelling over long sights and, since these same influences can be present during vertical angle measurement, some discussion of their magnitude in this context is appropriate.

Figure 4.18(a) shows the refracted line of sight between two stations A and B and the corresponding vertical angles, α and β, registered by theodolite. AD and BC are verticals through A and B respectively, such that angle c is subtended at the earth's centre by AC and BD, which are parallel, c is also the angle between those verticals. S_{AB} is the distance from A to B at mean sea level; h_A and h_B are the heights of A and B above mean sea level and R is the mean radius of the earth. To estimate the difference in level, BC, angles BAC and ABD are required.

In Fig. 4.18(a)

$$B\hat{A}C = \alpha - r + c/2 = \alpha + (c/2 - r)$$
$$A\hat{B}D = \beta + r - c/2 = \beta - (c/2 - r)$$

Therefore

$$B\hat{A}C + A\hat{B}D = 2B\hat{A}C = \alpha + \beta$$

Therefore

$$B\hat{A}C = \frac{\alpha + \beta}{2}$$

Note that

$$c = \frac{S_{AB}}{R} \text{ radians}$$

$$= \frac{S_{AB}}{R \sin 1''} \text{ seconds}$$

The coefficient of refraction can be written as r/c with $(c/2 - r)$ the combined correction for curvature and refraction: an average value of $15''$ of arc per km can be accepted for sights in excess of 5 km.

Also

$$BC = AC \frac{\sin B\hat{A}C}{\sin A\hat{B}C}$$

$$= AC \frac{\sin \dfrac{\alpha + \beta}{2}}{\sin \left(90 - \dfrac{\alpha + \beta}{2} - c/2\right)}$$

Fig. 4.18

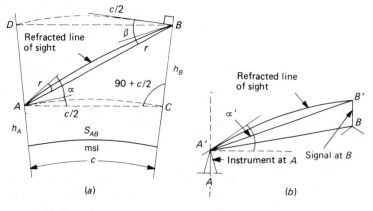

(a)

(b)

$$= AC \; \frac{\sin \dfrac{\alpha + \beta}{2}}{\cos \left(\dfrac{\alpha + \beta}{2} + c/2 \right)}$$

For practical purposes, this expression is usually written as

$$BC = AC \tan \left(\frac{\alpha + \beta}{2} \right)$$

Thus, by taking reciprocal observations α and β, the height difference BC can be determined.

Inspection of Fig. 4.18(a) reveals that

$$\begin{aligned} AC &= 2(R + h_A) \sin c/2 \\ &= 2(R + h_A) \; c/2 \sin 1'' \end{aligned}$$

where $c/2$ is a small value in seconds of arc.

Thus
$$AC = 2(R + h_A) \frac{S_{AB} \sin 1''}{2R \sin 1''}$$

$$= \left(1 + \frac{h_A}{R} \right) S_{AB}$$

Unless h_A is large $AC \simeq S_{AB}$.

It will be noted that the curvature correction $c/2$ depends upon distance AC and it is possible that, when BC is small compared to AC, α is an angle of depression although $B\hat{A}C$ is still overall an angle of elevation. A further point to be remembered is that the heights of the instruments and the heights of the corresponding signals need not be the same. Assuming that the signals are always set higher than the instrument axes above ground level, as shown in Fig. 4.18(b), the angle of elevation at A will be larger than it should be, and an 'eye and object' correction of $\dfrac{B'B - A'A}{AC}$ radians is deducted from α', which will have been measured with respect to B', to obtain α.

Example 4.1 Reciprocal vertical angles were observed between stations A and B (Fig. 4.18(a)), the following measurements being obtained:

Height of signal at A	= 2.75 m
Height of signal at B	= 2.46 m
Height of instrument at A	= 1.46 m
Height of instrument at B	= 1.37 m
Angle of elevation recorded at A	= 01° 19′ 36″

Estimate (a) the angle of depression recorded at B, given that the effective distance AC is 5950 m, and the difference in level between A and B is 139.2 m, and (b) the correction for refraction, if the correction for curvature is 01′ 38″.

(*a*) For the angular observations from A and B, respective 'eye and object' corrections of $\dfrac{2.46 - 1.46}{5950}$ radian (i.e. 35″) and

$\dfrac{2.75 - 1.37}{5950}$ radian (i.e. 48″) obtain.

Therefore α = the corrected angle of elevation at A
$$= 01° \ 19' \ 36'' - 35'' = 01° \ 19' \ 01''$$

Now $AC \tan\left(\dfrac{\alpha + \beta}{2}\right) = BC$

Therefore $\quad 5950 \tan\left(\dfrac{\alpha + \beta}{2}\right) = 139.2$

and $\quad (\alpha + \beta) = 02° \ 40' \ 50''$

Thus, β = the corrected angle of depression at B
$$= 02° \ 40' \ 50'' - 01° \ 19' \ 01'' = 01° \ 21' \ 49''.$$

Let β' be the recorded angle of depression

then $\quad\quad\quad\quad\quad \beta' + 48'' = \beta$

therefore $\quad\quad\quad\quad\quad \beta' = 01° \ 21' \ 49'' - 48''$

$$= \mathbf{01° \ 21' \ 01''}$$

(*b*) $\quad\quad\quad\quad$ Now $c/2 = 01' \ 38'' = 98''$

Hence $\quad \left(\dfrac{\alpha + \beta}{2}\right) = 01° \ 20' \ 25'' = \alpha + c/2 - r$

and $\quad -r = 01° \ 20' \ 25'' - (01° \ 19' \ 01'' + 98'')$

Therefore $\quad \mathbf{r = 14''}$

Slope length AB (Fig. 4.18(*a*)) can be measured directly from A with instruments described later in Chapter 5.

From Fig. 4.18(*a*) we have

$$\frac{BC}{\sin\ (\alpha + c/2 - r)} = \frac{AB}{\sin\ (90 + c/2)}$$

Therefore $BC = AB\ \dfrac{\{\sin \alpha \cos\ (c/2 - r) + \cos \alpha \sin\ (c/2 - r)\}}{\cos\ c/2}$

$$\simeq AB\ \{\sin \alpha + (c/2 - r)\ \cos \alpha\}$$

assuming $c/2$ and r are small.

Similarly

$$AC \simeq AB\ \{\cos \alpha - (c - r)\ \sin \alpha\}$$

These two expressions may be used to achieve the reduction of measured slope length AB in certain of those instruments, so that direct readouts of BC and AC are presented to the observer.

> **Example 4.2** The slant range BA was measured by EDM as 1877.48 m, whilst the angle of depression from B to A was found to be $1°57'12''$, eye and object corrections having been made. Determine the reduced level of A given that of B as 279.73 m. Take the radius of the earth to be 6 380 248 m and $r = c/14$.

With reference to Fig. 4.18, in triangle ABD,

$$ADB = 90° - c/2$$
$$DBA = \beta + r - c/2$$
$$DAB = 180° - 90° + c/2 - \beta - r + c/2$$
$$= 90° + c - \beta - r$$

in which β is the measured angle of depression from B to A. Thus,

$$\frac{AB}{\sin(90° - c/2)} = \frac{AD}{\sin(\beta + r - c/2)} = \frac{BD}{\sin(90° - \beta - r + c)}$$

whence for small values of c and r,

$$AD = \frac{AB \sin(\beta + r - c/2)}{\sin(90° - c/2)} = AB(\sin \beta - (c/2 - r)\cos \beta)$$

$$BD = \frac{AB \cos(\beta + r - c)}{\sin(90° - c/2)} = AB(\cos \beta + (c - r)\sin \beta)$$

Now
$$AB \sin \beta = 1877.48 \sin 1° 57' 12''$$
$$= 63.99 \text{ m (trial for height difference)}$$
$$AB \cos \beta = 1877.48 \cos 1° 57' 12''$$
$$= 1876.39 \text{ m (trial for chord } BD)$$

It is common practice to assume that BD (or AC when using α) is the length of the chord between the verticals at mid-height of A and B when calculating c. In this example, it could be estimated to be at a reduced level of $(279.73 - 63.99/2) = 247.73$ m AOD.

Hence $R \sin c/2 = 1876.39/2$, where $R = 6\ 380\ 248$ m

$$\sin c/2 = \frac{1876.39}{2} \times \frac{1}{6\ 380\ 248} = 0.000\ 147\ 0$$

$$c/2 = 30.3''$$
$$c = 60.6''$$
and
$$r = 4.3''$$

Thus $c/2 - r = 26.0'' = 0.000\ 126\ 1$ rad
and $c - r = 56.3'' = 0.000\ 272\ 9$ rad

Substituting into the expression for DA above, we get

$$DA = 1877.48(0.034\ 085\ 5 - 0.000\ 126\ 0)$$
$$= 63.76 \text{ m}$$
$$\text{RL of } A = 279.73 - 63.76$$
$$= \mathbf{215.97 \text{ m}}$$

Similarly if required,

$$\text{chord } DB = 1877.48 \ (0.999\ 418\ 7 + 0.000\ 009\ 3)$$
$$= 1876.41 \text{ m}$$

The small change in length will have no influence upon the value of angle c as derived.

Optical distance measurement

Not only can the theodolite be used to measure heights and angles as described previously, but it can also be used to measure distance. This is achieved by means of *stadia lines* engraved upon the telescope diaphragm or by sighting a bar of known length, the *subtense bar*. The term *Tacheometry* is applied to this process.

In the past, specialist optical and mechanical devices were manufactured to speed up the booking and reduction of tacheometric observations; these have largely been replaced by electromagnetic distance measurement techniques discussed in Chapter 5.

The stadia system

The invention of the stadia principle has been variously ascribed to the Englishman, William Green, who described in 1778 the method of stadia measurement using two fixed wires; to the Dane, Brander, who between 1764—73 constructed the first glass diaphragms with fine lines cut on them and applied them to the measurement of distances; and to the Scot, James Watt, who in 1771 used a tacheometer of his own construction in Scotland. There is reason to believe that they worked independently and evolved substantially similar methods of surveying, though it is interesting to note that working in the field of astronomy the great Dutch scientist, Huygens, had constructed a simple type of micrometer eyepiece as long ago as 1659.

In Fig. 4.19 the field of view through the telescope of a typical theodolite shows the two additional horizontal hairs (or lines) engraved on the diaphragm, called stadia lines, intersecting the image of a staff. In effect, these lines define a fixed angle and the measurement process involves observing the staff intercept subtended by this fixed angle over the required distance. In addition to the normal levelling staff, shown

Fig. 4.19 View of staff through a typical telescope

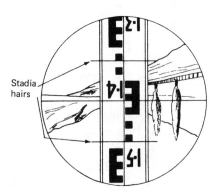

Stadia hairs

in Fig. 4.19, there are other types of stadia staves designed to give easier reading on long sights.

The basic principles are shown in Fig. 4.20 in which it is assumed that the telescope is level and the staff is vertical. Considering the rays which pass through the optical centre of the objective to form an image in the plane of the diaphragm, then by similar triangles

$$\frac{OX}{Ox} = \frac{u}{v} = \frac{AB}{ab} = \frac{s}{i}$$

Also, if $OF = f$ = focal length of objective,

$$\frac{1}{v} + \frac{1}{u} = \frac{1}{f}$$

whence
$$u = \frac{AB}{ab} \cdot f + f = \frac{f}{i}s + f$$

The horizontal distance D from the vertical axis of the tacheometer to the staff is obtained by adding the small distance c between the object glass and the vertical axis to u.

Therefore
$$D = \frac{f}{i} \cdot s + (f + c)$$

Thus if f, i and c are known and s is observed (i.e. $1.49 - 1.37 = 0.12$ in Fig. 4.19) D can be calculated.

The reduction of this formula would be simplified considerably if (a) the term f/i, the multiplying constant, is made some convenient figure, and (b) if the term $(f + c)$, the additive constant, can be made to vanish. Requirement (a) is solved by using values of f and i such that f/i equals 100. The second requirement was solved for external focusing telescopes by the Italian instrument-maker, Porro, in 1823 with a telescope incorporating an additional convex lens between the objective lens and the diaphragm. Note that this lens, which Porro called the *anallatic lens* (sometimes referred to as the anallactic lens), has not the same function as the concave lens used to focus the telescope in *internal*-focusing theodolites and levels. In this context, the term *anallatic point* can be encountered which may be defined as the point to which optical distance measurements are referenced. The above

Fig. 4.20

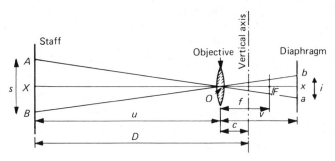

theory applies to those external focusing telescopes in which the eyepiece system moves. However, virtually no external focusing telescopes are in use nowadays, having been replaced by telescopes with internal focusing lenses. The theory developed above is not rigorous for these latter telescopes but the principles, and form of result for horizontal distance D, are the same. Reference can be made in this respect to the papers 'Telescopes anallactic or otherwise' by J.E. Jackson, and 'The design of anallactic surveying telescopes with a positive focusing lens' by S.B. Roy published in *Survey Review* and dated respectively April 1975 (No. 179) and July 1979 (No. 193).

With a modern internal focusing telescope set horizontally, horizontal distance D can be established using the straight-line formula

$$D = Cs + K$$

where C is the multiplying constant
  s is the staff interval intercepted by the stadia lines, and
  K is the additive constant.

The manufacturers usually arrange for C and K to be effectively 100 and 0 respectively in their instruments, the anallatic point being at the intersection of the three axes on a theodolite. In practice, K might be found to vary somewhat for short sights, e.g. less than 25 m, but unless great accuracy is required no correction is necessary.

There is no requirement to be able to move along the line being measured as with taping. All that is necessary is that the assistant, who carries the staff on which the instrument is sighted, shall be able to reach the various points to be surveyed and levelled, and that a clear line of sight exists between the instrument and the staff. For this reason, optical distance measurement is very useful in broken terrain and where obstacles such as rivers, roads and crops occur.

The fieldwork in tacheometry is rapid compared with direct measurement and levelling (the name derives from the Greek $T\alpha\chi\upsilon\sigma$ — swift and $\mu\epsilon\tau\rho\upsilon$ — a measure) and so it is still used to produce large scale contoured plans of areas for many engineering purposes.

Inclined sights

Although a stadia survey could be carried out with the telescope level, work would be tedious in broken and hilly terrain, and since it is on such ground that optical distance measurement comes into its own, we see that the basic formula $D = Cs + K$ must be modified to cover the general case when the line of sight is inclined to the horizontal.

From Fig. 4.21 where A, C and B are the readings given by the three lines with the staff held vertically, and A', C and B' are those which would be given if the staff were normal to the line of collimation,

$$D = C(A'B') + K$$
$$A'B' = AB \cos \theta \quad \text{(assuming } C\hat{A}'A = C\hat{B}'B = 90°\text{)}$$
$$= s \cos \theta$$

Therefore $D = Cs \cos \theta + K$
and $H = D \cos \theta$

Fig. 4.21

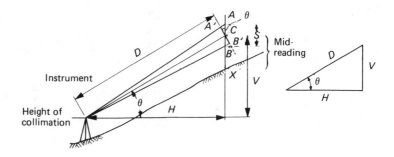

$$
\begin{aligned}
&= Cs \cos^2 \theta + K \cos \theta \\
V &= D \sin \theta \\
&= Cs \cos \theta \sin \theta + K \sin \theta \\
&= \tfrac{1}{2} Cs \sin 2\theta + K \sin \theta
\end{aligned}
$$

The additive constant in most modern instruments is essentially zero, so the above formulae reduce to

$$H = Cs \cos^2 \theta$$

and

$$V = \tfrac{1}{2} Cs \sin 2\theta$$

When booking the vertical angle θ, the following convention is used:

Elevation — i.e. sight uphill — θ positive;
Depression — i.e. sight downhill — θ negative.

The level at staff station X is given by:

(1) Sight uphill;
   Level at X = Height of inst. + V − mid-reading (CX)
(2) Sight downhill;
   Level at X = Height of inst. − V − mid-reading (CX)

The method of using the above formulae and booking readings is illustrated by the excerpt from the tacheometric book (Table 4.1).

The points, surveyed by the polar method of bearing and distance, i.e. radiation, Fig. 1.1(c), are plotted schematically in Fig. 4.22. At station A, the theodolite has been orientated by setting the horizontal circle to read zero whilst pointing at station B. A sight is then taken to a bench mark so that the height of the collimation axis and hence reduced levels can be calculated. The staff is then observed at all points of interest visible from station A with a final reading being taken on station B to check that the horizontal circle has not been accidentally disturbed. At this time, the reduced level of station B can be calculated and entered in the table so that heights may be determined from this point. Observations can now be taken at station B with station A as reference for both direction and height. The reference height of station A is found from the height of the collimation axis at A less the height of the instrument at station A ($27.38 − 1.56 = 25.82$ m).

As an example of the calculation, staff position 1 has been observed from A with $\theta = +3° 00'$ and $s = 0.75$ m so that

Fig. 4.22

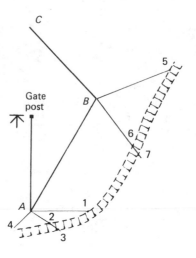

$$H = (100 \times 0.75) \cos^2 3° = 74.8 \text{ m}$$

and
$$V = (100 \times 0.75) \frac{\sin 6°}{2} = 3.92 \text{ m}$$

The reduced level of position 1 is therefore

$$27.38 + 3.92 - 1.41 = 29.89 \text{ m}$$

when we take the mid-reading to be 1.41 m.

Accuracy of stadia tacheometry

Three principal sources of error have to be considered: (1) staff readings, and hence staff intercept; (2) non-verticality of the staff; and (3) measurement of the vertical angle.

(1) Assuming a multiplying constant of 100, for each ± 1 mm uncertainty in the staff intercept there will be ± 100 mm uncertainty in the horizontal distance. As with levelling, the accuracy of staff reading decreases as the length of sight increases and for this reason the maximum length of sight should never exceed 150 m and for most purposes never more than 100 m. It is recommended that the lower reading be kept above 1 m to minimize the effects of differential refraction.

(2) The resultant error in the measured distance due to a tilt, δ, in the staff is dependent on both the inclination of the telescope and the length of sight. It is important that a circular target bubble be fitted to the staff to ensure that it is held as vertical as possible, but to keep any residual error within acceptable limits the inclination of the sight should never exceed ± 10° and the length of sight should be kept within the limits set above.

(3) Provided the normal observing procedure is followed, any error in the measured distance due to an error in the vertical angle will be small compared to other sources.

Table 4.1 Stadia Survey — Peel Park, Salford

Inst. stn. and Ht. of inst. axis	Staff stn.	Horizontal angle	Vertical angle θ	Stadia readings	$D = Cs + K = 100s$ (m)	Mid-reading	Horizontal distance $H = 100s\cos^2\theta$ (m)	Vertical distance $V = 100s\frac{\sin 2\theta}{2}$ (m)	$\pm V -$ Mid-reading (m)	Ht. of inst.	R.L. at staff	Remarks
A 1.56 m	B	00°00′00″	+ 0°20′00″	$\frac{2.425}{1.185}$	174.0	2.055	174.0	1.01	−1.04			
	$\overline{1}$	300°31′00″	0° 0′0″	$\frac{1.700}{0.475}$	122.5	1.085	122.5	—	−1.08	27.38	26.30	B.M. (n gatepost
	1	30°00′00″	+ 3° 0′0″	$\frac{1.780}{1.030}$	75.0	1.405	74.8	3.92	2.51		29.89	Bottom of bank
	2	65°00′00″	+ 6° 0′0″	$\frac{1.410}{1.135}$	27.5	1.270	27.2	2.86	1.59		28.97	Bottom of bank
	3	65°00′00″	+11° 0′0″	$\frac{1.455}{1.000}$	45.5	1.225	43.8	8.52	7.30		34.68	Top of bank
	4	176°30′00″	+ 5° 0′0″	$\frac{0.635}{0.260}$	37.5	0.450	37.2	3.26	2.81		30.19	Bottom of bank
	B	00°00′00″									26.34	
B 1.57 m	A	00°00′00″	0°00′0″	$\frac{2.965}{1.205}$	176.0	2.085	176.0	—	−2.09	27.91	25.82	
	5	188°45′00″	+ 4° 0′0″	$\frac{2.250}{1.240}$	101.0	1.745	100.5	6.96	5.24		33.15	Bottom of bank
	6	267°30′00″	0°0′0″	$\frac{1.200}{0.625}$	57.5	0.910	57.5	—	−0.91		27.00	Bottom of bank
	7	267°30′00″	+ 8°40′0″	$\frac{1.465}{0.700}$	76.5	1.080	74.8	11.40	10.32		38.23	Top of bank
	C	81°11′40″	+ 0°40′0″	$\frac{3.195}{1.660}$	153.5	2.425	153.5	1.78	−0.64		27.27	
	A	00°00′00″										

It is worth pointing out, as an example, that an observation error of 1° results in an error of 0.006 in $\cos^2\theta$ when $\theta = 10°$, i.e. a relative error of $0.6s$ when $C = 100$.

Stadia tacheometry is a very easy and relatively cheap method of optical distance measurement but, as has just been explained, its accuracy is severely limited. In normal use, it is generally accepted that a distance measured in this way will be good to approximately 1 in 500 and that height differences will be determined to the order of ± 40 mm.

Subtense tacheometry

Before EDM (Chapter 5) came into common usage it was possible to make accurate measurements of distance over rough terrain using a 1″ theodolite and a subtense bar. The technique using proprietary subtense bars is largely obsolete but is included here because the logic of the process could be used with any accurate horizontally mounted measuring bar should EDM be unavailable. It also has uses in industrial measurement systems (Chapter 6).

Fig. 4.23 Subtense bar (*Courtesy*: Leica UK Ltd)

Fig. 4.24

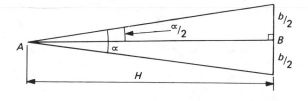

Figure 4.23 shows the Wild subtense bar with its levelling head mounted on a tripod. The outer steel casing is hinged at the middle and contains invar wires anchored there and tensioned by springs at the target ends: the target holders themselves are of brass.

Figure 4.24 shows the principle of the system. Length *AB* is required and accordingly the subtense bar has been set up normal to line *AB* using the directing telescope mounted on the bar near the hinge. A theodolite, preferably reading to one second of arc, has been set up at *A* and the horizontal angle subtended there by the targets on the bar at *B* is then measured. It will be appreciated that since the bar is mounted horizontally, refraction has an equal effect on both readings. Furthermore, since the angle subtended (α) is measured in the horizontal plane, the horizontal distance *H* obtains directly from

$$H = \frac{b}{2 \tan \alpha/2} \text{ where } b = \text{length of bar}$$

Thus, if the vertical angle θ is measured to the sighting mark on the hinge, the difference in levels is given by $V = H \tan \theta$.

Accuracy of subtense tacheometry

The accuracy with which the horizontal distance *H* may be computed depends on three factors, (*a*) the accuracy of the angle measurement, (*b*) the accuracy of the length of the subtense bar, and (*c*) the accuracy of the right angle at *B*; of these the influence of angle measurement is the most important factor.

$$H = \frac{b}{2 \tan \dfrac{\alpha}{2}} \simeq \frac{b}{\alpha} \text{ for small values of } \alpha$$

Therefore
$$dH = -\frac{b}{\alpha^2} \cdot d\alpha = -\frac{H^2}{b} \cdot d\alpha$$

the negative sign indicating that if α is measured larger than it should be, the distance H is reduced. Thus, if the error in angle measurement is $\pm 1''$, and the bar length is 2 m,

$$dH = \frac{\pm H^2}{2.000} \cdot \frac{1}{206\ 265}\ \text{m}$$

since $1'' = 1/206\ 265$ radian. The error in distance is proportional to the square of the distance, and so the best results are obtained at short distances, with a limiting minimum distance of 40 m say. At shorter distances than this, the error in angle measurement increases.

H (m)	40	50	75	100	125	150
dH (mm)	3.9	6.3	14.2	25.2	39.4	56.8
Fractional error	1/10 200	1/7940	1/5290	1/3970	1/3170	1/2640

To use the subtense bar for longer lengths involves either the subdivision of the line into bays of about 40 m, leapfrogging the bar and instrument at alternate ends of the bay or the use of an auxiliary base. In Fig. 4.25 an auxiliary base AB has been measured at right angles to the survey line AC. By measuring the angle β at C the length $AC = H$ can be computed.

Fig. 4.25

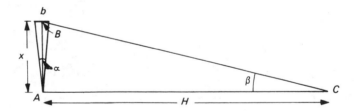

The length $AB(= x)$ should be about $\sqrt{2H}$, so that $\alpha = \beta$ or thereabouts; this gives the most favourable conditions for extending to AC, which may be up to 1000 metres long.

Errors are introduced in both the angle measurements, and if m_α and m_β are the mean errors in angle measurement, then, for small angles, writing

$$H = x/\beta$$

$$\text{error in } H = \pm \sqrt{\left(\frac{\partial H}{\partial x}\right)^2 (dx)^2 + \left(\frac{\partial H}{\partial \beta}\right)^2 (d\beta)^2} \text{ (see Chapter 8)}$$

$$= \pm \sqrt{\frac{1}{\beta^2} (dx)^2 + \left(\frac{H^2}{x}\right)^2 (d\beta)^2}$$

Due to errors m_α and m_β in angular measurement

$$dx = \pm \frac{x^2}{bK} m_\alpha \text{ (where } K = 206\ 265)$$

$$d\beta = \pm \frac{m_\beta}{K}$$

$$\text{error in } H = \pm \sqrt{\frac{H^2}{x^2} \frac{x^4}{b^2 K^2} m_\alpha{}^2 + \frac{H^4}{x^2 K^2} m_\beta{}^2}$$

$$= \pm \frac{H}{K} \sqrt{\frac{x^2}{b^2} m_\alpha{}^2 + \frac{H^2}{x^2} m_\beta{}^2}$$

Very long lines can be measured as in Fig. 4.25 but replacing the subtense bar by a series of auxiliary bases.

Example 4.3 An auxiliary base line *AB* was set out at right angles to a survey line *DAC* from its mid-point *A*. The angles subtended at *D* and *C* by that auxiliary base line were equal to the angle subtended at *A* by a 2 m subtense bar established at *B*. If the length of *DC* was then found to be 1800 m, determine the fractional error of the measurement for angular errors of $\pm 1''$.

Referring to the layout in Fig. 4.25, $AC = H = 900$ m

and
$$AB = x = \sqrt{2} H = \sqrt{2} \times 900 = 42.42 \text{ m}$$

$$m_{AC} = \pm \frac{900}{206\ 265} \sqrt{\left(\frac{42.42^2}{2^2} + \frac{900^2}{42.42^2} \right)}$$

$$= \pm 0.131 \text{ m} = m_{AD}$$

But $DAC = DA + AC$

Hence
$$m_{DC} = \pm \sqrt{[(0.131)^2 + (0.131)^2]}$$

$$= \pm 0.185 \text{ m}$$

The fractional error is thus $\dfrac{0.185}{1800} = \dfrac{1}{9735}$

---

1 (*a*) Describe a suitable procedure for setting up a theodolite over a ground station preparatory to observing horizontal and vertical angles, giving reasons for each step in the procedure.

(*b*) Give full details of two different routines for observing and booking horizontal angles and give examples of the sort of observations for which each routine might be appropriate.

2 A difference in level of 103.3 m was established by trigonometrical levelling between two stations, *A* and *B*. Assuming the radius of the earth to be 6367 km and taking the effect of refraction to be

one-seventh that of curvature, estimate (to the nearest second) the values of the angles measured at the two stations.

The effective distance *AB* was 10 090 m, the height of the instrument was 1.2 m at each station, and the heights of the signals at *A* and *B* were 3.8 m and 4.2 m respectively.

Answer: +33′ 53″, −36′ 39″.

3 (*a*) Derive an expression for the combined correction for curvature and refraction for non-reciprocal vertical angle observations.

(*b*) Vertical angle observations were made at a triangulation station *A* to a station of unknown height *B*. From the following data calculate the height of station *B*:

Height of station *A*: 252.45 m
Observed vertical angle at *A*: 1° 29′ 30″ elevation
Distance *A* to *B*: 12 712 m
Height of instrument at *A*: 1.56 m
Height of signal at *B*: 2.94 m
Coefficient of refraction: 0.07
Radius of earth: 6.356×10^6 m.

Answer: 593.1 m.

4 The following readings were taken by a theodolite set up at station *B* on to targets established at stations *A* and *C*.

	Horizontal circle		Vertical circle
Pointing	FL	FR	mean
A	25° 24′ 30″	204° 14′ 20″	+52° 12′
C	144° 43′ 50″	323° 33′ 30″	−20° 15′

Estimate the corrected readings of the horizontal circle in the face right case.

Answer: 204° 14′ 06″, 323° 33′ 21″.

5 In a tacheometric traverse, of which the first three stations are *A*, *B* and *C*, at station *B* it is found that the instrument height had not been measured above the peg *A*. Find from the tabulated readings the missing instrument height and the reduced level of the ground at *C*, if that at *A* is 83.44 m. The instrument constants are 100 and 0.

Station	Point	Vertical circle reading	Stadia readings	Height of instrument (m)
A	*B*	+5° 42′	2.43; 2.07; 1.71	?
B	*A*	−5° 24′	1.68; 1.34; 1.00	1.28
B	*C*	−5° 24′	1.68; 1.44; 1.20	1.28

Answer: 1.38 m; 85.2 m.

6 Observations were made from station *A*, reduced level 463.55 m, to station *B* and *C* as follows.

Instrument station	Station	Vertical angle	EDM slant range
A	B	$+2°\ 01'\ 16''$	2476.84 m
	C	$-1°\ 41'\ 37''$	1531.33 m

Eye and object corrections were applied to all measurements in the table. Determine the difference in level between B and C assuming $r = 0.07c$ and the radius of the earth to be 6370 km.
Answer: 132.42 m.

7 The gradient of the line joining two stations A and B was known to be 1 in 11. A conventional tacheometer having a multiplying constant of 100 and zero additive constant was set up at A and an angle of depression of $5°\ 00'$ was noted on the vertical circle when the reading of the lower stadia line on a staff held vertically at B was 1.000 m. If the height of the instrument axis at A was 1.210 m, estimate the other staff readings and deduce the horizontal distance from A to B.

(*Salford*)

Answer: 1.656, 2.312 m, 130.2 m.

8 The following readings were taken on a vertically held staff. Calculate the horizontal distance between the theodolite and the staff, and the elevation at the staff station, if the height of the instrument axis is 37.36 m above datum.

Vertical angle	Staff reading
$+4°\ 13'\ 30''$	1.00
$+5°\ 58'\ 20''$	3.00

Answer: 65.02 m, 41.16 m above datum.

9 (*a*) Describe a method of tacheometry using a horizontal 2-metre subtense bar and a theodolite reading directly to one second, giving (i) the merits of the system; (ii) the conditions under which the system would be used; and (iii) the factors affecting the accuracy of the system.

(*b*) How can a distance of 600 m be measured with an accuracy of about 1/10 000 when the error of angular measurement is $\pm 1''$?

(*c*) The length of a line $ABCDE$ was measured by establishing auxiliary base lines at B and D respectively. The lengths of these lines, which were perpendicular to $ABCDE$, were determined by means of a 2 m subtense bar, positioned such that the angles subtended by the auxiliary bases at A, C and E were equal to those subtended by the subtense bar at B and D. Given that $AB = BC = CD = DE$ and assuming that the standard error of angular measurement was $\pm 1''$, determine from first principles the fractional error in the deduced length of 4000 m for AE.
Answer: (*b*) Subdivide into 6 bays, (*c*) 1/13 072.

10 Three collinear ground control points X, Y and Z have the following co-ordinates:

	E (m)	N (m)
X	285.00	320.00
Y	295.00	360.00
Z	302.50	390.00

A tacheometer was set up at station Q, eastwards of XYZ, and the following information booked:

Station	Staff station	Horizontal angle	Vertical angle	Stadia	Readings	
Q	X	00° 00′	−02° 00′	1.000	x_1	x_2
	Y	19° 43′	00° 00′	1.000	y_1	y_2
	Z	42° 52′	+03° 00′	1.000	z_1	z_2

If the multiplying constant and the additive constant of the tacheometer were 100 and 0 respectively, estimate the values of $x_1, x_2, y_1, y_2, z_1, z_2$, obtained on a staff held vertically.
Answer: $x_1 = 1.526$ $x_2 = 2.053$
$y_1 = 1.390$ $y_2 = 1.780$
$z_1 = 1.340$ $z_2 = 1.679$

11 A 2 m long subtense bar was placed above station B and the angle subtended at station A was read as 2° 40′ 20″. Intermediate level information was later recorded using a theodolite, with tacheometric constants 100 and 0 at station C, and a staff held normal to the line of sight. The following data was recorded onto stations A and B:

Sighting	Horizontal circle	Vertical circle	Bottom	Mid	Top
$C \rightarrow A$	0° 0′ 00″	−5° 10′ 00″	1.459	1.649	1.839
$C \rightarrow B$	80° 24′ 20″	+10° 23′ 00″	—	1.235	—

What is the difference in level between A and B?

(*Salford*)

Answer: 8.879 m.

5 Electromagnetic distance measurement

Traditionally, accurate distance measurement was the most difficult part of a surveying operation but the introduction of Electromagnetic Distance Measurement (EDM) has completely revolutionized this. Nowadays most surveyors and civil engineers will work with electronic microprocessor-controlled instruments that can measure long distances to within a few millimetres at the press of a button. Two groups of instrument can be identified, namely the electronic, or microwave, types and the electro-optical types nowadays normally working with an infra-red beam. In each case the velocity of the electromagnetic wave in air must be known precisely for the accurate determination of distance.

In 1926 Michelson determined the velocity of light as 299 796 km/s by measuring the time taken for light to travel between two concave mirror systems. An eight-sided drum was positioned at the principal focus of one system, each side being a plane mirror, and the drum was rotated until a steady image of a source of light was seen in an eyepiece. This occurred at 528 revolutions per second thereby implying a travel time of 1/4224 s. The US Coast and Geodetic Survey had established the length of line over which the light travelled, and they appreciated that the method could be used in reverse to measure distance. However, it was never adopted directly for that purpose since rotational methods are limited by mechanical considerations.

The development of distance measuring instruments tended to follow closely developments in physics and electronics technology; for instance a much faster rate of interruption of light was obtained non-mechanically by means of the Kerr Cell. If two flat conducting electrodes are placed in nitrobenzine the liquid becomes doubly refracting when an electric potential difference is established between the electrodes. On placing the cell between crossed Nicol prisms forming a polarizer and analyser, some light will be transmitted when the potential difference is applied, since a component of vibration is now produced in the plane in which the analyser transmits, so that light emerges from the analyser. By means of an oscillator, a high rate of interruption can be applied to this emergent light. The use of this technique was suggested in 1929 for distance measurement, and later the Swedish physicist Dr E. Bergstrand used it for investigations into the velocity of light and also for the measurement of distance by the Geodimeter, manufactured by AGA of Stockholm. This was the first commercial instrument to be introduced and it belongs to the electro-optical group. The velocity of the electromagnetic wave *in vacuo* is accepted at the present time to be 299 792.5 km/s.

The Kerr cell has given way to the KDP-crystal which has a similar effect in that the plane of polarization of polarized plane light is rotated, whilst the development of the gallium arsenide diode has led to the introduction of many short-range instruments in the electro-optical group. Other significant changes have arisen with the introduction of the semiconductor and the microprocessor.

Electromagnetic distance measurement systems

Essentially the instruments mentioned in this chapter consist of a transmitter, set up at one end of the length to be measured, sending out a continuous wave, to the receiver at the other end. This wave, termed the carrier wave, is then modulated and the length determined as explained later.

Choice of frequency

This is a fundamental problem in EDM systems. The electromagnetic spectrum is continuous from visible light with frequencies of the order of 10^{14} Hz, corresponding to wavelengths of the order of 10^{-6} m, to long radio waves with frequencies of 10^4 or 10^5 Hz, corresponding to wavelengths of the order of 10^4 m. The relationship between frequency and wavelength is shown in Fig. 5.1.

It is convenient to divide the instruments in current use into three distinct categories depending on the frequency of the carrier signal:

(1) low frequency radio systems with carrier frequencies of the order of 10^5 to 10^6 Hz (wavelengths of the order of 10^3 or 10^2 m),
(2) microwave radio systems with carrier frequencies of the order of 10^{10} Hz (wavelengths of the order of 10^{-2} m),
(3) visible and infra-red light systems with carrier frequencies of the order of 10^{14} Hz (wavelengths of the order of 10^{-6} m).

Generally speaking it is found that the lower frequency signals provide greater range but require larger transmitters, and being affected by the atmosphere are therefore less accurate for EDM purposes than those of higher frequency. However, for marine and air navigation and for much hydrographic work long range is vital, accuracy

Fig. 5.1 The elctromagnetic spectrum

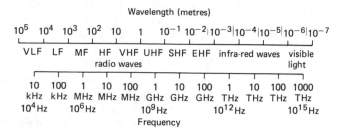

requirements are comparatively low and permanent or semi-permanent transmitters are appropriate. All these factors point to the use of low frequency signals and indeed many position-fixing systems operate in the low or medium frequency range.

For practical field instruments for engineering and land surveying the higher frequencies are most useful as the instruments can be made small and transportable and the propagation through the atmosphere is more stable. However, at these frequencies it is more difficult to measure the phase differences, as discussed later, and the wavelengths are so small that it is impractical to use directly the waves themselves for the measurements. The solution adopted is to modulate the high frequency carrier with a lower frequency wave and to use the modulated wave for measurement purposes.

Modulation is a process whereby certain characteristics of the carrier wave are varied or selected in accordance with another signal. The carrier signal does not have to be at a precisely determined frequency but it must be produced efficiently and in such a form that it can be modulated easily. The modulation signal, being the one used for the actual measurement, has to be an accurately controlled frequency.

Some instruments use an amplitude modulation, whilst others use a frequency modulation, the differences being indicated in Fig. 5.2, but the difference is unimportant from the point of view of the operator. In amplitude modulation the amplitude of the carrier wave is varied above and below its unmodulated value by an amount proportional to the amplitude of the modulation signal and the frequency of that signal. The amplitude of the carrier remains constant in frequency modulation

Fig. 5.2 Amplitude modulation and freqency modulation of a carrier wave

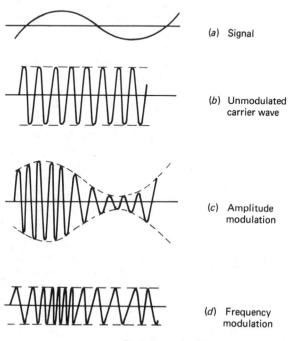

(*a*) Signal

(*b*) Unmodulated carrier wave

(*c*) Amplitude modulation

(*d*) Frequency modulation

but now its frequency is continuously varied by an amount proportional to the instantaneous amplitude of the modulating signal and at the frequency of that signal.

The term phase modulation may also be encountered. This can be described as an indirect method of frequency modulation, the phase of the carrier being modulated by a signal whilst the amplitude of the carrier remains constant.

Phase differences and distances

There are two options available in the use of electromagnetic waves for distance measurement, either pulse transit times or phase changes being measured. In the former case the unmodulated carrier is usually a pulse train or series of pulses, not the continuous wave referred to previously. When resolution of distance to 0.01 m or better is required the latter system is usually adopted, being based on the relationship between the transmitted and received signal.

Consider a transmitter sending out an oscillating signal at a constant frequency, f, to a receiver. Were the two touching, then the transmitted signal and received signal would be in phase, but as the receiver moves away from the transmitter the received signal will lag behind the transmitted signal due to the time of travel of that signal. Thus there will be a phase difference between the signals, and if the difference in phase between the signals at the transmitter and the receiver is measured the distance between them can be deduced. When that distance is equal to the wavelength the phase difference will be 2π and the signals will be in phase, as in fact they will be each time the distance apart is an integral wavelength. Therefore within an unknown distance, d, all that one measurement of phase difference will give is the residual part of d over and above an integral number of completed wavelengths.

It is not possible to compare instantaneously the phase of the signals at a transmitter and distant receiver. Therefore EDM systems adopt the technique of either retransmitting the signal back to the transmitter (microwaves) or reflecting the signal back to the transmitter (electro-optical) and making the phase comparison there. Thus it is always a double path which is measured.

The fundamental equation which relates slope distance to phase delay or phase difference may be written as

$$d = n\lambda + \frac{\phi\lambda}{2\pi} + a$$

where

d = double distance, i.e. total travel of wave

λ = modulation wavelength = $\dfrac{V_0}{\mu f}$

n = number of complete wavelengths within d

ϕ = phase difference between the outgoing and incoming signals

a = an additive constant related to geometrical and electrical eccentricities

Fig. 5.3

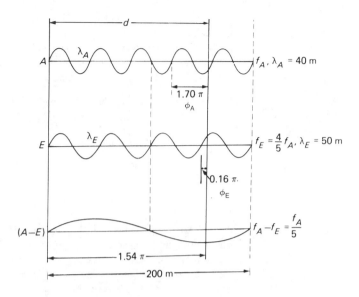

V_0 = velocity of the electromagnetic wave *in vacuo*
μ = refractive index
f = frequency

Generally n is unknown and d can be found by repeating the measurements of phase difference at frequencies differing from a 'fine' measuring frequency. In modern instruments this process is carried out automatically.

In Fig. 5.3 two modulation frequencies are shown with the phase delays then arising due to the distance d of travel. The wavelengths are such that five given by one frequency occupy the same length as four given by the other (lower) frequency. This particular length (200 m) is also covered by one whole wavelength whose frequency is equal to the difference of the two modulation frequencies. Moreover the difference in phase between the two is always equal to the phase of the wave given by the difference frequency. Thus, measuring the two phase delays consequent to travelling over a double distance up to 200 m is equivalent, on subtraction, to measuring the phase which would have been given had the difference frequency been applied. A phase difference of 2π applies to a distance of 200 m, and so if a phase difference of 1.54π be deduced on subtraction, a double distance of $1.54\pi \times 200/2\pi$ m is involved, i.e. 154 m. Naturally when a distance in excess of 200 m is being measured, this particular difference frequency gives an unknown number of whole lengths of 200 m plus a part length of 200 m. Thus a set of different frequencies has to be applied, which when related to the basic 'fine' frequency allows stages of double distance such as 2000 m, 20 000 m and 200 000 m to be evaluated without ambiguity.

In the measurement illustrated the phase difference with the lower wavelength was 1.70π and since this corresponds to $1.70\pi \cdot 40/2\pi$, i.e. 34.0 m, the distance measured must therefore have been $(n \times 40 + 34.0)$ where n is an integer. The measurement with the second wavelength was made primarily to enable n to be *identified*.

It will be seen from Fig. 5.3 that,

$$d = n_A \lambda_A + \phi_A/2\pi \times \lambda_A$$
$$= n_E \lambda_E + \phi_E/2\pi \times \lambda_E$$

The reader should also note that if

$$\phi_A > \phi_E \quad \text{then} \quad n_A = n_E, \text{ as in Fig. 5.3,}$$

whilst if

$$\phi_A < \phi_E \quad \text{then} \quad n_A = n_E + 1$$

Accordingly values of n_A can be determined in the corresponding stages of double distance.

The figures in the illustration have been arranged to be exact, and so the distance calculated using the difference frequency works out exactly as 154 m, i.e. $(3 \times 40 + 34)$. In practice there may be small inaccuracies in the measurements and the result from the difference frequency may not tally exactly with the first measurement, e.g. if the phase in the second measurement had been 0.15π the result would have been $1.55\pi \cdot 200/2\pi$, i.e. 155 m but the distance was known to be $(n \times 40 + 34.0)$ and the result is close enough to enable n to be *identified* as 3 giving the distance as 154.0 m as before. This is an important point which should be well understood as it is this which is the reason for the unique accuracy of EDM. If the wavelength is correct and the integral number of wavelengths is correctly *identified* then the major part of the distance is determined without error; the only part of the distance which is measured is the residual part over and above an integral number of wavelengths.

Effective wavelength

It has been mentioned above that the length of the double path is always measured by EDM systems.

To measure a distance D with a signal of wavelength λ and a double distance $d = 2D$ with a signal of wavelength 2λ gives an identical result in respect of both numbers of complete wavelengths and residual phase difference. Instead of determining the actual double distance travelled by the signal and then dividing it by two, it is convenient to use an *effective wavelength* of half the true wavelength and thence to calculate the single distance directly. By using effective wavelengths of $\lambda_A = 20$ m and $\lambda_E = 25$ m the same single distance of 77 m would have been derived.

In practice to achieve high precision λ_A is kept comparatively short, 10 m being a common value.

Precision is determined *also* by the accuracy of measurement of ϕ. Some instruments can resolve to only 1% phase, others to 0.1% or better:

Phase measurements may be achieved in various ways, i.e.

(1) by a resolver which consists of a stator and rotor, the latter being placed into an angular position with respect to the former corresponding to the phase difference between the transmitted and returns signals;
(2) by a variable light path;
(3) by a digital system in which the transmitted wave when passing through zero voltage activates a counter which counts pulses of a selected frequency until stopped by the returned wave.

It is recommended that reference be made to C.D. Burnside, *Electromagnetic Distance Measurement*, third edition (Granada Technical Books) for more detailed information on this and other aspects of the subject.

In modern instruments the phase difference between the outgoing and incoming signals is not measured at the operating frequencies but is transformed to a corresponding difference at much lower frequency. This greatly improves the accuracy of measurement of phase difference such that resolution to one thousandth part of a cycle can be readily obtained.

Microwave systems

The original instruments were developed in South Africa by Dr T.L. Wadley who gave a comprehensive account of the development in *Survey Review*, vol XIV, Nos 105/106. These systems held an important position in land surveying as they were used to measure distances from, say, 50 m to at least 50 km. However, the introduction of global positioning systems (GPS), as discussed in Chapter 7, has diminished that importance, nevertheless they are worthy of inclusion in view of the part they have played over many years.

As shown in Fig. 5.4, one instrument, the Master, is set up at one end of the distance to be measured and a second, the Remote, is established at the other end. An operator is required at each, inter-communication being possible by means of built-in radio telephones. A modulated signal is transmitted from the Master, received by the Remote and transmitted back to the Master where the phase difference between the transmitted and returned modulation signal is measured and displayed. The display is usually calibrated to read out directly in metres and in most instruments the operators carry out a simple sequence of switching, and successive parts of the distance are read and recorded.

The carrier wave is typically about 10 GHz (i.e. of wavelength say 0.03 m) and the measurement is carried out using a frequency modulation of about 7.5 or 15 MHz. The effective modulation wavelength is typically 10 m and, with a phase resolution of 0.1% phase this gives a basic precision of about 0.01 m.

The first model microwave Tellurometer, MRA1, was demonstrated in the UK in 1957 and, although new models have appeared at intervals since then, the same basic measuring principle is still in use. This is illustrated in Fig. 5.4 with modulation frequency f_1 being transmitted

Fig. 5.4

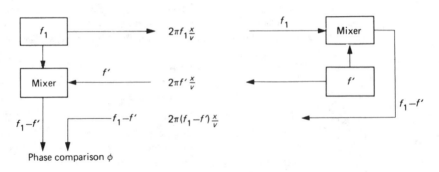

Phase comparison ϕ

on a carrier from the Master and modulation frequency f', 1 kHz lower, being returned from the Remote. If the instantaneous phases of f_1 and f' be given by ϕ_1 and ϕ' respectively, then it can be seen that the final phase comparison ϕ at $(f_1 - f')$ will be given by:

$$\phi = \left(\phi_1 + 2\pi f_1 \frac{x}{v} - \phi' + 2\pi (f_1 - f') \frac{x}{v} \right) -$$

$$\left(\phi_1 - \left(\phi' + 2\pi f' \frac{x}{v} \right) \right)$$

$$= 2\pi f_1 \frac{2x}{v}, \text{ where } v \text{ is the velocity of the signal.}$$

f' is eliminated from the equation and the system behaves identically to one in which the double distance phase delay is measured for f_1 transmitted from the Master to the Remote and back again.

MRA 101

One of the models of the Tellurometer which has been much used is the MRA 101 and, although superseded by later models, it may be instructive to consider some details of it in view of the previous discussion in this chapter. It uses a tunable klystron to generate a carrier wave with a frequency which can be varied from 10.05 to 10.45 GHz. Five different modulation frequencies (patterns) are applied to this carrier using the output from a crystal-controlled oscillator, each modulation being produced by a separate crystal. The crystals are in a thermostatically-controlled oven which has to be warmed up to its proper operating temperature before the oscillations are produced at the required frequency. The basic modulation frequency is 7.492377 MHz which gives an accurate 40 m wavelength at the design refractive index of 1.000325; this implies an effective wavelength of 20 m. The phase difference is measured in a rotary phase resolver and read on a graduated dial attached directly to the rotor. A vernier is provided which enables the reading to be made to 0.001 rotation, i.e. to 0.001 phase. By a simple mechanical device the first reading is arranged to be the sum of the phase differences measured using the direct modulation and a second measurement made with the modulation electrically reversed; this results in one rotation of the dial corresponding to 10 m so that the reading of the dial gives the residual part of the distance over and above integral multiples of 10 m and one unit of readout represents 0.001 rotation, i.e. 0.01 m.

The operating procedure is designed so that successive readouts are of the difference in the phase difference of the basic modulation and the phase difference of each subsidiary modulation. With an appropriate choice of frequency for the subsidiary modulations the readout is successively of the residual distance over and above integral multiples of 100 m, 1000 m, 10 000 m and 100 000 m and one digit of the readout represents 0.1 m, 1 m, 10 m and 100 m successively. As the maximum range of the instrument is some 50 km, the whole measured distance is completely determined. Each of the readouts is likely to contain errors because each one is arrived at by taking the small difference between two very similar quantities. However, as already explained, this does not matter because each readout is used to *identify successive digits* of the total distance and the results are sufficiently accurate to enable the *identification* to be made *without ambiguity*.

For example, when using the MRA 101 to measure a particular distance, the successive readouts on the dial were 892, 291, 630, 068 and 007.

Thus the first reading represented 892 $\times$ 0.01 = 8.92 m
second   291 $\times$ 0.1   = 29.1 m
third   630 $\times$ 1   = 630 m
fourth   068 $\times$ 10   = 0680 m
fifth   007 $\times$ 100 = 00700 m

The first reading gave the best value for the distance it represented, i.e. 8.92 m, and thus can be accepted for the last three figures of the distance. Allowing for slight inaccuracies in the second reading and adjusting it to conform to the first, the integer for the 'tens' figure in the distance must be 2, i.e. the last four figures of the distance are 28.92. Similarly the next integers are deduced as 6, 0 and 0 so that the final value for the measured distance is 00628.92. In all instruments where successive digits are evaluated the 'break-out' of the readings is made in this common sense fashion. It is customary for all readings after the first to record the first two digits of the readout only as these are sufficient for the identification of successive digits.

Once these readings, usually called the *Coarse Readings*, have been taken, a series of measurements of the first reading is taken so as to refine its value and these are called the *Fine Readings*.

Although the signals will penetrate mist or rain they are affected by atmospheric conditions and are particularly sensitive to changes in relative humidity. Meteorological observations are required for the best quality work so that the refractive index obtaining at the times of the observations is established: the instruments are calibrated to read distance for a given value of that index. Wet-and-dry-bulb temperatures and barometric pressures are obtained to revise that value to improve accuracy. Line-of-sight conditions are needed and grazing rays should be avoided since they lead to disturbed readings: accordingly station selection is important. The MRA 101 will give satisfactory observations under reasonably adverse atmospheric conditions but the best results are obtained in gentle sunshine with light breezes and low relative humidity.

As the signals are radio waves, they cannot be very highly collimated, so that a beam spread of 5° to 10° is normal. This is advantageous since there is some reception even when the Master and Remote are not exactly aligned. Once contact has been established the optimum orientation can be found by rotating the instruments to maximize the signals. Operation is also possible in high winds which cause the instruments to vibrate and create conditions which would be impossible for the more highly-collimated instruments.

On the other hand the spread of the beam and the wavelength of the carrier are such that stray reflections from the ground or other objects, i.e. vertical faces, smooth water and parked cars near to the measuring line, are common. These interact with the main signals and cause phase changes which give errors in measurement. This is referred to as ground swing, for which the remedy is to take a series of measurements of the distance using fine readings on slightly different carrier frequencies and then to mean the results. The spread of such readings indicates the magnitude of the error.

Although the MRA 101 is a robust instrument, functioning well in tough conditions, the above factors and its high power consumption make it tedious to use. A good working figure for its accuracy is $\pm D \times 10^{-5}$ for distances of 5 km or more.

Electro-optical instruments

These are the instruments most likely to be used by the civil engineer. They measure lengths from a few metres to 1 km or more, and some in fact can measure up to 60 km. As with the microwave instruments, all need line-of-sight conditions.

The main components of instruments in this group are (1) a light source, visible light, produced by a tungsten lamp, Xenon flash tube or laser light, or infra-red light, (2) a light modulator, (3) optical parts for transmitting and receiving the modulated light, (4) photomultiplier and phase meter, and (5) readout unit. In addition a passive reflector system is needed, usually a retro-directive prism, at the Remote station.

The daylight range of instruments having tungsten light-sources is much reduced when compared to the microwave instruments, since their radiation has to compete with that of the sun. Consequently, instruments achieving the longer ranges require a laser as the light source. In addition the laser can be projected in narrow beams, eliminating 'ground swing' whilst its light is monochromatic and coherent. The laser can be taken as analogous to the oscillator of a radio transmitter, except that the oscillations are produced at optical frequencies which are much higher than radio frequencies.

Most short-range instruments were introduced after the development of the luminescent gallium arsenide diode. This diode emits an infra-

Fig. 5.5 Prism on a detail pole

red carrier beam which can be amplitude modulated directly since the intensity of light emitted is proportional to the current fed to the diode. The infra-red beam is reflected at the Remote station, normally by a prism (Fig. 5.5), and returned to a photodiode embodied in the instrument at the Master station where it is processed into an electrical signal. The instruments are small and light and some are designed to be attached to theodolites so that angles and distances may be measured with the same instrumental set-up. Such instruments are designed as one unit, and in most cases they cannot be used without their appropriate theodolite although the theodolite can be used without the EDM.

Some instruments have coaxial transmitters and receivers (Fig. 5.6), whilst in others these two components are side by side or above and below the telescope. For those with coaxial systems the reflectors used are standard corner cubes, but for side-by-side systems special reflectors which separate the transmitted and reflected signals by an appropriate distance may be necessary. For all instruments the range achievable using one prism is limited and to measure up to the full range an assembly of several prisms is needed. As far as possible a particular instrument should always be used with the same set of prisms since different prisms can introduce different corrections to the measured distances, as discussed later.

The reader should be aware that all of the EDM systems described here measure the slope distance (S) along the carrier wave. If the instrument displays horizontal distance (H) this is because it has measured the vertical angle (θ) and performed the calculation

$$H = S \cos \theta$$

In a similar manner, the vertical difference in level (V) comes from the calculation

$$V = S \sin \theta$$

where the value of V is the difference between the height of collimation

Fig. 5.6 Principles of the MA100 coaxial optical system (*Courtesy*: Telumat UK Ltd)

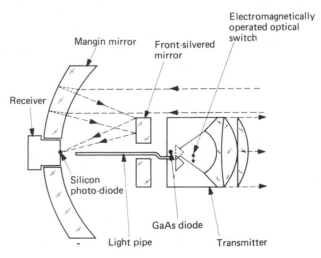

Mangin mirror

Front-silvered mirror

Electromagnetically operated optical switch

Receiver

Silicon photo-diode

GaAs diode

Light pipe

Transmitter

of the instrument and the height of the prism. It is common practice, when detail surveying or setting out, with the prism on an adjustable detail pole (Fig. 5.5), to set the detail pole to the height of the instrument so that V is the direct difference in ground level. However, if this is not done then the following correction calculations will be necessary.

> **Example 5.1** A slope distance of 204.834 m was determined by EDM, the instrument and prism having been set up 1.645 m and 1.873 m respectively above their stations. The vertical angle between the stations was later measured as $2°\ 17'\ 40''$, the instrument station being the higher of the two. Calculate the horizontal length of the line.

Fig. 5.7

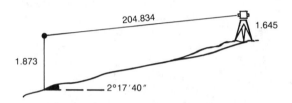

In Fig. 5.7, had the height of the instrument and prism above the ground stations been the same then the measured slope distance would have been parallel to the line between the stations. However, since the prism height is greater than the instrument height the line of sight along which the slope distance was measured is inclined upwards from the line connecting the stations. Thus,

$$\text{angle of inclination} \ = \ \frac{1.873 - 1.645}{204.834} \ \text{radians}$$

$$= 0.001131 \ \text{radians}$$
$$= 03'\ 50''$$

Hence the line of sight was inclined at an angle of $(2°\ 17'\ 40'' - 03'\ 50'')$ to the horizontal, i.e. $2°\ 13'\ 50''$.

Thus, the reduced horizontal distance between the stations is

$$204.834 \cos 2°\ 13'\ 50''$$

$$= \textbf{204.679 m}$$

Many types of EDM can be used over substantial distances where the effect of the curvature of the earth and refraction cannot be ignored. This topic is treated in greater detail in Chapter 4 under trigonometrical levelling.

The waves used by these instruments are not so seriously affected by the refraction of the atmosphere as microwave instruments and for many purposes it is unnecessary to measure the atmospheric conditions. However, the signals are greatly affected by mist or rain and in poor visibility the operating range is greatly reduced. Moreover there is also the problem of vibration of the more 'highly-collimated' instruments due to wind etc., causing vibration.

There is a wide range and variety of infra-red instruments on the market. New models are frequently introduced and it is impossible in a book of this nature to attempt to describe all the instruments currently available. The main features of typical infra-red instruments will be described and salient details of the Kern Mekometer will be given since the latter exhibits a higher order of accuracy than most other available instruments.

'Add-on' EDM

The Geodimeter 220 (Fig. 5.8) is a small lightweight instrument which is normally fixed to a simple mount on the telescope of a theodolite. (Most theodolite telescopes can be fitted so that the mount does not affect the performance of the theodolite in any way.) The instrument has its transmitter and receiver side by side but is used with standard circular corner cube prisms, and the manufacturers claim a range of 2.3 km using one prism under conditions of clear visibility. The basic modulation frequency is 14.986 MHz which implies an effective wavelength of 10 m at the design refractive index of 1.000273.

Fig. 5.8(a) The Geodimeter 220 and Geodat 124 (*Courtesy*: Geotronics (UK) Ltd)

A built in sensor measures the vertical angle and inputs it to the instrument's microprocessor so that slant distance, horizontal distance and difference in height between the instrument and reflector can be displayed at will. A special optional accessory, known as the Tracklight, is available (Fig. 5.8(b)). This can be attached underneath the instrument and it provides a visible guide to the staff man to assist him to position the reflector within the spread of the beam. Once on line a tracking mode on the instrument updates horizontal distance every 0.4 seconds so that the instrument man is continuously aware of distance to the reflector even when it is being moved. Such a feature is of great value for setting out, but the instrument has a further important aid in this respect. A built-in microphone converts speech transmission into signals which are carried on the infra-red beam to the reflector, where a suitable receiver is mounted on the detail pole allowing the staff man to receive direct instructions from the instrument man.

Figure 5.8(a) shows the Geodimeter 220 as viewed by the instrument man, with a display selector and mode selector on the left and right respectively. A set option allows the adjustment of prism constants and atmospheric corrections, corresponding values being presented on the liquid crystal display. The ROE option (Remote Object Evaluation)

Fig. 5.8(b) The Geodimeter 220 showing Tracklight and 'Unicom' speech converter. (*Courtesy*: Geotronics (UK) Ltd)

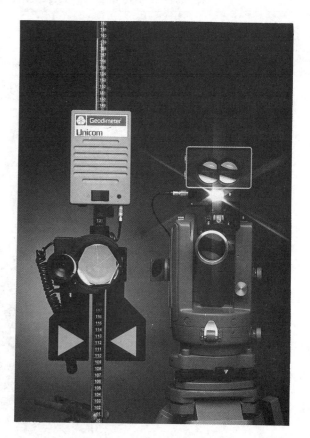

allows the setting out or measurement of heights of objects with a continuously varying display as the instrument is tilted. A distance accuracy of ±5 mm + 5 ppm is claimed for this instrument when measuring to a static reflector.

Kern Mekometer

Figure 5.9 shows the optical principles of the Kern Mekometer 3000 which, until being superseded by the Mekometer 5000, was the most accurate EDM instrument. It was designed at the National Physical Laboratory and the principles of the design are set out in articles by the designers, K. D. Froome and R. M. Bradsell in *Survey Review*, XXIII, No. 179. The commercial form, engineered and marketed by Kern, is a comparatively large and heavy instrument which has to be used as an entity, i.e. it cannot be attached to a theodolite.

It resolved distances to 0.1 mm and is mentioned in this section because its movable prism system illustrates the second method of phase measurement mentioned previously.

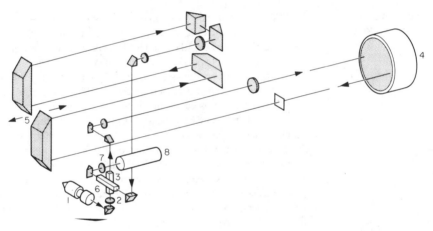

Fig. 5.9 Optical principles of Kern Mekometer ME3000 (*Courtesy*: Leica UK Ltd)

The carrier signal is produced by a xenon flash tube (1) and the modulation is produced by passing the signal through a polarizing filter (2) and a KDP crystal (3). The modulation frequency applied to the crystal has a frequency of 449.5103 MHz which gives a basic wavelength of 0.6000 m (effective wavelength 0.3000 m) under stipulated meteorological conditions. This modulation frequency is derived from a quartz cavity resonator which is connected to the atmosphere through soft bellows and this partially compensates for meteorological changes. The phase comparison is achieved by passing the returned signal (4) through a second KDP crystal (6) and polarizing filter (7). This second crystal is modulated by the same signal as the first and the second filter is crossed with respect to the first. As a result, if the return signal is exactly in phase with the transmitted signal it is completely demodulated and this condition can be accurately identified by a photomultiplier cell and a null meter. The condition is established by passing the return signal through a variable light path

(5) and changing the path length to achieve a null reading; the change in the path length is measured directly. Effectively this allows the residual part of the distance over and above an integral number of complete wavelengths to be measured directly on a linear scale.

An internal accuracy of 0.2 mm $+ D \times 10^{-6}$ was claimed for the ME 3000, and accuracies of that order were achieved, a major contributor being the fact that part distance was measured directly. However, the maximum slope distance, D, was of the order of 2.5 km with three prisms.

Now that model ME 5000 has been introduced an accuracy of 0.2 mm $+ 0.2 \times D \times 10^{-6}$ is claimed by the manufacturer and, since the xenon tube has been replaced by a helium-neon laser, distances of 8 km can be measured. The carrier wave is continuously modulated within a frequency range of 460 MHz to 510 MHz and distance is diaplayed to 0.1 mm. The emission power of the laser when leaving the instrument is stated as 0.3 mW which gives it a class 2 rating to BS 4803: Part 1: 1983 *Radiation Safety of Laser Products and Systems*.

Electronic tacheometers

An EDM mounted on a theodolite is a very convenient system for making tacheometric observations since distances and heights can be measured directly by the EDM and horizontal angles can be measured by theodolite, but most major instrument manufacturers now produce instruments which have been specifically designed for this type of work. One of the main problems in tacheometry is taking and recording the mass of readings involved, so these electronic tacheometers all have circles which are read electronically. In operation the observer points the instrument at the reflector and then at the touch of a button the circle readings and distances are recorded automatically. It is necessary for a keyboard to be provided so that identification data can be input with the readings, and most instruments also incorporate a microprocessor so that some reduction of the readings can be carried out on the spot if required. All these instruments enable a large number of points to be surveyed and the readings recorded in a short time with a small survey team. It is common practice to use the instruments with a portable data recorder like the Geodat 124, illustrated in Fig. 5.8a, so that data can be loaded directly into a computer for analysis, plotting, etc.

Total Station instruments

The most common type of electronic instruments now available are termed Total Station instruments. These incorporate a theodolite with electronic circles and an EDM. The EDM normally works concentric with the telescope eyepiece and is generally housed in a casing that forms part of the telescope. Figure 5.10 shows a Sokkisha Set 4C instrument. On this model the theodolite circles are accurate to 5" although other instruments in the same series can be graduated to 1" or 10" accuracy. They are automatically indexed by passing the telescope through horizontal after the instrument has been switched on. The horizontal circle can be switched to measure either clockwise or anticlockwise and the scales give a near continuous display as the instrument is turned. The theodolite incorporates a dual-axis liquid compensator that allows tilt up to 3' to be corrected before the angle is displayed. At larger tilts an error message appears on the liquid crystal display panel indicating that the instrument has been disturbed.

The EDM uses an infra-red carrier wave with two modulation frequencies to give a minimum range of 1 km with one prism, increasing to nearly 3 km with a bank of prisms. The manufacturer quotes the accuracy of distance measurement as ± 5 mm $+ 3$ ppm. The instrument can be used in a tracking mode for setting out where the distance is updated every 0.4 seconds with a reduced accuracy. The EDM measures slope distance and a microprocessor can calculate and display horizontal distance and difference in level. It also has a number of inbuilt trigonometric routines allowing calculation of co-ordinates, remote elevation and missing lines.

In addition to the functions contained in the onboard microprocessor, the instrument can interface with data loggers or other computers either to store survey data or to retrieve previously prepared setting out data. However, one of the new features of the latest generation of total station instruments is the provision of a removable memory card. This is about the size of a credit card and slots into the side of the instrument; typically each card has 32 K bytes of memory which can store or supply the data for about 500 points. This is a very convenient method of transporting and temporarily storing data. A memory card reader is needed to interface the card with a personal computer.

Fig. 5.10 Sokkisha Set 4C Total Station Instrument (*Courtesy*: Sokkisha (UK) Ltd)

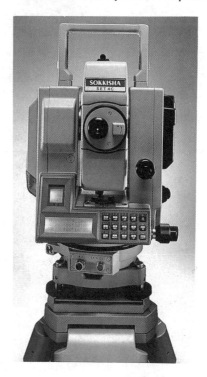

The first automatic instruments, Geodimeter 460 and Geodimeter 4000, use servo motors to facilitate pointing of the telescope. This is a major development in instrument technology and has allowed the Geodimeter 4000 to be marketed as the first commercial single operator instrument.

In these instruments, as in other Geodimeter total stations, the angles are measured by virtue of a magnetic field on solid state circles. Geotronics claim that the relevant measuring devices are of a completely new design and can be termed surface averaging electromagnetic high frequency devices, because the high frequency field is integrated over the complete circle. The measurement is such that eccentricity errors are eliminated and a standard deviation of less than 2 seconds of arc is quoted for any direction of orientation.

Fig. 5.11 Geodimeter
460 (*Courtesy*:
Geotronics (UK) Ltd)

Positioning of the line of sight of the Geodimeter 460, Fig. 5.11, in both the horizontal and vertical planes is controlled by precise servo motors, with friction clamps, so that not only can the surveyor obtain a precise pointing but the line of sight can be aligned in a calculated direction. A microprocessor is in control of the various functions and both an internal memory and an external memory (e.g. Geodat 400) may be utilized to store or recall up to 5100 data points.

Measurement of angles when either traversing or observing rounds is simplified because after one coincidence on the targets the servo motors take over on both faces for further pointings thus giving the surveyor precise confidence in the relocation of the target. A dual-axis compensator corrects for level errors and reveals them on the display; collimation and trunnion axis errors can be measured and stored in the memory for later record and analysis.

For setting out purposes a special version, Geodimeter 400CDS, is available. This has on-board software and an internal memory which can be loaded with either the X, Y and Z co-ordinates, or bearing and distance, of the point to be set out. Once on site the engineer calls up the point reference and the instrument automatically turns to the correct bearing. After the staff man has aligned the reflector with the instrument the difference between his position and the proposed point is shown on the display. The Geodimeter 400CDS has a wide measuring beam and includes the Tracklight and communication facility, detailed earlier for the Geodimeter 220.

Fig. 5.12 Remote Positioning Unit for the Geodimeter 4400 (*Courtesy*: Geotronics (UK) Ltd)

The Geodimeter System 4400 is an entirely new concept in that only one man is required to cover the whole survey or setting out operation. This man does not remain with the instrument at its station but moves as required with the reflector unit (the Remote Positioning Unit, RPU), (Fig. 5.12).

The station unit, Geodimeter 4400, is somewhat similar to the Geodimeter 460 described above, since its characteristics include servo controls for automatic pointing and the dual-axis compensator for automatic correction of trunnion axis tilt and collimation error. It can be used as a measuring instrument, like the Geodimeter 460, when it has a maximum range of 2300 m onto a single prism. However, it is equipped with an automatic search and aim function which can be switched on to locate the RPU.

Figure 5.13 shows the Geodimeter 4400 on the left together with the RPU, and it will be noted that there is telecommunication between the two. Contact between them is established immediately the RPU is switched on and all functions can be controlled from the remote unit

Fig. 5.13 Geodimeter
4400 and RPU
(*Courtesy*: Geotronics
(UK) Ltd)

thereafter. In addition to the telecommunication unit, the RPU consists
of a measuring rod, a prism, an internal battery and a computer recorder
with alpha numeric keyboard, which forms the control unit. Each unit
can have extended internal memory sufficient to store 2700 points or
a variety of computer programs for data analysis.

By following a predetermined search route the station unit is able
to find and lock onto the remote unit whose movement it will then track
automatically. When the RPU has been located the station unit can
take measurements under control from the RPU. All data is passed
by the telecommunication connection to the RPU computer where it
is stored and displayed. The display is mirrored at the station unit.
For accurate work the RPU must be carefully positioned vertically
above the point being measured.

There are two key uses for this type of system, detail surveying and
setting out. For the former the advantage to the surveyor is that he
can take direct control of the location of the RPU and enter a clear
description of the point using the keyboard, thus eliminating location
and identification errors. For the latter the location of the point is
quicker because all the engineer needs to do is scrutinize the display
on the RPU to determine the relative position of the unit to the intended
point. It is possible to set the instrument so that this task is performed
in a 'zeroing' manner in the X, Y and Z directions for a 'fix'.

The automatic search and aim function has a range of about 500 m,
effectively giving a covering of 785 000 m^2 though there is some
dependence upon atmospheric conditions and background radiation.
Geotronics quote an accuracy of ± 6 mm for a range of 100 m for
standard measurements, rising to ± 15 mm when fast tracking is being
carried out. The search function can be carried out at night and this
has been noted to be a distinct advantage for work in built-up areas
or near main roads where the level of background interference is higher.

The 'one-man' concept has operational and manpower cost advantages but problems may arise because the station unit is unmanned. This makes the instrument vulnerable to theft or damage by passing traffic, especially on construction sites.

Zeiss Eldi rangefinder

Figure 5.14 shows a self-contained instrument, the Zeiss Eldi 4, mounted on a theodolite. This is typical of a number of simple EDM units available on the market. The infra-red carrier signal has a wavelength of 860 nm and it will be noted that the transmitting and receiving optics are coaxial. A microprocessor (Fig. 5.15) controls the rangefinder signal, computation of mean distance values and, if connected to an electronic theodolite such as the Zeiss Eth 3 of Fig. 5.14, data transfer to allow reduction to horizontal distance and difference in level.

With a single prism the instrument will measure up to 1 km with a claimed accuracy of ± 5 mm + 3 ppm. With a bank of prisms a range of 2 km can be achieved. The instrument is designed as a low cost accessory to a theodolite for the routine measurement of distance where some of the fancier features of a total station instrument are not justified.

Fig. 5.14 Zeiss Eldi 4 Rangefinder (*Courtesy*: Carl Zeiss)

Fig. 5.15 Zeiss Eldi 4: detail of microprocessor (*Courtesy*: Carl Zeiss)

Time pulse instruments

There is an alternative type of EDM which works by a different principle of distance measurement. An infra-red laser carrier wave is transmitted from the instrument to the receiver and back again in the normal way, but instead of a continuous modulation and measurement of phase, the carrier is pulsed and the time that the pulse takes to travel out and back along the carrier is determined. Knowing the pulse velocity the time can be converted into a distance. The Zeiss Eldi 10 rangefinder and Wild Distomat DI3000 are examples of this type of equipment. Both are designed to fix on top of a theodolite and can work up to 7 km with a single prism and 16 km to a bank of prisms. One of the key advantages of this equipment is that it can measure to moving targets and could, for example, be used to control the position of machinery such as cranes and earth movers. In a tracking mode the Zeiss Eldi 10 can measure at less than 0.2 second intervals to an accuracy of ± 5 mm + 3 ppm. Its transmitter and receiver are mounted side by side, the former being a gallium arsenide infra-red laser diode and the latter a photodiode.

The Wild DIOR 3002 instrument (Fig. 5.16) is a special version of the DI3000 series, in that not only will it measure to a prism in the conventional manner but it can also be used over limited distance without a reflector. The operating range depends upon the reflectance characteristics of the target surface, together with its angle to the pulse train and the ambient light. A maximum range of 200 m is suggested by the manufacturer for normal light conditions when observing smooth, dry, light coloured surfaces. In the best conditions an accuracy between 5 to 10 mm can be expected.

Fig. 5.16 Wild DIOR 3002 reflectorless EDM (*Courtesy*: Leica UK Ltd)

For ease of location a small helium-neon laser can be fitted to the instrument as shown in Fig. 5.16. This projects co-axially with the infra-red laser and will thus mark the measurement spot.

The instrument has many uses, for example in the survey of tunnel profiles, rock or quarry faces, stock piles or complex structural shapes like cooling towers. It can also be used to measure moving surfaces like changing water levels for hydrographic surveys. (*See* Spence and MacPherson, 'Surveying rock slopes in difficult conditions', *Highways and Transportation*, May 1991).

Data recorders

It is now normal practice to record data in the field electronically for direct offloading into a computer for analysis and plotting. In the same manner, setting out data can be prepared in the office and fed directly into the instrument on site using the data recorder. Most manufacturers produce devices which range from memory cards to fully portable computers that can drive printers and plotters. Figure 5.8(a) shows the Geodimeter Geodat 124, which is primarily a data storage unit with basic computational software. Figure 5.17 shows a Zeiss REC 500 portable computer; this is a version of the Husky Hunter computer for which a selection of commercial software can be obtained. As well as the data logging function, a computer of this type is capable of full data analysis with printing, plotting or disc storage in the field using suitable portable peripheral devices. Information can also be transmitted to a computer in the office using telephone lines and suitable modems (Fig. 5.18).

Fig. 5.17 Zeiss REC 500 portable computer (*Courtesy*: Carl Zeiss)

Fig. 5.18

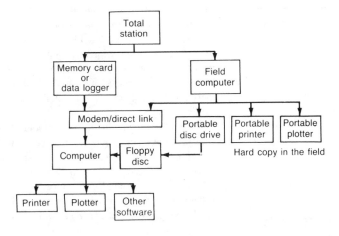

Sources of error   EDM measures the distance travelled by the electromagnetic wave from the transmitter to the reflector and the expression for this distance x established by the phase difference ϕ can be written as

$$x = K\phi + a$$

where K and a are constants. In general this is a slope distance and for most survey work reduction to the horizontal is required. Some instruments have a built-in facility to achieve this directly, otherwise one measures either the vertical angle or the height difference between the terminal stations. Either of the two measurements can be used to calculate a correction to be applied to reduce the measured slope distance to the horizontal (*see* Chapter 2 and Example 5.1).

Over long distances, such as might be measured using microwave instruments, the vertical angle should be determined by simultaneous reciprocal observations from both ends of the line, as single-ended observations are affected by the curvature of the earth and the refraction of the atmosphere.

Once the measurement has been reduced to the horizontal it is exactly equivalent to any other measured distance, and if necessary the distance should be corrected to sea level (Chapter 2) and also multiplied by the projection scale factor (Chapter 7).

The reader might also be referred to a correction required for the curvature of the ray path and the curvature of the earth. This correction should certainly be taken into account in geodetic surveying, but it does not amount to 1 part in 10^6 until the distance is nearly 24 km and for all ordinary purposes it can be ignored.

Apart from the above, the readings themselves are subject to error and certain other corrections may be needed. The principal sources of error are:

(1) error in instrumental constant
(2) error in the reflector constant
(3) error in frequency
(4) cyclic error

Instrumental and reflector constants   In most EDM instruments the signals travel over some distance internally during both transmission and return. The point from which the signal can be considered to be transmitted, the electronic centre, may differ from the geometric centre referred to when locating the instrument over a station. This gives rise to a constant, which must be applied to all distances measured with that instrument. But in addition there can be a discrepancy between the optical and geometrical centres of the reflectors.

It is preferable that the instrument and its associated reflector(s) be considered as a single entity so that a combined instrumental constant, a, can be evaluated. Its value can be specified by the manufacturer or may be adjusted by him to be zero.

The electronic components are very stable but even so, particularly if the instruments are used in extreme conditions, some might exhibit small changes with time, and accordingly it is advisable to check the instrumental constant periodically by taking careful measurements over known distances or alternatively using a series of collinear points as mentioned below. In any case the constant should be checked if the reflectors are changed: also it is good practice to check when very precise measurements are being made, e.g. for deformation studies. Since even reflectors from the same manufacturer can have slightly different corrections, the instrument should always be calibrated with the actual reflectors to be used.

Triangulation stations (Chapter 6) whose co-ordinates are known, or obtainable, can be used to provide a known distance. When comparing this distance all possible corrections must be applied to compare like with like, and it is suggested that to minimize refraction uncertainties calibration should be carried out over shorter lengths if available.

An alternative approach is to set out collinear points and then measure the difference between them. If these points be denoted as $ABCDEFG$, one possible set of observations would be:

$$(AG + a_1) = (AB + a_1) + (BG + a_1)$$

so that

$$a_1 = AG - (AB + BG)$$

a_1 is the value of the additive constant derived from those measures. Four further values can be obtained using the other intermediate points and so a mean value of the constant could be found. Alternatively AB, BC, etc. can be measured such that for n bays

$$a = \frac{AG - \Sigma AB}{n - 1}$$

Great care must be taken when setting up the instruments and reflectors over the respective stations for this method to give good results. Any change in the known constant leads to a zero error in distance and the infra-red instruments can be rather suspect in this respect. The electrical and optical signals are not coherent over the active area of the diodes so that there is a variation of phase between the input and output at points over those areas. Moreover the transmitted beam tends to spread so that the reflector only samples part of the radiation and that particular portion depends upon the alignment of instrument and reflector. Thus small phase errors and, consequently, distance errors can be produced. Also at the photo-diode phase delay between optical input and electrical output can differ, depending upon the angles of incidence of the returned light through the receiver lens on to the active area. If the whole of the lens is effective the average of the phase delays is constant and can be eliminated but if the lens is partially blocked asymmetrically, or perhaps if a non-standard reflector is used, then distance errors can arise.

Frequency error and atmospheric correction

The constant K depends upon the modulation frequencies and the refractive index. Usually f is precisely controlled but some drift may occur over time. Should any error be found it is suggested that a correction be applied to measured distance rather than attempting instrument adjustments. Note that distance is determined in terms of wavelength λ which is inversely proportional to μf, where μ is the refractive index and f the frequency. Thus distance is inversely proportional to f assuming μ remains constant.

Modulation frequency and refractive index can be selected for a direct readout of distance. For instance, if the basic modulation frequency is 14.985520 MHz, this implies an effective wavelength of 10 m at the design refractive index of 1.000274, corresponding to a pressure of 1013.25 mb and 20°C temperature. If other pressures and temperatures obtain, a proportional error will arise unless the value of the index is corrected accordingly.

Microwave instruments are affected more than electro-optical instruments by changes in μ. Since two operators are involved in the former system, wet-and-dry bulb temperature and barometric pressure are read at both stations, the correction being derived from the mean readings. In the shorter-range systems the readings need only be taken at the instrument station.

The respective expressions for refractive index μ for microwaves and light waves are given by the expressions

$$(\mu - 1)10^6 = 77.6 \frac{p}{T} - 13 \frac{e}{T} + 3.8 \frac{e \times 10^5}{T^2}$$

and

$$(\mu - 1) = 0.2696 \frac{(\mu_0 - 1)}{T} (p - 0.14 \, e)$$

in which p is the total pressure in millibars, e is the partial pressure of water vapour (in mb) and T is the temperature in degrees Kelvin (K); e is established from the difference in readings of the wet (T_w) and dry (T) bulbs such that

$$e = e' - 0.5 \, (T - T_w) \frac{p}{1006}$$

e' being the saturated pressure at T_w. Since T_w is difficult to determine exactly, it presents a source of uncertainty in precise distance measurement and it has been suggested that it is of major influence in the determination of the magnitude of the standard errors of distance measurements.

The refractive index of light is given by a different expression to that quoted for microwaves since the velocity of light is not independent of wavelength as is the case assumed, say, for microwaves of length 30 mm. The dry air index μ_0 increases from 1.0002908 to 1.0002982 as we progress from red to violet through the electromagnetic spectrum, T and p being 273 K and 1013.25 mb respectively. A group refractive index μ_g is taken for the electro-optical instruments in lieu of μ (see Bomford's *Geodesy*) and μ_{g0} replaces μ_0, such that $(\mu_{g0} - 1)10^6 = 287.604 + 4.8664/\lambda^2 + 0.068/\lambda^4$ in which λ is the carrier wavelength (in μm). Thus if $\lambda = 900$ nm (i.e. $0.900 \, \mu$m) $\mu_{g0} = 1.0002940$ and, on substitution in the first expression $\mu_g = 1.000274$ when $T = 293$ K, $p = 1013.25$ mb and e is zero. A change of 1 part per million in μ_g will induce a change of 1 part per million in $\nu (= V_0/\mu_g)$ of opposite sense, and this will be reflected in λ when f is fixed.

Cyclic error

A cyclic error is one whose magnitude depends on the actual phase difference ϕ which is being measured, i.e. on the residual part of the distance over and above an integral number of complete effective wavelengths. It is more likely to occur in microwave instruments and it is likely to be small. A possible source is that of the contamination between the two $(f - f')$ signals shown in Fig. 5.4. The mean of forward and reversed observations mentioned for the MRA 101 should eliminate it.

To test for the error it is necessary to set up a calibration base with a fixed position at one end and a movable one at the other. The latter position should be capable of being moved along the actual line being measured through a range of just over one complete wavelength. The instrument or the reflector, or in the case of microwave equipment either instrument, can be mounted at the movable position but for the sake of convenience in this discussion the fixed position will be referred to as the instrument position and the movable one as the reflector position.

The procedure is to warm up the instrument and then to measure a succession of distances with the reflector moved through a defined distance for each successive measurement, e.g. if $\lambda = 10$ m the reflector might be mounted on a rail 10.1 m long, and moved by 0.1 m between readings. A suitable substitute for the rail might be the top of a flat wall but with a little ingenuity a suitable base can always be found. The readings are carefully recorded but the higher multiples of the basic wavelength are ignored as they will be the same for all readings. One reading is now treated as a reference — it is convenient to take the reading of the shortest distance — and for each successive distance the increase in the measured distance is calculated. But the increase in the true distance is known as it is the distance moved by the reflector and so the 'error' (recorded increase less the known increase), is calculated. This is then plotted against the part of the reading which represents the residual part of the path to give a graph of the cyclic error.

A constant can always be added to or subtracted from the graph so as to 'reference' it at any desired residual distance. As the cyclic error correction must always be applied before evaluating the instrumental and reflector constant, this shift in the graph has no effect on the measured distance.

Accuracy of EDM

It will have been noted that typical accuracies of measurement have been given for various instruments in the form

$$\pm e \text{ mm } \pm p \text{ mm/km}$$

These are both standard errors and the overall standard error of a measurement is given by

$$s_D^2 = e^2 + [D \cdot p \cdot 10^{-6}]^2$$

where D is the distance measured.

For short distances such as 100 m it is clear that e is more important and p can be ignored, but when D is say 10 km then p becomes more important, e is largely dependent on the sensitivity of the phase resolver and other electrical components, and is therefore a function of instrument design. p is dependent on the accuracy of the modulation and on the accuracy of the value used for the refractive index μ. With good calibration f should be correct to better than 1 ppm but as indicated earlier the accuracy of the determination of μ is more difficult to assess. With microwave systems a change of 1 mb in the water vapour pressure gives an error of $+5.1$ ppm in distance, whereas with visible light systems a change of 1 mb in the water vapour pressure in the atmosphere can be ignored.

Thus over short ranges it is the value of e which determines the accuracy in the latter case and the atmospheric correction can be ignored. Even over longer distances or for the highest precision the measurements of temperature and pressure can be comparatively crude. If this is done the stated values for e and p can be used to estimate the precision of measurement of distance.

The same sort of considerations apply to microwave measurements but in this case the wet and dry bulb temperatures in particular must be measured to better than 1 °C so as to make a good estimate of the water vapour pressure. Even then a value of 5 is about the best which can be reasonably anticipated for p.

The influences of pressure and temperature become negligible when a 'two-colour' system of measurement is adopted, and although humidity can still have some effect it is likely to be less than 1 part per million. This principle has been used in the development of Georan 1 which embodies an argon-ion laser as the light source. Wavelengths of 458 nm (blue) and 514 nm (green) are used in the measurement of distance, there being one measurement system for each colour, these being transmitted and received together. The modulation method mentioned for the Mekometer has been adopted in this instrument (i.e. using KDP crystals) and the fundamental frequency of modulation is of the order of 500 MHz.

If d is the measured distance and μ is the refractive index we may write

$$x = \frac{d}{\mu}$$

where x is the required distance.

For the two colours

$$x = \frac{d_g}{\mu_g} = \frac{d_b}{\mu_b}$$

Put
$$x = x \left(\mu_g - \frac{(\mu_g - 1)}{(\mu_b - \mu_g)} (\mu_b - \mu_g) \right)$$

and then
$$x = d_g - K_g(d_b - d_g)$$
$$= d_b - (K_g + 1)(d_b - d_g)$$

in which
$$K_g = \frac{\mu_g - 1}{\mu_b - \mu_g}$$

d_g and d_b are the approximate distances measured using green and blue colours respectively. K_g is independent of atmospheric density and is weakly dependent upon the composition of the atmosphere, so it can be accurately determined from readings of temperature and pressure. It has a value of about 57 and accordingly d_b and d_g need to be determined precisely. Reference should be made to papers by G. Shipley and R.H. Bradsell in *Survey Review*, No. 179, for further details of the principles of this instrument. Another instrument, the Terrameter, operating on the same principle has been in use in America for some years, and has produced results consistent to better than 1 part in 10^7. Details are given in a paper by E.N. Hernandez and G.R. Huggett in *Proceedings of the 41st Annual Meeting of the American Congress on Surveying and Mapping*, 1981.

Exercises 5

1 Show that distance d in Fig. 5.3 can be determined by the expression

$$(\alpha_A - \alpha_E) \frac{\lambda_A \lambda_E}{\lambda_E - \lambda_A}$$

where α_A and α_E are the respective fractional parts of phase difference.

2 The standard errors of a single measurement of length by an instrument are given as ± 1.5 mm and ± 3 mm/km. What is the standard error of a measurement of 1500 m?
Answer: ± 4.7 mm.

3 Determine the additive constant of a certain electro-optical instrument from the following measurements of length.

Line	Length (m)	Line	Length (m)
AB	191.733	BG	313.466
AC	222.247	CG	282.953
AD	252.740	DG	252.461
AE	283.262	EG	221.940
AF	313.673	FG	191.526

ABCDEFG are collinear, and *AG* was measured as 505.091 m.
Answer: -0.109 m.

4 The carrier of a certain electro-optical instrument has a wavelength of 830 nm. Single distance to 10 m increments is displayed at standard conditions of 15°C and 1013.25 mb. Estimate the relevant modulation frequency, and the correction to be applied to a measured distance of 2736.84 m when the temperature is 18°C and

the pressure 746.5 mmHg. Assume dry air conditions.
Answer: 14.98544 MHz, +0.02 m.

5   The instrument whose behaviour was illustrated in Fig. 5.3 was
used to measure the lengths of lines XY and XZ, a third frequency
f_B being employed in addition to frequencies f_A and f_E. Calculate
the length of YZ given that $f_B = 49/50 \times f_A$ and that $XYZ = 59°$
41′ 30″.

Phase difference readings

Line	f_A	f_E	f_B
XY	1.535π	0.028π	1.384π
XZ	0.208π	1.367π	1.324π

Answer: 43.2 m.

6   An EDM whose carrier wavelength is 0.860 μm has a modulation
wavelength of 20.000 m, corresponding to a frequency of
14.98552 MHz at a specified pressure of 1013.25 mb. What is the
specified temperature?
Answer: 239°K or 20°C.

6 Survey methods

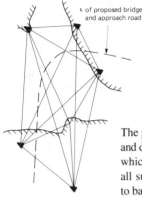

⟨ of proposed bridge and approach road

The previous chapters have described a variety of methods for angular and distance measurement and now in this chapter the methods of survey which employ these techniques are described. They are adaptable for all survey tasks whether it be the production of a site plan on which to base engineering design work, the setting out of roads and buildings or the control of aerial photography for photogrammetric plotting.

Co-ordinate systems

In all but the simplest surveys covering small areas, the relative positions of the control points are calculated in a co-ordinate system rather than directly plotted by scale and protractor. This is the most satisfactory method because:

(1) it enables errors to be assessed and adjusted,
(2) each station is plotted independently from precise calculations,
(3) it is not dependent on any angle-measuring device.

In plane surveying a system of plane rectangular cartesian co-ordinates is used to define the positions of points in plan. Naturally the axes of such a system have to be defined and it is usual in practice to adopt north and east directions for this purpose. The north–south axis is the principal direction, or reference meridian, to which bearings are related. This axis can be chosen from one of the following:

(a) The true meridian, or true north, i.e. the trace which a plane containing the north and south poles and the origin of the axes makes on the earth's surface. It may be located by methods described in Chapter 7.
(b) Magnetic north, which may be east or west of true north, again as mentioned in Chapter 7.
(c) National Grid north which is related to true north as explained in Chapter 7.
(d) An arbitrary direction, e.g. one selected survey line which is in a convenient direction.

Grid co-ordinates

The grid co-ordinates of a point P within this system are given by the perpendicular distances E_P and N_P from the two principal axes, at whose intersection the origin O of the system is located (Fig. 6.1). These are referred to as eastings and northings respectively, the former being quoted conventionally before the latter, but in (*d*) above *x* and *y* co-ordinates may be adopted if there is any likelihood of confusion.

The plan of a large area may take a number of individual sheets and, for continuity and convenience, a network of eastings and northings lines may be drawn on the plan at selected intervals and the whole system becomes a grid system.

The origin is usually located at the extreme south and west of the area so that all co-ordinates are positive. A convenient way of ensuring this, and one often used, is to allot large positive co-ordinates to one of the survey stations.

Fig. 6.1   **Fig. 6.2**

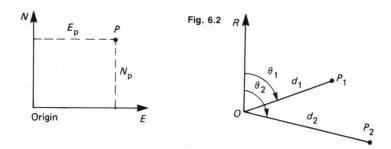

Polar co-ordinates

In Fig. 6.2, if *O* is an origin chosen at a convenient position and *OR* is a chosen reference direction, then P_1 can be located by its polar co-ordinates d_1 and θ_1, where d_1 is the distance from the origin and θ_1 is the clockwise angle between *OR* and OP_1. Co-ordinates of this type are often used in survey computations when fixing and plotting detail and for setting out work using total station instruments.

Geographical co-ordinates

The geographical co-ordinates of latitude and longitude are never used in plane surveying but they are sometimes needed in geodetic work. Even then they are usually converted to plane rectangular co-ordinates for computation purposes. Maps often show selected meridians and parallels and the network of lines produced is referred to as a graticule. It is seldom square or rectangular and must not be confused with a grid.

Bearings

Bearings may be of two types: (1) whole-circle bearings, and (2) quadrantal (reduced) bearings.

Whole-circle bearings

This is the standard way of defining a bearing in surveying practice. The whole-circle bearing (α_{AB}) of a line *AB* is defined as the clockwise angle from 0° to 360° at *A* between the direction to north and the direction to *B*.

Clearly, the bearing of the line *AB*, i.e. the bearing of *B* from *A* (Fig. 6.3(*a*)), differs by 180° from the bearing of the line *BA* (α_{BA}), i.e. the bearing of *A* from *B* (Fig. 6.3(*b*)). Thus

Fig. 6.3

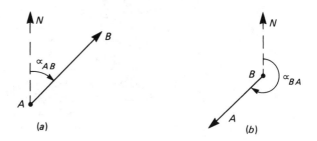

$$\alpha_{BA} = \alpha_{AB} \pm 180°$$

If at any time in the course of computations the value of a bearing is calculated as being greater than 360°, then the appropriate multiple of 360° should be subtracted from it, and similarly, if it becomes negative, the appropriate multiple should be added.

Quadrantal bearings

The quadrantal or reduced bearing of a line is defined as the angle lying between 0° and 90°, between the direction to north or south and the direction of the line. Note that the east and west directions are never used as reference lines but nevertheless are included since they indicate the direction, either east or west, of the line from grid north.

The use of quadrantal bearings requires some care because, as can be seen from Fig. 6.4, they increase in clockwise and in anticlockwise directions in alternate quadrants and formulae involving them are not universal. This can be a common source of error to the beginner, but the fact that the angles never exceed 90° is attractive because these

Fig. 6.4

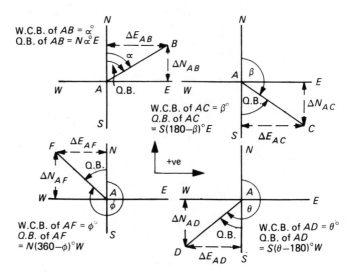

are the angles output by pocket calculators after trigonometric calculations. If they are used, it becomes more important than ever to draw diagrams for all but the most simple calculations.

Figure 6.4 illustrates the connection between quadrantal bearings and whole-circle bearings. It will be noted that in the four cases $\Delta E = l \sin (WCB)$ and $\Delta N = l \cos (WCB)$, l being the length of line. The signs of ΔE and ΔN with respect to the positive directions are given automatically by trigonometrical formulae and conventions. It will be obvious to the reader that if the grid co-ordinates of A are known, those of B etc. can be deduced once ΔE and ΔN have been determined.

Example 6.1 What is the whole-circle bearing and the quadrantal bearing of a line joining point A (1023.25 mE, 2134.32 mN) to point C (1141.71 mE, 2012.89 mN)?

These co-ordinates refer to line AC on Fig. 6.4.

$$Q.B. = \tan^{-1} \frac{(E_C - E_A)}{(N_C - N_A)}$$

$$= \tan^{-1} \frac{(1141.71 - 1023.25)}{(2012.89 - 2134.32)}$$

$$= -44° \ 17' \ 26'' \quad \text{(by pocket calculator)}$$

Quadrantal bearing = **S 44° 17′ 26″ E**

$$W.C.B. = 180° - 44° \ 17' \ 26''$$
$$= \mathbf{135° \ 42' \ 34''}$$

Example 6.2 In Fig. 6.4, if AF has a length of 91.88 m and a whole-circle bearing of 295° 25′ 45″, what are the co-ordinates of F if those of A are 1932.76 mE, 770.47 mN?

If the co-ordinates of a third point G are 2016.29 mE, 952.18 mN, what is the length of FG?

The easting of a line $= l \sin W.C.B.$

Thus
$$\Delta E_{AF} = 91.88 \sin 295° \ 25' \ 45''$$
$$= -82.98 \ m$$

The northing of a line $= l \cos W.C.B.$

Thus
$$\Delta N_{AF} = 91.88 \cos 295° \ 25' \ 45''$$
$$= +39.45 \ m$$

Co-ordinates of F are

$$E_F = 1932.76 - 82.98 \quad = \mathbf{1849.78 \ mE}$$
$$N_F = \ \ 770.47 + 39.45 \quad = \ \ \mathbf{809.92 \ mN}$$

For line FG

$$\Delta E_{FG} = 2016.29 - 1849.78 = 166.51 \text{ m}$$
$$\Delta N_{FG} = 952.18 - 809.92 = 142.26 \text{ m}$$

From Pythagoras' Theorem

$$l_{FG} = \sqrt{\Delta E_{FG}^2 + \Delta N_{FG}^2}$$
$$= \sqrt{166.51^2 + 142.26^2}$$
$$= \textbf{219.01 m}$$

Horizontal control

We now consider two different methods of control survey: traverse and triangulation. As implied previously, in a control survey the relative positions of a framework of reference points are determined to act as a base for detail survey or setting out works. The measurements should be taken to a degree of accuracy such that the unavoidable random errors that occur in all measurements do not sum up to exceed the limits set down by the specification. If this is done the measurements for the location of detail or setting out can be taken to a lower order of accuracy since the errors in these will be contained by the control framework.

In Chapter 2 the idea of a control framework was introduced when describing tape and offset surveying. To survey a small area, lines were laid down and detail located with respect to these by measuring distance along and perpendicular to the line: the series of triangles of survey lines was the control framework. Shape and scale were, in this case, determined by distance measurement and the position and orientation were selected arbitrarily. Control frameworks can be formed in many different ways using a variety of observing techniques and the position and orientation fixed with respect to other data, e.g. the national co-ordinate system, magnetic north.

Since there is a limit to the length of the line that sensibly can be measured in tape and offset surveying, there is a limit to the size of each triangle and so to control a large area demands many adjoining triangles. Small errors can accumulate rapidly when building a control framework in this way but this problem can be overcome by ensuring that the control framework connects into a 'simple' one which covers the whole area and has been surveyed to a higher accuracy. In this way, many different levels of control can exist.

Traversing

Traversing is a method of control surveying. A series of *control points* (stations), each one being intervisible with its adjacent stations, will be chosen to fulfil the demands of the survey, the lines joining these stations being the *traverse lines*. The survey then consists of the measurement of (*a*) angles between successive lines (or bearings of each line), and (*b*) the length of each line. Given the co-ordinates of the first station and the bearing of the first line the co-ordinates of all successive points can be calculated.

If the figure formed by the lines closes at a station, i.e. if they form a polygon, or it starts and finishes at points of known co-ordinates, then a *closed traverse* has been obtained, the two being distinguished as a closed loop traverse (Fig. 6.5(a)) and a closed line traverse (Fig. 6.5(b)). In the latter case, if the survey is connected to the National Grid the measured lengths need to be adjusted by an appropriate scale factor as discussed in Chapter 7, after corrections have been applied to those measurements.

A traverse starting at, say, station A and ending at E which has not been co-ordinated previously, is called an *unclosed traverse* (Fig. 6.5(c)). Each type has its particular uses, but the closed traverse is the more satisfactory figure since it is the easiest one to which to apply corrections for the errors which invariably occur. The unclosed traverse survey can be carried out when the survey is comparatively long and narrow, such as that required for a trunk sewer, pipeline, main trunk road or rail construction (though where the length is great, consideration should always be given to the possibility of tying in to Ordnance trigonometrical stations).

A closed traverse survey may be used for the framework of surveys for housing or factory sites, and determination of the perimeters of lakes, etc. They may also have to be undertaken when setting out shafts to tunnels which are being driven under built-up areas: in such cases, of course, it may be impossible to set out a surface line directly above the tunnel. The closed line traverse has the advantage over the closed loop traverse in that mistakes in the 'finishing' co-ordinates and bearings and in the scale of the distance measurement should be revealed.

Traverse types are often identified by either the equipment used or their accuracy. A first-order traverse might have leg lengths of up to 50 km measured by microwave EDM and angles measured by a precise geodetic theodolite, e.g. Wild T3. For such a traverse a gross error of one part in 100 000 might be expected (the total error in the traverse expressed as a fraction of the total length on the traverse).

A precise control traverse is often used for the original mapping and for the setting out of linear engineering works such as roads and railways. Leg lengths up to 1 km can be readily measured with 'infra-red' EDM and the angles measured with a theodolite reading to 1 second of arc, e.g. Wild T2.

On small sites, or in urban areas where visibility is greatly restricted, leg lengths may be up to 250 m and measurement could be by EDM or steel tape. The angles of the traverse might well be measured with a theodolite reading to 20 seconds.

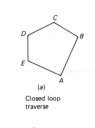

(a)
Closed loop
traverse

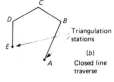

Triangulation
stations
(b)
Closed line
traverse

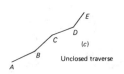

(c)
Unclosed traverse

Fig. 6.5 Types of traverse

Choice of stations and procedure

On the surveys which form a fair part of the duties of an engineering assistant in this country, the party consists of the surveyor himself and two 'chainmen' or helpers. The angles will be read by the surveyor and booked by him, while the helpers will often be stationed near the signals at the adjacent stations to make any adjustments to their verticality (and sometimes, particularly in traffic, to protect them). On surveys of great extent, several parties under the control of one chief surveyor may function simultaneously. The chief surveyor will carry out the reconnaissance for positioning the stations, and will generally co-ordinate the work.

The stations should be chosen with the requirements of the survey in mind, aiming for good visibility between stations and bearing in mind any subsequent setting out. When surveying land for a housing site, for instance, the traverse lines will be used for picking up much of the detail to be plotted, so that they will follow the perimeter of the site. Cross lines, sited to serve as check lines, will also be arranged, if possible, to enable the collection of other detail. For setting out purposes (Chapter 10), the nature of the district may fix the stations. This is often the case in unclosed traverses, where bends in roads etc. dictate the precise pattern. When there is no such restriction and the framework is for control purposes only, then long sights are preferable, and to avoid refraction the lines of sight should be well clear of the ground. The legs should be of approximately equal length and it is suggested that no traverse should contain more than ten legs before closing, whenever possible.

Stations when chosen should be placed in such a way that there will be no displacement, since some time may elapse between angular and linear measurements, plotting the survey, and actual construction work, if any. The stations may serve for the survey, or for control of levelling or contouring operations over the site, and also for setting out. On roads, etc. short heavy nails may be driven in and located by ties to nearby permanent features, while in fields, stout wooden pegs with small-headed nails driven in their tops are often favoured. It is preferable to set these in concrete, if possible, and to locate them by ties from nearby features. Alternatively, it is possible to purchase proprietary forms of plastic or metallic marking systems for purposes of monumentation. Areas liable to flooding and settlement are always suspect, and it must be remembered that although a station position may be very acceptable insofar as the lines of sight from adjacent stations are acceptable, the station itself will have to be occupied and should therefore be suitable for (*a*) setting up the instrument, and (*b*) the surveyor to be able to read the angle(s) subtended there.

Linear measurement

Traverse lines will normally be measured by EDM instruments with direct correction to the horizontal. Where this is not possible, measurements can be made by steel band applying the full range of standardization corrections detailed in Chapter 2.

Fig. 6.6

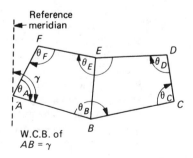

Reference meridian

W.C.B. of $AB = \gamma$

If internal angles are being read, it is usual to proceed from station to station round the traverse in an anti-clockwise direction. Starting at A, Fig. 6.6, the instrument will be directed to F, the back station, and then wheeled to B, the fore station, as previously described. The next station to be occupied will be B, where the telescope is directed first on A and then on C. The station E would perhaps be observed here as a check. It is advisable to change face and zeros at each station, a suitable observing sequence being:

> Observe back station, *Face Left*
> Observe fore station, *Face Left*
> Observe fore station, *Face Right*
> Observe back station, *Face Right*

This comprises one set and the observer can now change the zero setting and repeat the procedure as many times as required. The angles may be booked in the field book on separate pages or, probably, at most, two sets to the page. For example, when using a glass-circle theodolite one might book at stations A and B:

Station A	Face Left	Face Right	Angle FL	Angle FR	Mean Angle
AB	93° 03′ 20″	273° 03′ 20″	93° 01′ 40″	93° 01′ 20″	93° 01′ 30″
AF	00° 01′ 40″	180° 02′ 00″			
AB	183° 06′ 20″	03° 07′ 00″	93° 01′ 20″	93° 01′ 40″	93° 01′ 30″
AF	90° 05′ 00″	270° 05′ 20″			

$$F\hat{A}B = 93° \, 01′ \, 30″$$

Station B

BC	163° 43′ 30″	343° 43′ 30″	162° 41′ 10″	162° 41′ 20″	162° 41′ 15″
BE	84° 56′ 40″	264° 56′ 30″	83° 54′ 20″	83° 54′ 20″	83° 54′ 20″
BA	01° 02′ 20″	181° 02′ 10″			
BC	252° 55′ 50″	72° 55′ 30″	162° 41′ 10″	162° 41′ 00″	162° 41′ 05″
BE	174° 08′ 40″	354° 09′ 10″	83° 54′ 00″	83° 54′ 40″	83° 54′ 20″
BA	90° 14′ 40″	270° 14′ 30″			

$$A\hat{B}E = 83° \, 54′ \, 20″ \qquad\qquad A\hat{B}C = 162° \, 41′ \, 10″$$

If the external angles are observed then one should occupy the stations in a clockwise direction. Naturally, it is essential that the data be consistent in any one traverse, either all internal or all external angles being measured.

When all the internal angles have been measured, it is possible to make a check, since the sum of the angles should equal $(2n - 4)$ right angles, where n is the number of sides. Similarly, the sum of the external angles should be $(2n + 4)$ right angles.

Should there be a large error in the sum of the observed angles, then the angles must be re-checked. The value of the check sight BE is now apparent, since the error mentioned above may be isolated into one of the two quadrilaterals by balancing their values as separate units. There will usually be a small error and this may be apportioned equally among the angles.

The adjustment and computation of a traverse is most conveniently explained by reference to an example. The first example is the closed loop traverse shown schematically in Fig. 6.7 and an abstract of the data is given in Table 6.1.

Fig. 6.7

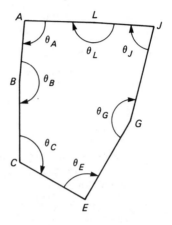

Table 6.1

Line		Mean included angle	Length (m)
AB	(θ_A)	94° 10′ 00″	103.40
BC	(θ_B)	178° 19′ 00″	157.25
CE	(θ_C)	118° 21′ 45″	143.36
EG	(θ_E)	94° 42′ 25″	169.08
GJ	(θ_G)	158° 07′ 30″	176.74
JL	(θ_J)	89° 03′ 55″	110.60
LA	(θ_L)	167° 15′ 50″	140.83

The whole-circle bearing of *AB* is 187°22′20″

Having observed the lengths of the lines and angles of a closed traverse, the unavoidable errors that occur in the data must be determined to find if they are acceptable and, if so, the total error (misclosure) must be distributed between the observations. This distribution process, usually called *adjustment*, must cause as little change to the observed data as possible, and obviously if the misclosure is unacceptable some data will have to be reobserved. Finally, the co-ordinates of the traverse stations are calculated.

Angular misclosure

The internal angles of a closed loop traverse should sum to $(2n - 4) \times 90°$ where n is the number of stations. Table 6.2 shows that the sum of the seven angles in the traverse sum to 900° 00′ 25″ whereas their sum should be $(2 \times 7 - 4) \times 90° = 900° 00′ 00″$. The traverse has an angular misclosure of 25″, which lies within the acceptable limits (see page 176) so that this misclosure can be distributed

Fig. 6.8

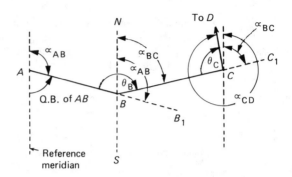

and corrections applied as shown in Table 6.2. All the angles were measured with the same accuracy and under the same conditions. It is considered that they are all equally liable to error and so the overall misclosure has been removed by applying a correction of $-3''$ or $-4''$ to alternate angles.

Calculation of bearings

Starting with the known or assumed bearings of one line, the whole-circle bearings of all other lines must be determined.

Referring to Fig. 6.6, the mean internal angles are found to be θ_A, θ_B, etc. while the whole-circle bearing of AB has been determined as α_{AB}.

Conditions at B, Fig. 6.6, are reproduced in Fig. 6.8, the dotted line through B being the north−south meridian NBS.

The whole-circle bearing required is $N\hat{B}C = \alpha_{BC}$. The whole-circle bearing of BB_1, which equals that of AB, $= \alpha_{AB}$.

$$N\hat{B}B_1 = A\hat{B}S \text{ (vertically opposite)}$$
$$= \alpha_{AB}$$

Therefore
$$\alpha_{BC} = S\hat{B}A + A\hat{B}C - N\hat{B}S$$
$$= \alpha_{AB} + \theta_B - 180°$$

i.e. the whole-circle bearing of BC is given by the sum of the whole-circle bearing of AB and the internal angle at B *minus* $180°$.

Inspection of C shows that the whole-circle bearing of CD, which equals α_{CD} is given by the sum of the whole-circle bearing of $BC(\alpha_{BC})$ and the internal angle at $C(\theta_C)$ *plus* $180°$.

To summarize, then, for the general case. To determine the whole-circle bearing of a line at a station:

(1) Add the included angle at the station to the whole-circle bearing of the previous line.
(2) If the sum obtained is below $180°$, then add $180°$ to it (i.e. as for line CD).
(3) If the sum exceeds $180°$, then deduct $180°$ from it (i.e. as for line BC).

In Fig. 6.7, $\alpha_{AB} = 187° \; 22' \; 20''$

$$\theta_A = \quad 94° \ 09' \ 56'' \qquad \theta_E = \quad 94° \ 42' \ 22''$$
$$\theta_B = 178° \ 18' \ 57'' \qquad \theta_G = 158° \ 07' \ 26''$$
$$\theta_C = 118° \ 21' \ 41'' \qquad \theta_J = \quad 89° \ 03' \ 52''$$
$$\theta_L = 167° \ 15' \ 46''$$

$$\text{W.C.B. of } AB = 187° \ 22' \ 20''$$
$$\theta_B = \underline{178° \ 18' \ 57''}$$

$$365° \ 41' \ 17''$$
$$\text{Ddt. } \underline{180° \ 00' \ 00''}$$

Therefore
$$\text{W.C.B. of } BC = 185° \ 41' \ 17''$$
$$\theta_C = \underline{118° \ 21' \ 41''}$$

$$304° \ 02' \ 58''$$
$$\text{Ddt. } \underline{180° \ 00' \ 00''}$$

Therefore
$$\text{W.C.B. of } CE = 124° \ 02' \ 58''$$
$$\theta_E = \underline{\quad 94° \ 42' \ 22''}$$

$$218° \ 45' \ 20''$$
$$\text{Ddt. } \underline{180° \ 00' \ 00''}$$

Therefore
$$\text{W.C.B. of } EG = \quad 38° \ 45' \ 20''$$
$$\theta_G = \underline{158° \ 07' \ 26''}$$

$$196° \ 52' \ 46''$$
$$\text{Ddt. } \underline{180° \ 00' \ 00''}$$

Therefore
$$\text{W.C.B. of } GJ = \quad 16° \ 52' \ 46''$$
$$\theta_J = \underline{\quad 89° \ 03' \ 52''}$$

$$105° \ 56' \ 38''$$
$$\text{Add } \underline{180° \ 00' \ 00''}$$

$$\text{W.C.B. of } JL = 285° \ 56' \ 38''$$
$$\theta_L = \underline{167° \ 15' \ 46''}$$

$$453° \ 12' \ 24''$$
$$\text{Ddt. } \underline{180° \ 00' \ 00''}$$

$$\text{W.C.B. of } LA = 273° \ 12' \ 24''$$
$$\theta_A = \underline{\quad 94° \ 09' \ 56''}$$

$$367° \ 22' \ 20''$$
$$\text{Ddt. } \underline{180° \ 00' \ 00''}$$

$$\text{W.C.B. of } AB = \underline{187° \ 22' \ 20''} \ \text{Check}$$

Easting and northing differences   In the position now reached the lengths of the lines are known, the internal angles have been measured and adjusted, and whole-circle bearings have been calculated. As mentioned at the beginning of the chapter, it is possible at this stage to plot the survey using a scale and protractor and, in fact, this may be done in order (a) to determine

Table 6.2

Line	Length	Included angle	Corrected included angle	W.C.B.	Easting difference +	Easting difference −	Northing difference +	Northing difference −
AB	103.40	94° 10' 00"	94° 9' 56"	187° 22' 20"		13.27		102.55
BC	157.25	178° 19' 00"	178° 18' 57"	185° 41' 17"		15.59		156.48
CE	143.36	118° 21' 45"	118° 21' 41"	124° 2' 58"	118.78			80.27
EG	169.08	94° 42' 25"	94° 42' 22"	38° 45' 20"	105.85		131.85	
GJ	176.74	158° 7' 30"	158° 7' 26"	16° 52' 46"	51.32		169.13	
JL	110.60	89° 3' 55"	89° 3' 52"	285° 56' 38"		106.35	30.37	
LA	140.83	167° 15' 50"	167° 15' 46"	273° 12' 24"		140.61	7.88	
	1001.26	900° 00' 25"	900° 00' 00"		275.95	275.82	339.23	339.30
					275.82		339.23	339.23
					+0.13			Error −0.07

Table 6.3

Line	Correction to ΔE	Corrected ΔE	Correction to ΔN	Corrected ΔN	Co-ordinates		
AB	−0.01	− 13.28	+0.01	−102.54	A	1000.00 E	2000.00 N
BC	−0.02	− 15.61	+0.01	−156.47	B	986.72 E	1897.46 N
CE	−0.02	+118.76	+0.01	− 80.26	C	971.11 E	1740.99 N
EG	−0.02	+105.83	+0.01	+131.86	E	1089.87 E	1660.73 N
GJ	−0.02	+ 51.30	+0.01	+169.14	G	1195.70 E	1792.59 N
JL	−0.02	−106.37	+0.01	+ 30.38	J	1247.00 E	1961.73 N
LA	−0.02	−140.63	+0.01	+ 7.89	L	1140.63 E	1992.11 N
	−0.13		+0.07				

a suitable scale for plotting (if this has not already been decided by, for example, statutory requirement), or (*b*) to obtain a suitable layout scheme on several sheets, if the survey is a large one.

However, for the reasons given, it is preferable to use *co-ordinates* for fixing the positions of the traverse stations.

The co-ordinates are derived from easting and northing differences, which are effectively projections of the lines as shown in Fig. 6.4. Thus, the next step is to calculate the easting and northing differences for each line of the traverse. As mentioned previously

$$\Delta E = l \sin (\text{W.C.B.})$$

and

$$\Delta N = l \cos (\text{W.C.B.})$$

Great care must be taken with the signs of the differences since some will be positive and some negative. Provided that whole-circle bearings are used and the correct sign is associated with the trigonometrical function, ΔE and ΔN will automatically have the correct signs. For example, in the case shown in Fig. 6.9 where the traverse leg *LA* has a bearing of 273° 12′ 24″ (4th quadrant), the sine, and hence ΔE, is negative whereas the cosine, and hence ΔN, will be positive.

Fig. 6.9

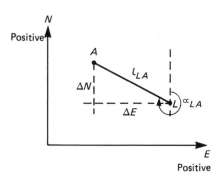

In Table 6.2, the easting and northing differences for all traverse legs have been calculated. Since this traverse is in the form of a closed loop, the algebraic sum of all the easting differences and all the northing differences should be zero, i.e. the traverse should finish where it started. It can be seen that this is not the case, the errors *dE* and *dN* in the easting and northing directions being +0.13 m and −0.07 m respectively.

(1) Note that in Tables 6.1 and 6.2, *AB* indicates not only the line in question but also the included angle at *A*; similarly, *BC* gives the included angle at *B*, and so on.

(2) If the respective differences of each line be plotted, a point A_1 would result, near *A*, but 0.07 m south and 0.13 m east of *A*. A_1A is the closing error of the survey.

$$\text{Closing error} = \sqrt{(dE^2 + dN^2)} = \sqrt{0.13^2 + 0.07^2} = 0.15 \text{ m}$$

at a bearing of $\tan^{-1}(+0.13/-0.07)$ (or, $\tan^{-1}(dE/dN)$), which is approximately 118° 18′.

The closing error, or positional error, of 0.15 m deduced from the measurements obtained during the traverse implies a fractional linear error of 0.15/1001.26 or, say, 1 in 6700 in horizontal distance.

Detection of error in linear measurement

The size of the closing error can indicate the presence of some error which is not admissible for balancing out by methods now to be discussed.

Should there be a large error in linear measurement, i.e. failure to book one tape length, then so long as only one such error has been made, it will be found that the bearing of the closing error will be similar to the bearing of the line in error.

Determination of co-ordinates

The co-ordinates of the traverse stations can be determined after the closing error is eliminated and three methods for doing this are now described: (a) Bowditch's method, (b) Transit rule, and (c) the method of unaltered bearings.

(a) *Bowditch's method* This method assumes that the error in the bearing of a line due to some inaccuracy in angular measurement gives a displacement of one end of that line, relative to the other end, equal and at right angles to that displacement, due to the error in measurement of its length. The error in length (l) is taken to be proportional to $\sqrt{l}$.

Fig. 6.10

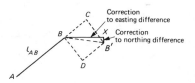

Thus, in Fig. 6.10 the displacement of B due to the error in linear measurement is BC (and is proportional to $\sqrt{l}$), and the displacement BD due to an error in bearing is equal to BC, with a result that B has a final position of B'. The total error BB' thus equals $\sqrt{BC^2 + BD^2}$.

The method of least squares (Chapter 8) states that the sum of the weighted errors squared will be a minimum, and after weighting the lines inversely as the square of the error of each line (i.e. $\sqrt{(l)}^2 = l$), then $\Sigma(\sqrt{BC^2 + BD^2}/l)$ is to be a minimum.

This leads to the following conclusions:

(1) Correction to an easting difference ΔE_{AB}

$$= dE \times \frac{\text{length of side } AB}{\text{perimeter of the traverse}}$$

(2) Correction to a northing difference ΔN_{AB}

$$= dN \times \frac{\text{length of side } AB}{\text{perimeter of the traverse}}$$

i.e. point B moves to B' over a distance of

$$\sqrt{\left(\frac{l_1}{\Sigma l}\right)^2 (dE)^2 + \left(\frac{l_1}{\Sigma l}\right)^2 (dN)^2} = \frac{l_1}{\Sigma l} \sqrt{dE^2 + dN^2}$$

The quadrantal bearing of the movement is $\tan^{-1} \dfrac{BX}{B'X}$

where

$$\tan^{-1} \frac{BX}{B'X} = \frac{\dfrac{l_1}{\Sigma l} dE}{\dfrac{l_1}{\Sigma l} dN} = \frac{dE}{dN}$$

and this is the bearing of the closing error.

Thus, generally speaking, the bearing of each line of the traverse will be altered after applying Bowditch's method of correction. This method is popular in practice for the average type of engineering survey, since (i) it is easy to apply, and (ii) the corrections, although affecting the bearings of the lines, do not affect the plotting to a noticeable extent. Table 6.3 illustrates typical corrections to the traverse *ABCEGJLA* (Table 6.2).

Notes on Table 6.3

(i) Correction to an easting difference

$$AB = \frac{103.40}{1001.26} \times (-0.13) = -0.01 \text{ m}$$

Correction to a northing difference

$$AB = \frac{103.40}{1001.26} \times (+0.07) = +0.01 \text{ m}$$

and similarly for other lines.

The corrections are equal to the errors, but are opposite in sign and so in the example given the correction to ΔE_{AB} will be negative and the correction to ΔN_{AB} will be positive,

i.e. Corrected $\Delta E_{AB} = -13.27 - 0.01 = -13.28$ m
Corrected $\Delta N_{AB} = -102.55 + 0.01 = -102.54$ m

(ii) It was assumed that the co-ordinates of A were 1000.00 E, 2000.00 N. Then, co-ordinates of B are

$$1000.00 - 13.28 = 986.72 \text{ E}$$

and

$$2000.00 - 102.54 = 1897.46 \text{ N}$$

Co-ordinates of C are

$$986.72 - 15.61 = 971.11 \text{ E}$$

and

$$1897.46 - 156.47 = 1740.99 \text{ N}$$

As a check, the co-ordinates of A should be derived from those of L, proceeding in a cyclic manner round the traverse.

co-ordinates of L 1140.63 E   1992.11 N
apply corrected ΔE_{LA} -140.63 corrected ΔN_{LA} $+7.89$

Hence co-ordinates of A $+1000.00$ E $+2000.00$ N

which are those from which all the others were deduced.

(*b*) *Transit rule* In the previous method all lines will have some correction made in both easting and northing, as called for in the totals of dE and dN (Table 6.2). The Transit rule or method has no mathematical background, and lengths of lines do not enter into the calculations. The rule is:

$$\text{Correction to } \Delta E_{AB} = dE \cdot \frac{\Delta E_{AB}}{\Sigma \Delta E_{AB}}$$

$$\text{Correction to } \Delta N_{AB} = dN \cdot \frac{\Delta N_{AB}}{\Sigma \Delta N_{AB}}$$

Table 6.4

Line	AB	BC	CE	EG	GJ	JL	LA
Correction to ΔE	0.00	0.00	-0.03	-0.03	-0.01	-0.03	-0.03
Correction to ΔN	$+0.01$	$+0.02$	$+0.01$	$+0.01$	$+0.02$	0.00	0.00

and Table 6.4 indicates corrections by this method for the traverse in Table 6.1.

Hence, if a line has no easting difference, then it will not have an easting correction and, similarly, if it has no northing difference, then there will be no northing correction. So long as the line is running parallel to one of the co-ordinate axes, then the corrections applied will not alter the bearing of the line (compare with Bowditch's method), but when the line is inclined to the axes, corrections will be made which do affect the bearings.

Comparison of Tables 6.3 and 6.4 indicates the effects of the two methods. Note that line *LA* by Bowditch's rule will have some alteration in bearing, whereas by the Transit rule the alteration is virtually zero.

(*c*) *Unaltered bearings* A further method which may be employed is one which leaves the bearings of lines unaltered after the corrections have been made. Corrections are applied to each line, such that

$$\frac{\text{Correction to } \Delta E}{\text{Correction to } \Delta N} = \frac{\Delta E}{\Delta N}$$

Much more calculation is involved in this method, and it is likely that the work involved would not be justified in the usual type of traverse unless a computer program was available. The main value of this method lies in its application to the correction of precise traverses.

Computation of closed line traverses

The computation procedure described for the closed loop traverse has to be modified slightly for closed line traverses and the changes will be explained again by reference to an example. Figure 6.11 shows a line traverse to fix control stations 1, 2 and 3 lying between stations A and B whose co-ordinates are known. It will be noted that the angles measured and listed in the data cited in Table 6.5 lie on the same side of the line of the traverse. This is in the interests of consistency but in practice the angles on the other side of the line could be observed for checking purposes. The orientation of the scheme is controlled by observing angles at A and B between the traverse legs and lines of known bearing AR_1 and BR_2 respectively.

Fig. 6.11

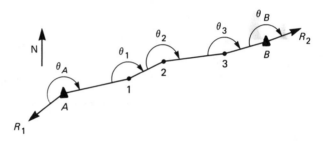

In the previous example, the computations were presented in a number of tables so that the various steps could be illustrated. The computations and adjustment of the traverse are shown in Table 6.6 together with the field observations, the input data being shown in bold type. The format is suitable for all traverse surveys and it facilitates independent checking. The procedure is largely self-explanatory, merely differing from that of the loop traverse in the way the angular and co-ordinate difference misclosures are assessed. Although the angles are no longer those of a closed polygon, their sum can be used for correction purposes quite readily from the knowledge of the bearings of lines AR_1 and BR_2.

Fig. 6.12

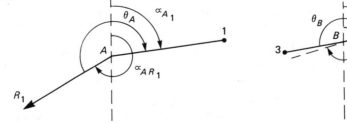

Table 6.5 Abstract of traverse data

Observed horizontal clockwise angles		Reduced horizontal distances (m)	
R_1–$\hat{A}$–1	203° 41′ 28″	A–1	703.28
A–$\hat{1}$–2	162 37 21	1–2	473.29
1–$\hat{2}$–3	193 18 06	2–3	687.48
2–$\hat{3}$–B	170 08 49	3–B	202.31
3–$\hat{B}$–R_2	189 35 52		

Known co-ordinates	A	3208.49 m E	4375.29 m N
	B	5074.49 m E	5227.47 m N

Known bearings	A–R_1	228° 27′ 30 ″
	B–R_2	67° 48′ 48 ″

Referring to Fig. 6.12, it will be apparent that the forward bearing of $A1$ is

$$(\alpha_{AR_1} + \theta_A - 360°)$$

and hence the back bearing of $A1$ is

$$(\alpha_{AR_1} + \theta_A - 180°)$$

Applying the same argument in respect of the other stations of the traverse

$$\alpha_{AR_1} + (\theta_A + \theta_1 + \theta_2 + \theta_3 + \theta_B) - 5 \times 180° = \alpha_{R_2 B}$$

Now

$$\begin{aligned}
\alpha_{AR_1} &= 228° \ 27' \ 30'' \\
\Sigma\theta_1 &= 919° \ 21' \ 36'' \\
&\ \underline{} \\
&\ 1147° \ 49' \ 06'' \\
\text{Ddt.} &\ 900° \ 00' \ 00'' \\
&\ \underline{}
\end{aligned}$$

$$\begin{aligned}
\text{Apparent bearing } \alpha_{R_2 B} &= 247° \ 49' \ 06'' \\
\text{Ddt.} &\ 180° \ 00' \ 00'' \\
&\ \underline{}
\end{aligned}$$

$$\begin{aligned}
\text{Apparent bearing } \alpha_{BR_2} &= 67° \ 49' \ 06'' \\
\text{Given bearing } \alpha_{BR_2} &= 67° \ 48' \ 48'' \\
&\ \underline{}
\end{aligned}$$

$$\begin{aligned}
\text{Angular error} &= + 18'' \\
\text{Total correction} &= - 18''
\end{aligned}$$

In addition, for this line traverse the sum of the easting co-ordinate differences and the northing co-ordinate differences should be equal to the co-ordinate differences between stations A and B, i.e.

$$\Delta E_{AB} = E_B - E_A = (5074.49 - 3208.49) = 1866.00 \text{ m}$$
$$\Delta N_{AB} = N_B - N_A = (5227.47 - 4375.29) = 852.18 \text{ m}$$

Table 6.6

Station	Obs angle / Correction / Corr angle	Back Brg / Corr angle / For'd Brg (α)	Horizontal distance (l)	Sin Brg / Cos Brg	ΔE Correction Corr ΔE		ΔN Correction Corr ΔN		Corr ΔE Final E	Corr ΔN Final N	Station
					+	−	+	−			
A	**203° 41' 28"**								3208.49	4375.29	A
	−04"	**228° 27' 30"**	703.28	0.951853	+669.42		+215.59		+669.51	+215.68	
	203° 41' 24"	203° 41' 24"		0.306554	+ 0.09		+ 0.09		3878.00	4590.97	1
	162° 37' 21"	72° 08' 54"			+669.51		+215.68				
1	162° 37' 18"	**252° 08' 54"**	473.29	0.816843	+386.60		+273.02		+386.66	+273.08	
	162° 37' 18"	162° 37' 18"		0.576860	+ 0.06		+ 0.06		4264.66	4864.05	2
	193° 18' 06"	54° 46' 12"			+386.66		+273.08				
2	193° 18' 02"	**234° 46' 12"**	687.48	0.927644	+637.74		+256.75		+637.83	+256.84	
	193° 18' 02"	193° 18' 02"		0.373465	+ 0.09		+ 0.09		4902.49	5120.89	3
	170° 08' 49"	68° 04' 14"			+637.83		+256.84				
3	170° 08' 46"	**248° 04' 14"**	202.31	0.850046	+171.97		+106.56		+172.00	+106.58	
	170° 08' 46"	170° 08' 46"		0.526709	+ 0.03		+ 0.02		5074.49	5227.47	B
	189° 35' 52"	58° 13' 00"			+172.00		+106.58				
B	−04"	**238° 13' 00"**									
	189° 35' 48"	189° 35' 48"									
	67° 48' 48"										
	919° 21' 36"	Sum	2066.36	Totals	+1865.73		+851.92		Positional misclosure		
	919° 21' 18"	Required	= Σl						$[(-0.27)^2 + (-0.26)^2]^{1/2}$		
	+18"	Misclosure							= 0.37 m		
	Angular correction			Actual ΔE	+1865.73		ΔN	+851.92	Fractional linear error		
	2 (α −3")			Required ΔE	+1866.00		ΔN	+852.18	1/5585		
	3 (α −4")			Misclosure dE		−0.27	dN	−0.26			

Three-tripod traversing equipment

Errors can arise due to the inaccurate centring of the theodolite, angle target or aiming mark, EDM and EDM reflector prism system at the various ground stations. These centring errors arise especially when short survey lines are used, as often occurs in town surveys. Positional errors accumulate along the traverse, and these can be minimized or even avoided by using forced centring equipment. Essentially, this consists of a number of tripods and tribrachs (detachable levelling bases) which can be mounted thereon, together with targets and optical plummets. The theodolites used in this system also have such detachable bases into which they and the targets, reflectors, etc., can be mounted so that their centre is always in the same position. Many tribrachs have bubble tubes and optical plummets built in and so can be levelled and centred over the ground mark. Alternatively, special optical plummets with sensitive bubbles can be inserted into the forced centring tribrachs if the highest precision in levelling and centring is required. The minimum number of tripods and tribrachs required is three — hence the name of the equipment. With targets set at stations A and 2, observations can be taken at 1 (Fig. 6.11). The tripod and tribrach with target is moved from A to 3 and the theodolite (or EDM) at 1 interchanged with the target at 2. Observations can now be taken at 2 to 1 and 3 and this leap-frog process can continue all along the traverse. Note that any particular tripod and tribrach assembly is not moved until all observations at and to that station have been completed.

If a loop traverse is being observed, four tripods and tribrachs are really required, one being left at the starting station so that final observations can be taken on a target mounted there. Additional tripods, tribrachs, targets and reflectors, over and above the basic set, are useful anyway to make the process of 'three' tripod traversing more efficient since fewer movements of equipment are required. In effect, the traverse is between the centre of the tribrachs and not the ground marks, eliminating centring error from being carried along the traverse.

The accuracy of the whole traverse is influenced by the exactness with which the vertical axis of the theodolite can be placed in the lines previously occupied by the vertical axes of the targets. Centring errors must be of low magnitude, and alternative methods of centring have been developed, one frequently encountered being the stub axis system. Such axes, on theodolites or targets, etc., fit closely into recesses in the tribrachs mounted on the tripod heads; locking devices are provided in the tribrachs. British theodolites have shallow stub axes of relatively large diameter and these are located by two studs in the tribrach and then locked by a third.

A mean square error of ± 0.1 mm has been mentioned by some writers in respect of centring errors, but a German specification demands a value of ± 0.03 mm for the stub axis system. Over a sighting distance of 10 m, which may not be excessive in certain circumstances underground, the former value suggests an error of ± 2 sec. The lateral shift in an unclosed traverse is given by the expression

$$\pm \frac{s_{\bar{x}} L \sqrt{n/3}}{206\ 265}$$

where L is the length of traverse, n the number of angles read, and $s_{\bar{x}}$ the mean square error of angular observation, including centring effects. The importance of restricting centring errors to low magnitude can be appreciated if the case of a survey of length 2 km and sighting distances 100 m with $s_{\bar{x}} = \pm 5''$ be considered: a lateral shift of ± 0.124 m can occur. The Hungarian Optical Works manufacture a forced centring device enclosed in a rotating sleeve. This allows the stub axis system to be rotated through $180°$ in order that eccentricity errors can be eliminated by taking the means of pairs of observations.

Accuracy

The angular and linear (or positional) misclosures should be due solely to random error in the measurement, any significant systematic effects having been removed: they are therefore proportional to the respective standard errors $s_{\bar{x}A}$ and $s_{\bar{x}d}$ in the measurements (see Chapter 8).

In so far as misclosure of a closed traverse is concerned, the standard errors in angular measurement and distance measurement can be expressed in the form of $\pm s_{\bar{x}A}\sqrt{n}$ and $\pm s_{\bar{x}d}\sqrt{n}$ respectively, where n is the number of stations.

In the majority of traverse surveys insufficient repeat observations are taken to establish standard errors $s_{\bar{x}A}$ and $s_{\bar{x}d}$. However, the manufacturers of EDM equipment, for example, quote the standard error that can be achieved with their equipment under specified conditions and representative values were given in Chapter 5. In practice, the linear misclosure is based upon the length of the traverse rather than the number of stations. Very often the standard error per angle is assumed to be the reading accuracy of the instrument when a relatively small number of readings is taken.

Table 6.7

Order	Unadjusted horizontal distance	Unadjusted horizontal angles	Typical task
First	1 in 25 000	$2\sqrt{N}$ sec	Traversing for control of (i) aerial mapping (ii) expansive engineering project
Second	1 in 10 000	$10\sqrt{N}$ sec	Engineering surveys,
Third	1 in 5 000	$30\sqrt{N}$ sec	including setting out.
Fourth	1 in 2 000	$60\sqrt{N}$ sec	Small site surveys

Misclosure values suggested by Prior in the paper 'Accuracy of Highway Surveys', given to the 11th Congress of the International Society for Photogrammetry, together with typical tasks are shown in Table 6.7.

As an example, consider the misclosure of the loop traverse computed earlier. For this traverse, the angles were the mean of two rounds with a 20″ theodolite and so the standard error per angle could be taken as ±20″: the distances were measured with a steel band with an estimated standard error of, say, ± 0.1 m. There were seven legs and so the theoretical misclosures were ±20″ $\sqrt{7}$ = ±50″ and ±0.1 $\sqrt{7}$ m = ±0.26 m respectively.

The actual values of 25″ and 0.15 m were less than these values and were tending towards second-order accuracy according to Prior's values. His values serve as a guide to the specification required for the various tasks. Reference may also be made to the *Code of Practice for Accuracy in Building* BS 5606: 1978, which implies maximum probable differences between lengths measured between traverse stations and those calculated from the station co-ordinates, after adjustment, of ± 15 mm in 100 m and ± 20 mm in 200 m from lower limits of ± 5 mm, with a linear accuracy of 1:20 000 for lengths in excess of 500 m.

However, should the instruments read direct to one second and be fitted with optical plummets, the Standard states that the maximum value could well be improved to ± 7 mm to ± 10 m in 100 m. Appropriate plotting scales of 1:100 or 1:200 apply in this case compared to 1:500 for the previous case, in which a plummet need not be fitted to the theodolite. In each case, however, it is assumed the linear measurement will be by steel tape or band under standard pull with temperature and slope corrections applied. A Swedish standard quotes $0.75\sqrt{L}$ mm over L m when 'one second' theodolites are used and distances are measured twice by steel tape or band.

Triangulation

Triangulation is a method of control surveying. In the simplest form of triangulation the area is divided into a series of standard geometrical

Fig. 6.13

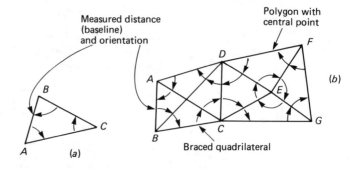

figures, such as braced quadrilaterals or polygons with central points, the corners of which form a series of accurately located control stations. The distinguishing features are shown in Fig. 6.13(a) and (b). If angles A, B and C are measured in triangle ABC (Fig. 6.13(a)) and the length of side AB is measured, then the lengths of sides BC and CA can be calculated. If the co-ordinates of A, say, are known, together with the orientation of AB, the co-ordinates of B and C can be calculated. In Fig. 6.13(b) the angles marked are measured and the lengths of all the other sides of the whole triangulation and the co-ordinates of the stations are worked out from these observed angles (after making necessary adjustments), the length of the initial side (conventionally known as the *base line*), the co-ordinates of A and the orientation of AB. Measurement of the one line provides the scale of the survey, whilst the angular observations define the shape.

If all the twenty angles were measured, rather more information would be available than required, but it is characteristic of these surveys that additional (or redundant) measurements are taken both to check the data and by adjustments to improve the precision of the final result.

Schemes of the simple design discussed so far are extremely useful when EDM instruments and calculating aids are not available because distance measurement and calculation can be kept to a minimum. Distance measurement by tape is difficult and tedious when carried out to the same order of accuracy as good angular measurements. However, when EDM equipment is available then more than one distance would be measured and the layout of the control scheme would not be restricted to braced quadrilaterals and centre-point polygons. In fact, we could measure lengths only, thereby producing a trilateration framework.

Most modern control schemes involved both angular measurement and the measurement of selected, or all, sides and so should not be called simply triangulation or trilateration surveys but, by convention, the name triangulation generally applies. In addition, the angular measurements of a triangulation scheme define the shape of the framework more efficiently than the distances of a trilateration survey. However, the details of planning, station marking and computing described in the following sections can be applied to a trilateration scheme.

Choice of stations   The area to be covered by a triangulation scheme must be carefully reconnoitred to select the most suitable positions for the control stations. Existing maps, especially if contoured, can be of great value since the size and shape of the triangles formed by the stations can be difficult to visualize in the field.

When planning the scheme, certain considerations should be kept in mind which may be summarized as follows:

(a) Every station should be visible from the adjacent stations. Rays passing close to either the ground or to an obstacle (grazing rays) should be avoided since they can be refracted due to air temperature differences (see comments on the Channel Tunnel in Chapter 10).

(b) The triangles formed thereby should be well-conditioned, that is to say, as nearly equilateral as possible. No angles should be less than 30°, if at all possible. The scheme should be kept as simple as possible, but with sufficient redundant observations to provide the necessary checks and to increase precision.

(c) The size of the triangles will depend on the configuration of the land, but they should normally be as large as possible compatible with the distinct bisection of signals, having regard to the type of theodolite used.

Short sights, in which the signal cannot be accurately bisected, should be avoided. In hilly terrain, longer sights may be obtained by using the higher points as stations; in flat country, consideration must be given to the elevation of the instrument and/or the signal by means of towers or staging (see Figs 6.14 and 6.16). The positionings of instrument and signal can be assessed using the expression $0.067 \, D^2$ given in Chapter 3 to estimate distance D over which the line of sight becomes tangential to mean sea level from a given height.

Fig. 6.14

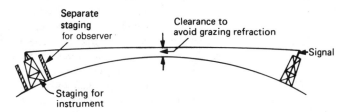

Separate staging for observer

Clearance to avoid grazing refraction

Signal

Staging for instrument

The cost of such staging adds considerably to the cost of a survey. The height must be such that the line of sight is at least 1.5 m clear of all obstructions.

(d) The end purpose of the triangulation scheme must be kept in mind. Where choice of station sites exists, the ones most suitable for connection to subsequent traverse and detail surveys should be used.

Once the framework has been designed, the station positions must be marked and beaconed. The ground work can vary in quality from a nail in a stout wooden peg to a mark in a metal plug set in the upper surface of a concrete pillar on which instruments can be mounted directly. Pillars are normally only used for the permanent marking of national schemes or for specialist work where the highest accuracy is required, e.g. monitoring the relatively small movements of a dam.

To make a station visible to a distant theodolite some form of beacon, which serves as a signal, is required. A simple target would be a pole, with a flag, set over the mark and guyed for stability. If distance measurement is proposed then there must be provision for the EDM reflector, often a bank of prisms if long sights are involved with infra-red instruments. The use of very tall tripods, with brightly coloured conical tops, will allow the instrument to be set up over the station whilst maintaining visibility in all directions. Luminous beacons are available, and these are either acetylene or electric lamps, or heliographs which use the sun. The former cater for night observations or for hazy conditions.

Angular and distance measurement

For improving the accuracy of measurement of the horizontal angles the best observing procedure is 'the method of rounds' providing that equal and good all-round visibility exists from the theodolite station.

To observe a 'round' or 'zero' at a station, set up and level the instrument at station O. The optical micrometer is set to read approximately zero in the face left position and a pointing is made on the reference station A, which will usually be the clearest observable beacon (Fig. 6.15). Then, swinging right, the instrument is pointed on B and the reading on the horizontal circle noted; stations C, D, E and, finally, A, are then bisected, ensuring in all cases that the direction of swing to the right is preserved. The face of the instrument may then be changed and pointings made on to stations E, D, C and B, always swinging to the left. The final reading on A should now differ from the first by $180°$; this completes one set of readings. The zero setting is now altered to use other parts of the scale and a similar procedure followed, the number of zeros required depending on the degree of precision required. Sixteen to thirty-two zeros may be taken for primary triangulations, in which side lengths up to 50 km apply, and eight to sixteen zeros for secondary triangulations having side lengths up to 15 km.

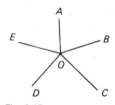

Fig. 6.15

Individual angles are deduced from the readings in the usual way and if the difference between any one measurement and the mean for that particular angle exceeds, say, 4 seconds, it should be repeated before leaving the station. A glass arc theodolite, such as the Wild T3 reading directly to 0.1 second, should be used for first-order work. For third-order work with side lengths of a few kilometres, a theodolite reading to 1 second should be used to observe from four to eight zeros.

Of the other alternatives, Schreiber's method may be mentioned in which, as well as measuring the angles subtended at the theodolite station by successive stations, the angles between each station and all the other stations are also measured independently. It will be seen that this means a considerable increase in the amount of work at each theodolite station, but has the advantage that even when all-round visibility is not available, work may be possible in the direction in which conditions are satisfactory.

The side lengths of the triangulation framework will normally be measured with an EDM instrument, the particular type chosen depending on the length of line and precision required, as mentioned in Chapter 5. The observing procedures and the necessary corrections to apply are also described in that chapter. For larger triangles GPS (Global Positioning Systems) may be used to obtain the co-ordinates of the stations, as mentioned in Chapter 7.

For simple schemes, although only one distance can be used in the computations, more than one should be measured, as additional data provide excellent checks on the quality of the work. Where no restriction is forced by the method of computation, up to 20% of the lines should be measured since it has been shown that this is advantageous from the point of view of the scaling and the shaping of the framework.

For a fuller treatment of this subject, and indeed the whole study of geodetic surveying, the student is referred to *Geodesy* by Brigadier Bomford (OUP).

Before the advent of EDM, at least one side length had to be measured by taping. For most accurate work this was achieved by repeated measurements with an invar band hung in catenary, the actual support being provided by wires of small diameter attached to the brass rings at each end of the tape. These wires passed over the pulleys of straining devices, being tensioned by weights attached to their other ends. The tensioning weight, which gave the field tension, was not to be less than twenty times the weight of the tape and where this was not possible it would be supported at mid-span or third points, the supports to be at constant gradient.

For the measurement of one bay the distance was measured between two tripods fitted with special measuring heads (index heads), the invar band being suspended just clear between the straining trestles.

Readings were taken of the positions of reference marks on the index heads relative to the graduated scales at each end of the bands. To avoid errors due to pulley friction, the tape was displaced by small amounts, and perhaps ten readings taken.

Unless protecting screens were erected, work was not possible on windy days, owing to tape movement. Temperature was measured at a number of points along the tape while the scale readings were being made, and the mean temperature obtained. Invar was used for the tapes since it has a lower coefficient of linear expansion than other possible materials and hence is less susceptible to temperature effects. One base line, of course, consisted of many bays in which the tripod heads were accurately aligned. The field data was corrected using the equations presented in Chapter 2 working to 0.0001 m and accuracies of the order of 1:500 000 were achieved by these methods. However, due to the advent of EDM systems and GPS, work of this order is not likely to be undertaken nowadays. For an authoritative (and pre-eminently readable) account of typical equipment and its use in base measurements, the student is recommended to read *The Measurement of the Ridge Way and Caithness Bases 1951−1952* by Major M.H. Cobb,

MA, ARICS, RE, OS Professional Papers, New Series, No. 18 (HMSO). The Ridge Way Base was 11.06 km long and the Caithness Base was 24.83 km long (see Fig. 6.16).

The re-triangulation of Great Britain

Until 1986 the Ordnance Survey maintained a triangulation network covering Great Britain, Fig. 6.16, but the introduction of GPS (Chapter 7) means that stations no longer have to be intervisible and many of the primary stations on remote hill tops are no longer used. In the re-triangulation surveys carried out up to 1986, most of the sides of the primary triangles had lengths between 40 km and 70 km whilst the largest, whose corners were Scafell Pike, Snowdon and Slieve Donard (Ireland), had side lengths in excess of 150 km. Nowadays, the 2000 or so concrete trig points that marked the corners of secondary, third- and fourth-order triangles within the primary network are co-ordinated by GPS and used directly for the collection of detail survey data and the control of other works.

Not all countries have a primary network consisting of intersecting quadrilaterals or polygons which covers the whole country. If the country is large, such a procedure would be excessively lengthy and costly and, instead, chains of figures may run along the meridians of longitude and the parallels of latitude. Triangulation networks may then be run between the various stations on the chains.

Similarly, if the country is long and narrow, it is sufficient to have a backbone of quadrilaterals as in Fig. 6.17, which shows the triangulation of the Nile Valley from Cairo to Beba. The figure also shows how short base lines can be connected by extension into the larger sides of the quadrilaterals.

Accuracy of triangulation

The primary measure of the precision of triangulation is the *average triangular error*, i.e. the average deviation of the sum of the measured angles in the triangles from 180° after correction for curvature.

In a small triangle, whose sides are of the order of two kilometres, the curvature of the earth may be considered negligible, and the three measured angles should sum to 180°. In practice, there will be a difference of a small number of seconds known as the *triangular error*.

Spherical excess

In larger triangles, an error arises from the fact that, though the angles are measured in the horizontal plane, the curvature of the earth throws these planes out of parallel with one another. The three angles of the triangle should now add up to more than 180°, and the excess, known as the *spherical excess*, is easily calculated from the area of the triangle.

Then $\Sigma(\text{measured angles}) - (180 + E) = \epsilon$

where E is the spherical excess, and ϵ is the triangular error whose value and sign are to be estimated.

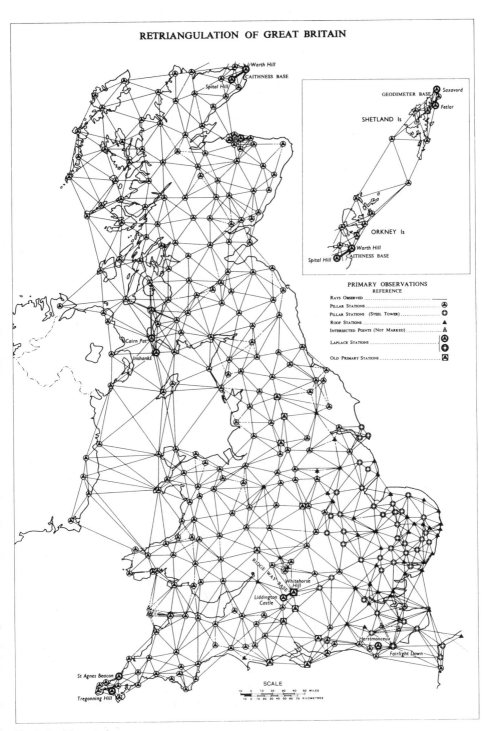

RETRIANGULATION OF GREAT BRITAIN

Warth Hill
CAITHNESS BASE
Spital Hill

GEODIMETER BASE — Saxavord
Fetlar
SHETLAND Is

ORKNEY Is

Warth Hill
CAITHNESS BASE
Spital Hill

Cairn Pat
Inshanks

RIDGE WAY BASE
Whitehorse Hill
Liddington Castle

Herstmonceux
Fairlight Down

St Agnes Beacon

Tregonning Hill

PRIMARY OBSERVATIONS
REFERENCE

Rays Observed ....
Pillar Stations ....
Pillar Stations (Steel Tower) ....
Roof Stations ....
Intersected Points (Not Marked) ....
Laplace Stations ....
Old Primary Stations ....

SCALE

10   0   10   20   30   40   50 MILES
10   0   10   20   30   40   50   60   70 KILOMETRES

Fig. 6.16 Intersected stations are fixed by means of intersecting sights from adjacent stations

Survey methods **183**

Fig. 6.17

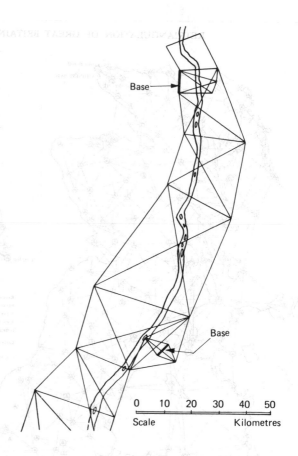

Base

Base

```
0    10   20   30   40   50
Scale                Kilometres
```

The spherical excess is calculated from

$$E = \frac{A}{R^2 \sin 1''}$$

where A = area of triangle, and R = mean radius of the earth.

For all but the most accurate work the area of the triangle can be estimated as if the area were plane, so that

$$E = \frac{\text{area of triangle in sq km}}{1000} \times 5.09''$$

In geodetic work of the highest accuracy, the average value of ϵ should be less than one second of arc.

Adjustment

The basic formulae for computing the co-ordinates of points from observed angles and distances were given earlier in the chapter. With triangulation, however, the problem arises that since too much data has been observed, some adjustment is required to get the best unique set of co-ordinates for the control stations from all the possible combinations of data. In the more complex figures, several triangles are obtained whose angles are interdependent upon one another, and these interdependent angles can be corrected simultaneously by the method of least squares or by the method of variation of co-ordinates which is eminently suitable for computer treatment. The principles are indicated in Chapter 8 but reference can be made to *Geodesy* (*op. cit.*) and to *Professional Paper 24* (New Series) published by the Ordnance Survey for further information.

Application of triangulation

Now that EDM is widely available, traversing has replaced triangulation for the provision of horizontal control for the majority of survey work. Although triangulation, trilateration or, as mentioned previously, a blend of the two, are normally selected for:

(1) the establishment of accurately located control points for surveys of large areas;
(2) the accurate location of engineering works such as (i) centre lines, terminal points and shafts for long tunnels, (ii) centre lines and abutments for bridges of long span, and (iii) complex highway interchanges;
(3) the establishment of accurately located control points in connection with aerial surveying;
(4) measurement of deformations of structures such as dams.

Figure 6.18 shows the control survey for the Channel Tunnel linking England to France. The survey contains a polygon with central point in England and a quadrilateral and chain of triangles in France, linked together by a series of intersecting quadrilaterals across the sea, and is based upon the Ordnance Survey trig points in Fig. 6.16.

Fig. 6.18 Control network for the Channel Tunnel (*Courtesy*: Editor, *New Civil Engineer*)

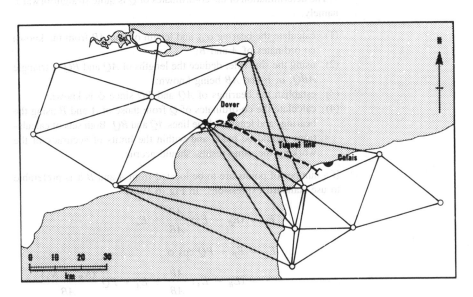

Examples of geodetic triangulations were given in a previous section, and an example of the second type of application is given in the diagram at the beginning of this chapter. This shows a triangulation scheme, whose average side length was about 5 km, to provide horizontal control for the major project to construct a bridge and approach roads.

We have also pointed out previously that pillars should be provided at the stations of important surveys. It is essential that centring errors be a minimum, and to this end various methods are employed such as grouting a centring plate to the top of the pillar, or employing special ball-tipped bolts at the base of the instrument to fit into a hollow cylinder cast into the pillar. When measuring dam deformation, one set of pillars should be provided outside the pressure zone of the dam in order to 'vet' the positions of the observation pillars established closer to the targets on the dam. An instrument reading to the order of 0.1 sec would be used in the measurement of angles and an instrument such as the Mekometer would be used to measure distances.

Intersection

This is the process of locating and co-ordinating a point from at least two existing control stations by observing horizontal directions at the control points. It has many uses apart from co-ordinating new control points, e.g. curve ranging (Chapter 11), and surveying detail in inaccessible positions including that high up on buildings. In this latter use, it is especially useful for co-ordinating existing conspicuous points such as church spires, flag-poles and radio aerials since no observations have to be taken at the points in question.

In Fig. 6.19(a), to fix the position of point Q from the control stations A and B, angles a and b can be observed. If the stations A and B are not intervisible then angular observations a' can be taken on to other known directions as in Fig. 6.19(b).

The determination of the co-ordinates of Q is quite straightforward, namely:

(1) calculate the bearing α_{AB} and length of line AB from the known co-ordinates of A and B;
(2) using the sine rule, deduce the lengths of AQ and BQ in triangle ABQ, a, b and AB being known;
(3) establish the bearings of AQ and BQ since ϕ is known;
(4) calculate the co-ordinates of Q from those of A and B using the bearings and lengths of the lines AQ and BQ. Both sets of calculations should be made and within the limits of accuracy of the working the same results should accrue.

If a number of points are to be located from A and B it is preferable to use standard expressions. In Fig. 6.19(a)

$$E_P = (E_B - E_A)\,\frac{AP}{AB} + E_A$$

$$E_Q = E_P + PQ\,\cos\phi$$

$$= (E_B - E_A)\,\frac{AP}{AB} + E_A + PQ\,\frac{(N_B - N_A)}{AB}$$

Fig. 6.19

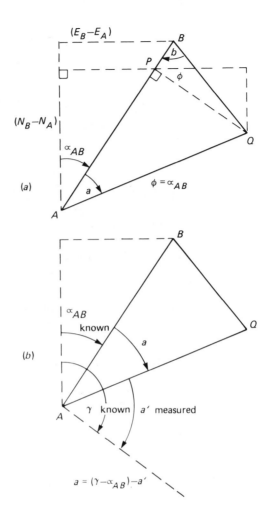

(a)

(b)

But $\quad AP = PQ \cot a$, and $AB = PQ \cot a + PQ \cot b$

Therefore $\quad E_Q = \dfrac{E_A \cot b + E_B \cot a + N_B - N_A}{\cot a + \cot b}$

Similarly $\quad N_Q = \dfrac{N_A \cot b + N_B \cot a + E_A - E_B}{\cot a + \cot b}$

Note that ABQ has been 'lettered' in a clockwise direction. Great care must be taken to ensure that the data are correctly presented when using the above expressions. If required, the expressions can be developed further by using the bearings of AQ and BQ in lieu of a and b:

$$E_Q = \frac{E_A \cot \alpha_{AQ} - E_B \cot \alpha_{BQ} - N_A + N_B}{\cot \alpha_{AQ} - \cot \alpha_{BQ}}$$

$$N_Q = N_A + (E_P - E_A) \cot \alpha_{AQ}$$

Fig. 6.20

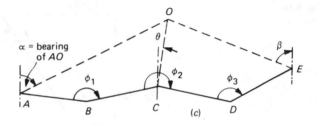

One application of the technique is the checking of an open traverse, bearings being taken from occasional stations, on to a salient point O, as shown in Fig. 6.20.

The co-ordinates of O may be determined from the observations within $ABCO$. From the observations in $OCDE$, the co-ordinates of E may be deduced and checked with those obtained directly from the angular and linear measurements along $ABCDE$.

> **Example 6.3** A new control station, F, is to be established from existing co-ordinated points T and D (Fig. 6.21). The horizontal clockwise angles at T and D have been observed as $D\hat{T}F = 44°\ 52'\ 36''$ and $T\hat{D}F = 284°\ 26'\ 38''$ respectively. Calculate the co-ordinates of station F given that the co-ordinates of T are 3931.82 mE, 7491.98 mN and of D are 2959.39 mE, 7487.09 mN.

Fig. 6.21

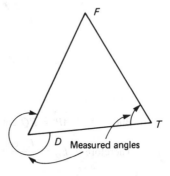

Measured angles

The co-ordinates of station F will be calculated directly using the standard expressions

$$E_F = \frac{E_T \cot \hat{D} + E_D \cot \hat{T} - N_T + N_D}{\cot \hat{T} + \cot \hat{D}}$$

$$N_F = \frac{N_T \cot \hat{D} + N_D \cot \hat{T} + E_T - E_D}{\cot \hat{T} + \cot \hat{D}}$$

in which $\hat{T}$ = 44° 52′ 36″ and $\hat{D}$ = 75° 33′ 22″.

Now cot $\hat{T}$ = 1.0043144 and cot $\hat{D}$ = 0.2575730

Therefore

$E_F =$

$$\frac{3931.82 \cdot 0.2575730 + 2959.39 \cdot 1.0043144 - 7491.98 + 7487.09}{1.0043144 + 0.2575730}$$

$$= 3154.00 \text{ m}$$

and

$N_F =$

$$\frac{7491.98 \cdot 0.2575730 + 7487.09 \cdot 1.0043144 + 3931.82 - 2959.39}{1.0043144 + 0.2575730}$$

$$= 8258.70 \text{ m}$$

It is suggested that the reader checks the values E_F and N_F by the alternative approaches.

Industrial measurement systems

One of the applications of the technique of intersection is in industrial measurement systems. Here, two electronic theodolites are used to accurately set up pieces of equipment and machinery or to align complex products such as an aircraft fuselage or the hull of a yacht. Figure 6.22 illustrates a typical configuration in which the X, Y and Z co-ordinates of point P are obtained from the measurement of the four angles indicated and a knowledge of length, b, and difference in height between the collimation axes of the instruments, h. With precise one-second theodolites, very accurate positioning or measurement of P can be obtained provided that b and h have been accurately measured. The system is normally computer controlled, with the two theodolites being linked directly to the computer allowing instantaneous inspection of the co-ordinates.

Fig. 6.22

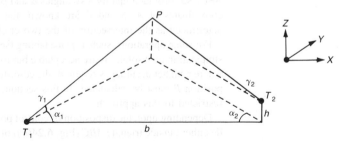

In Chapter 4 we discounted the subtense bar as a means of the measurement of survey lines, the role having been replaced by EDM. However, for the measurement of the relatively short distance between the theodolites (typically between 3 m and 20 m), EDM may not be accurate enough. In this context, a subtense bar set up approximately midway between the theodolites is an ideal method of determining b.

The overall accuracy of the position fix for P is dependent upon a number of factors. Zeiss quote the following 95% confidence values for their ITh 2 system:

(a) uncertainty of horizontal angle measurement, 0.06 mm per 10 m sighting distance;
(b) uncertainty of vertical angle measurement, 0.09 mm per 10 m sighting distance;
(c) length uncertainty of the base line, 0.02 mm at 3 m increasing to 0.22 mm at 20 m (using their subtense bar);
(d) position uncertainty, which depends upon the intersection angle of the lines of sight at P. With a 90° intersection at 10 m the uncertainty would be 0.12 mm but for 30° or 150° intersection this rises to 2.7 mm.

Overall accuracies of 0.05 mm could be achieved with a short base line and the best possible set up, and better than 0.5 mm with the worst practical arrangements.

The same equipment, and working method, can be used in the construction industry, for example, to survey the façade of existing buildings or accurately perform setting out tasks.

Resection

Resection is a method of determining the position of a point by observing horizontal directions from it to at least three points of known position. It is an ideal method of placing additional control points around a survey area in positions eminently suitable for detailing or setting out. Provided conspicuous points such as spires, aerials, beacons, etc., have been located by triangulation, traversing or intersection, any point at which three of these are visible can be co-ordinated.

The measurement of angle α (Fig. 6.23(a)) at P between A and B defines P to lie on the circle APB, whilst the measurement of angle β (Fig. 6.23(b)) at P between B and C defines P to lie on the circle BPC. So, provided that the two angles α and β are measured and the co-ordinates of A, B and C are known, the position of P can be determined as the intersection of the two circles APB and BPC.

For some operations, such as plane tabling (see later in this chapter), simple graphical solutions are acceptable but for the majority of work where a higher accuracy is required, the co-ordinates of the observer's position P must be calculated. In this section, consideration will be restricted to this approach.

Depending upon the disposition of control points A, B and C, P can lie either outside triangle ABC (Fig. 6.24(a)) or inside (Fig. 6.24(b)).

In each case, PB is a side common to two triangles in which A, B and C have known co-ordinates.

$$\frac{BP}{\sin \theta_1} = \frac{AB}{\sin \alpha} \quad \text{and} \quad \frac{BP}{\sin \theta_2} = \frac{BC}{\sin \beta}$$

Therefore

$$\frac{AB \sin \theta_1}{\sin \alpha} = \frac{BC \sin \theta_2}{\sin \beta}$$

Fig. 6.23

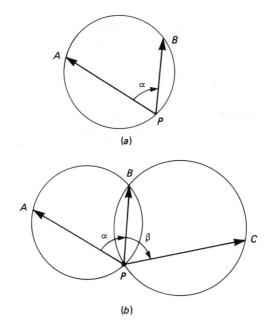

(a)

(b)

Fig. 6.24

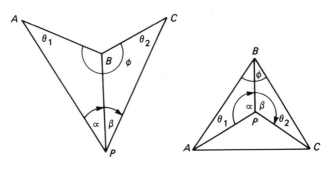

(a) Observer outside $\triangle ABC$ (b) Observer inside $\triangle ABC$

Also $\qquad\qquad \theta_1 + \theta_2 = 360° - \phi - \alpha - \beta$

Since AB, BC and ϕ can be found from the co-ordinate data, whilst α and β have been measured, θ_1 and θ_2 can be calculated.

The co-ordinates of P now follow since

(1) the lengths of AP, BP and PC can be calculated using the sine rule in triangles ABP, BPC;

(2) the bearings of those lines can be deduced from the bearings of AB and BC.

An alternative, and more rigorous, approach is given in Chapter 8.

Note that in two cases the determination of the co-ordinates of a point by resection can fail: firstly, when the three existing control points and the new point lie on, or very near to, the circumference of the same circle, and secondly, when all the points lie on, or very near to, the same straight line. Obviously, care must be taken when selecting the control points to ensure that neither of these distributions occurs.

Tienstra formula   If several such points have to be located it is preferable to use a standard formula such as that attributable to Tienstra. The method of station and angle identification must be adhered to without deviation.

Fig. 6.25

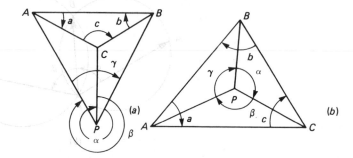

Note that the stations have been 'lettered' in a clockwise direction in Fig. 6.25.

α is the clockwise angle between directions PB and PC
β is the clockwise angle between directions PC and PA
γ is the clockwise angle between directions PA and PB

Write $K_1 (\cot a - \cot \alpha) = 1$
$K_2 (\cot b - \cot \beta) = 1$
$K_3 (\cot c - \cot \gamma) = 1$

Then

$$E_P = \frac{K_1 E_A + K_2 E_B + K_3 E_C}{K_1 + K_2 + K_3}$$

and

$$N_P = \frac{K_1 N_A + K_2 N_B + K_3 N_C}{K_1 + K_2 + K_3}$$

Example 6.4   Angles have been observed at station F (Fig. 6.26) between the co-ordinated control points P, W and D so that the position of the station can be determined by resection. Calculate the co-ordinates of F, given

Co-ordinates P 2876.24 mE   8754.11 mN
W 3810.80   7997.25
D 2959.39   7487.09

Tienstra's method will be used.
Bearings of PW, PD and DW are found as follows

Fig. 6.26

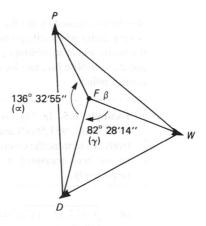

$$\alpha_{PW} = \tan^{-1} \frac{3810.80 - 2876.24}{7997.25 - 8754.11} = 129°\ 00'\ 09''$$

$$\alpha_{PD} = \tan^{-1} \frac{2959.39 - 2876.24}{7487.09 - 8754.11} = 176°\ 14'\ 43''$$

$$\alpha_{DW} = \tan^{-1} \frac{3810.80 - 2959.39}{7997.25 - 7487.09} = 59°\ 04'\ 13''$$

Thus in Fig. 6.26:

From co-ordinates			From data

$\alpha_{PW} = 129°00'09''$ Hence $P = \quad 47°14'34''$ $\quad \gamma = \quad 82°28'14''$

$\alpha_{WD} = 239°04'13''$ $\qquad W = \quad 69°55'56''$ $\quad \alpha = 136°32'55''$

$\alpha_{DP} = 356°14'43''$ $\qquad D = \underline{\quad 62°49'30''}$ Thus $\beta = \underline{140°58'51''}$

$\qquad\qquad\qquad\qquad\qquad\qquad 180°00'00''$ $\qquad\qquad\qquad 360°00'00''$

whence

$$K_1 = 1/(\cot D - \cot \beta) \qquad = 0.5722685$$
$$K_2 = 1/(\cot P - \cot \gamma) \qquad = 1.2619109$$
$$K_3 = 1/(\cot W - \cot \alpha) \qquad = \underline{0.7037878}$$
$$\qquad\qquad\qquad\qquad\qquad\qquad = 2.5379672$$

Hence $\qquad E_F = \dfrac{K_1 E_D + K_2 E_P + K_3 E_W}{K_1 + K_2 + K_3} = 3154.15 \text{ m}$

and $\qquad N_F = \dfrac{K_1 N_D + K_2 N_P + K_3 N_W}{K_1 + K_2 + K_3} = 8258.54 \text{ m}$

Free station

Many total station instruments, such as the Zeiss Elta range, and nearly all data loggers, contain software that will give the co-ordinates of points such as P in Fig. 6.25 following observation of distances and angle on to reflectors established at two stations, A and B. This is known as 'free station'. The co-ordinates of P are derived from the values of A and B previously entered into the system and the level of P can also be ascertained by height measurements with respect to A. This is very useful when setting out, since it allows the engineer to set up the instrument at an arbitrary point close to where he wishes to work, and then establish his exact location from observations onto two known control points.

Example 6.5 In Fig. 6.25, A and B have co-ordinates 1870.74mE, 953.56mN and 2013.53mE, 1020.23mN respectively. Determine the co-ordinates of P if, distances AP and BP having been measured by EDM as 143.94m and 178.08m respectively.

$$AB = \sqrt{(2013.53 - 1870.74)^2 + (1020.23 - 953.56)^2}$$
$$= 157.59 \text{ m}$$

Since $AP = 143.94$ m and $BP = 178.08$ m, from cosine rule

$$157.59^2 = 143.94^2 + 178.08^2 - 2 \times 143.94 \times 178.08 \cos P$$

Therefore
$$\cos P = 0.5383200$$
$$\hat{P} = 57°25'50''$$

From sine rule

$$\sin A = \frac{178.08 \times \sin 57°25'50''}{157.59}$$

$$= 0.9523145$$
Therefore $\quad \hat{A} = 72°14'05''$

Also $\quad$ W.C.$B_{AB} = \tan^{-1} \dfrac{(2013.53 - 1870.74)}{(1020.23 - 953.56)}$

$$= 64°58'18''$$

Whence $\quad$ W.C.$B_{AP} = 64°58'18'' + 72°14'05''$
$$= 137°12'23''$$

$E_p = 1870.74 + 143.94 \quad \sin 137°12'23''$
$$= \textbf{1968.53 mE}$$

$N_p = 953.56 + 143.94 \quad \cos 137°12'23''$
$$= \textbf{847.94 mN}$$

Although a position fix can be obtained by observations onto two stations, as shown above, for accurate work the surveyor should sight a third station for a check on the data.

The satellite station problem

As mentioned in the section on intersection, a church spire or similar tall feature can serve as an excellent signal and, hence, can be a useful control point. It is sometimes necessary to take angular observations from such points, however, and it would be extremely difficult to set up the instrument over this particular signal. In such a case the

Fig. 6.27

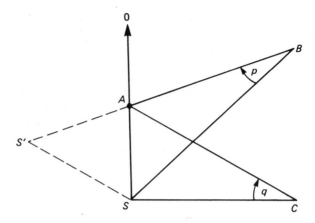

instrument is set up near the signal (which is at A in Fig. 6.27) at a nearby satellite station, or eccentric station, S, and the angle BSC measured at S is corrected to give the reading $B\hat{A}C$ as though the instrument were in fact set up at A.

BC is known, having been calculated from previuos data, and angles ABC and ACB will have been included in the programmes of angular measurement at stations B and C. A value of angle BAC is then estimated: $B\hat{A}C \approx 180° - (A\hat{B}C + A\hat{C}B)$. From this, approximate lengths of AB and AC are calculated using the sine rule. The distance AS and angles ASB and BSC are now measured as described previously to the required accuracy; the problem is then to obtain the best value for angle BAC.

In triangle ABS, $\sin p = \dfrac{AS}{AB}. \sin A\hat{S}B$

Also in triangle ASC, $\sin q = \dfrac{AS}{AC}. \sin (A\hat{S}B + B\hat{S}C)$

For small angles,

$$p'' = \frac{AS}{AB} \cdot \frac{\sin A\hat{S}B}{\sin 1''} \text{ and } q'' = \frac{AS}{AC} \cdot \frac{\sin (A\hat{S}B + B\hat{S}C)}{\sin 1''}$$

Now $O\hat{A}B = A\hat{S}B + p$
$O\hat{A}C = (A\hat{S}B + B\hat{S}C) + q$

Hence $B\hat{A}C = O\hat{A}C - O\hat{A}B = B\hat{S}C + (q - p)$

and the value reduced to true centre A is obtained. The angles in triangle ABC may now be summed up, and the triangular error determined and distributed.

It must be emphasized that in practice, accuracy of measurement of angle $B\hat{S}C$ must be of a high order when using a satellite station if serious errors and distortion of the framework are not to ensue. Bomford in *Geodesy* states that if AS/AB exceeds $1/1000$, significant error is likely to occur in the satellite correction.

The length AS would normally be of the order of, say, 10−15 m, and to retain an accuracy of 0.1 sec, that length must be correct to, say, 5 mm.

When observing angle $A\hat{S}B$ a pointing has to be made on a closely positioned elevated signal. Even with a diagonal eyepiece on the theodolite it will be difficult to measure $A\hat{S}B$ with the same precision as $B\hat{S}C$. Provided, however, that distance AS is small an error of one or two minutes of arc can be tolerated in $A\hat{S}B$ in so far as $(q-p)$ is concerned. Measuring distance AS to a high signal can be difficult. If a second station S' (Fig. 6.27) be established on the ground and distance SS' be determined then AS can be calculated after measuring angles $A\hat{S}'S$ and $S'\hat{S}A$.

Example 6.6 It is required to find the bearings of two lines TA and TB from a spire station, T, which has been co-ordinated by intersection. A satellite station S was set up 4.70 m from T in an approximately S−E direction, and from it the theodolite angles were measured:

Telescope pointing on *A* 00°00′00″
Telescope pointing on *B* 72°40′00″
Telescope pointing on *T* 305°00′00″

If point A lies N 10°04′32″E from station S, what are the true bearings of A and B from T?
AT is 15 km; BT is 16.7 km; log sin 1″ = $\overline{6}.68557$.

(*I.C.E.*)

$$T\hat{S}A \;=\; 360°00′00″ - 305°00′00″ \;=\; 55°00′00″$$
$$T\hat{S}A+A\hat{S}B \;=\; 55°00′00″ + 72°40′00″ \;=\; 127°40′00″$$

Fig. 6.28

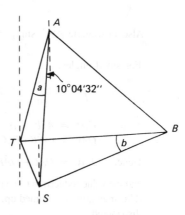

Using the notation of Fig. 6.28 in the formulae already given,

$$a \;=\; \frac{ST}{AT}\cdot\frac{\sin T\hat{S}A}{\sin 1″}$$

$$= \frac{4.70}{15\ 000} \cdot \frac{\sin 55^\circ 00' 00''}{\sin 1''}$$

$$= 53'' \text{ say}$$

Similarly
$$b = \frac{ST}{BT} \cdot \frac{\sin (T\hat{S}A + A\hat{S}B)}{\sin 1''}$$

$$= \frac{4.70}{16\ 700} \cdot \frac{\sin 127^\circ 40' 00''}{\sin 1''}$$

$$= 46'' \text{ say}$$

Therefore $A\hat{T}B$
$$\begin{aligned} &= A\hat{S}B + (b - a) \\ &= 72^\circ 40' 00'' + 46'' - 53'' \\ &= 72^\circ 39' 53'' \end{aligned}$$

Bearing TA
$$\begin{aligned} &= \text{bearing } SA + a \\ &= 10^\circ 04' 32'' + 53'' \\ &= \mathbf{10^\circ 05' 25''.} \end{aligned}$$

Bearing TB
$$\begin{aligned} &= \text{bearing } TA + A\hat{T}B \\ &= 10^\circ 05' 25'' + 72^\circ 39' 53'' \\ &= \mathbf{82^\circ 45' 18''.} \end{aligned}$$

Detail surveying

The approach to all surveys, whether it be to produce a topographic map or the plan for an engineering project, is to establish in the first instance a control framework as described previously, and then to fix detail or engineering features by measurements with respect to this framework.

As indicated in Chapter 1 there are four basic ways in which detail can be fixed with respect to a known line (Fig. 1.1), i.e. offsetting, tie lines, radiation and intersection. The procedure chosen for a particular job will of course depend on the personnel, the availability of equipment and its applicability to the task in hand; the accuracy required must also be considered.

It was suggested in Chapter 2 that measurements be taken to 10 mm rather than the limits implied by the scale of the finished plan and the specification but, in addition, distance and angular reading precisions can be matched. For example, if a point of detail has to be plotted within 0.25 mm at a scale of 1:100 then over a distance of 80 m this is equivalent to an angle of 60 seconds of arc. Thus the angles could be observed to that limit.

In general, it is always better to record measurements to a higher accuracy than required for plotting in case they are required for other purposes later.

Distinction is often made between 'hard' and 'soft' detail. The former implies well-defined points of detail such as buldings and walls, i.e. man-made features, whilst the latter generally refers to natural features such as river banks, hedgerows, etc., which are not so easy to define. Great care must always be taken when locating 'hard' detail, but the specification can sometimes be relaxed for 'soft' detail.

The processes of *offsetting* and measuring *tie lines* have been described in Chapter 2. On small traverse surveys the tape may be used to measure the leg lengths and the detail is normally collected during the course of measurement. It is preferable to split the work, as mentioned, so that the leg lengths are measured first and the detail collected later. In this way some check is made on the primary measurement.

Radiation is very suitable for detail collection. The process consists of measuring (i) the angles at a control station between the directions to another control station and the points of detail, (ii) the respective distances from the control station to the points of detail and (iii) the respective vertical angles subtended at the control station by these points (the vertical angles are used to reduce the distances to the horizontal and to calculate height difference). The use of stadia tacheometry in this context has been described in Chapter 4. Naturally the theodolite and tape can be used in combination to position detail by radiation but the detail needs to lie within one tape length of the instrument and steep slopes and/or 'cluttered' areas should be avoided. Height differences can be determined by measuring vertical angles to a mark on a ranging rod held at the detail point: this mark should be at the same height above the ground as the trunnion axis of the instrument.

The combination of the theodolite and electromagnetic distance measuring instrument is nowadays increasingly used to position detail by radiation, the reflector now being mounted on a pole at the matching height above the ground. In the simplest form, on each point of detail the horizontal and vertical circles of the theodolites are observed and the slope distance measured, and then the position and height of the detail point can be deduced. As explained in Chapter 5, Total Station instruments are available which automatically measure the horizontal and vertical circles and have the computing ability to give direct readouts of horizontal distance and height difference. Many of these instruments, as well as the electronic theodolites mentioned in Chapter 4, have the facility to download to a data logger or to record the data on magnetic cards. Especially with automatic recording, the measurements can be taken so quickly that two or even three assistants marking the points of detail are required to keep the instrument in continuous use. The range over which these units can be used is limited effectively by visibility, EDM range and the sensitivity of the theodolite.

The fourth method, *intersection*, is also used, but great care has to be taken when recording the observations to ensure that the direction from one station is matched with the corresponding direction from the other, since the two are usually observed at different times. The main use of this method is for the location of well-defined detail that is inaccessible from the control stations because neither staff nor target are required.

Plane-table surveying methods

The plane table differs from other surveying instruments in that the map or plan can be prepared in the field without the direct measurement of any angles and, except when contouring, without calculation. Maps can be produced directly with the plane table and its accessories since complete networks of control points may be arranged with it and the whole of the detail filled in. It may be put to this use, for instance, in the preparation of maps for exploratory surveys and where speed is required. Moreover, after control points have been fixed by a traverse or by triangulation the plane table can be used for the filling in of detail or the revision of plans. In the field it has much to commend it when compared with tape and offset surveying, particularly in urban areas, and it can be readily used for the location of contour-lines.

The equipment is simple, the most important piece being a drawing-board of convenient size mounted on a tripod as shown in Fig. 6.29(a); the plotting document should preferably be fastened to the underside of the board rather than damage the smooth upper surface.

Connection of the board to the tripod may be by means of a single screw, levelling being carried out by the legs, but Fig. 6.29(a) shows an adjustable head provided for this purpose, together with a spirit level. Although for most work the centring of the table above the ground station is not critical, a plumbing fork does allow more accurate location.

One important accessory is the alidade, Fig. 6.29(b), which, conventionally, is a sighting rule of either boxwood or metal and having folding sights which can be turned up at each end. One sight contains a narrow vertical slit or holes of small diameter, while the other consists of a vertical wire stretched across an open frame.

A more sophisticated alidade consists of a telescope, pillar-mounted on the base rule, which can be levelled independently of the table. Not only are clearer sights possible through the telescope, but distances and levels may be measured tacheometrically.

The basic procedure in plane tabling consists of taking sights on to an object with the table, correctly oriented, at two separate stations, Rays drawn on the board along these sight lines must intersect at the plotted position of the object. In Fig. 6.30 two stations A and B have been established on the ground at a known distance apart.

Line ab is plotted to the desired scale on the board which is set up at A (with a as nearly over A as possible). With the alidade laid along ab, the board is turned until a sight can be obtained on station B, and then clamped: by means of another accessory, the trough compass, the direction of magnetic north is found and marked on the plotting document, which nowadays is of matt-surfaced plastic material so that work can be carried out in damp conditions. With the board thus oriented, rays are drawn as shown taking care to identify them, after which the board is set up at B and re-oriented by sighting back on station A with the alidade held along ba. A check is obtained by means of the trough compass. Another set of rays is drawn as required, and the intersections locate the points of detail. New control stations, i.e. C, can be positioned relative to AB in a similar manner.

The choice of stations A, B, etc., will be governed by the nature of the country, but in general,

Fig. 6.29 (*a*) Plane table with tripod, spirit level and plumbing fork; (*b*) Typical alidade

(*a*)

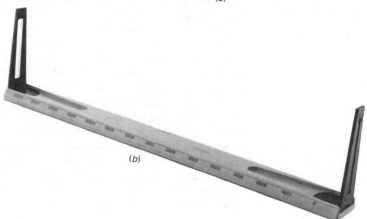

(*b*)

Fig. 6.30

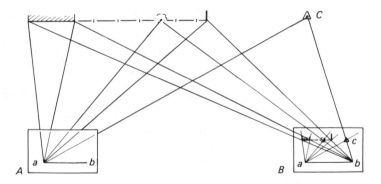

(1) they must be in conspicuous positions,
(2) they must be clearly defined and capable of being occupied in their turn,
(3) they must give well-conditioned triangles.

Detail can also be collected by *radiation*.

Refer to Fig. 6.31 in which the board is set up at a convenient station X and a series of rays drawn through x towards the points to be surveyed. The lengths XA, XB, etc., are now measured and the distances, reduced to the horizontal, are scaled off along the rays xa, xb, etc. The telescopic alidade mentioned previously is of great value in this method, which is mainly suitable for surveys of limited extent.

Fig. 6.31

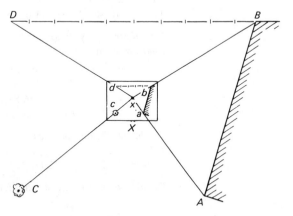

Exercises 6

1 A theodolite traverse carried out between two points A and B on the line of a proposed tunnel resulted in the following observations being obtained:

Line	Bearing	Length (m)
A–(i)	086°37′	128.88
(i)–(ii)	165°18′	208.56
(ii)–(iii)	223°15′	96.54
(iii)–B	159°53′	145.05

Calculate the bearing and distance of point *B* from point *A*.
Answer: 157°35′; 433.41 m

2 Using the data of a closed traverse *ABCDEA*, calculate the lengths of the lines *BC* and *CD*.

Line	Length	Bearing	ΔE	ΔN
AB	104.85 m	14°31′	+26.29	+101.50
BC		319°42′		
CD		347°15′		
DE	91.44	5°16′	+8.39	+91.06
EA	596.80	168°12′	+122.05	−584.21

(*I.C.E.*)

Answer: *DC* = 289.15 m; *BC* = 143.73 m.

3 The following table purports to give uncorrected lengths and bearings of the legs of a closed traverse, but it contains an error in transcription of one of the values of length.

Leg	AB	BC	CD
Length (m)	210.67	433.67	126.00
Bearing	20°31′30″	357°16′00″	120°04′00″
Leg	DE	EA	
Length (m)	294.33	223.00	
Bearing	188°28′30″	213°31′00″	

Find the error.   (*London*)
Answer: *BC* should be 343.67 m long.

4 *J* and *K* are survey stations and a mark is to be placed at *L*. Their co-ordinates are:

	E (m)	N (m)
J	120.38	66.07
K	26.83	77.48
L	60.10	27.05

Calculate the distances and bearings from *J* and *K* respectively to *L*.
Answer: 71.81 m, 60.42 m, 320°07′50″, 49°37′56″.

5 The following data were obtained for a closed traverse *ABCDEFA*:

	Length	ΔE	ΔN
AB	183.79	+183.79	0
BC	160.02	+98.05	+128.72
CD	226.77	−140.85	+177.76
DE	172.52	−154.44	−76.66
EF	177.09	0	−177.09
FA	53.95	+13.08	−52.43

Adjust the traverse by the Bowditch method. (*London*)
Answer: Corrections for *DE*, −0.05 m, +0.07 m.

6   The co-ordinates of three control survey stations are as follows:

Station 1	2487.19 mE	3547.37 mN
Station 2	3594.88 mE	1923.26 mN
Station 3	3602.72 mE	1917.12 mN

A theodolite was set up at Station 4 and the following horizontal circle readings were noted, the alidade having been rotated clockwise:

Pointing	Circle reading
Station 1	336°14′01″
Station 2	27°28′30″
Station 3	64°01′44″

Estimate the difference between the grid bearing of line 4−1 as derived from the above data and the recorded horizontal circle reading for the station. (*Salford*)
Answer: 10°09′46″.

7   The National Grid co-ordinates of two primary control stations, *R* and *B*, of a precise engineering survey network are

$$R: 324\ 022.039\ (E)\quad 342\ 846.959\ (N)$$
$$B: 324\ 967.822\ (E)\quad 341\ 829.829\ (N)$$

Horizontal angle measurements from two secondary control stations, *T* and *S*, to the South-West generally of *R* and *B* have been observed and the following mean values recorded:

$$R\hat{T}B = 43°39′15″ \quad T\hat{S}R = 33°59′41″$$
$$B\hat{T}S = 61°40′11″ \quad R\hat{S}B = 34°34′55″$$

The *approximate* azimuth of *ST* is 334°. Compute the National Grid co-ordinates of *T* and *S* and the azimuth and length of the line *S* to *T*.

(*Eng. Council*)
Answer: 322 958.452 mE, 341 914.592 mN; 324 280.398 mE, 341 083.822 mN; 334°5′7″; 1561.286 m.

8   (a) Describe the meaning of the following terms when used in the context of a survey control traverse:
     (i)   forward and back bearings;
     (ii) angular misclosure;
     (iii) linear and proportional misclosures; and
     (iv) Bowditch adjustment.
    (b) Explain how a singe blunder in either a measured angle or distance of a loop traverse may be isolated and suggest how you would correct for any mistake so located.
    (c) Figure 6.32 gives the observed angles for a horizontal traverse. Compute the unadjusted co-ordinates of the control

Fig. 6.32

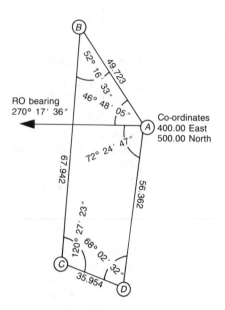

RO bearing
270° 17′ 36″

Co-ordinates
400.00 East
500.00 North

stations and confirm that the traverse misclosure is less than 1 part in 20000. *(London)*

Answer; c 355.077 mE   469.397 mN

9   In a triangulation a tall pointed spire, referred to as station *C*, is clearly visible from stations *A* and *B*. The horizontal distances *CA* and *CB* have been calculated from the unadjusted triangulation to be 3911 m and 3034 m respectively. A fourth station *S* is set up close to *C* so that *A, B* and *C* can be seen from it. The horizontal distance between *S* and a further station *T* is measured and found to be 22.10 m. The following mean horizontal circle readings are obtained from a theodolite set up at *S* and *T*:

Instrument at	Station sighted	Mean horizontal circle reading
S	C	038°08′46″
	A	174°52′20″
	B	232°01′06″
	T	347°23′34″
T	C	190°13′53″
	S	255°40′18″

Calculate the magnitude of the angle *ACB*.   *(I.C.E.)*

Answer: 56°49′14″.

10   In triangle *ABC*, station *C* could not be occupied and a satellite station *S* was established south of *C*. The following angles were then recorded on a theodolite set up at *S*:

Pointing on	Horizontal circle reading
B	$00°00'00''$
A	$71°54'30''$
C	$135°42'20''$

The lengths of *AC* and *BC* were estimated to be 16 480 m and 21 725 m respectively.

If the value of angle *ACB* was later deduced as $71°53'15''$, calculate the distance *S* from *C*. *(Salford)*

Answer: 16.30 m.

11   In triangle *ABC*, station *C* could not be occupied and so a satellite station *D* was selected at a distance of 9.22 m from *C* and lying inside that triangle. The following information was then obtained:

$$ABC = 64°48'40''$$
$$BAC = 53°11'20''$$
$$ADC = 141°27'20''$$
$$BDC = 156°29'30''$$
$$AB = 10\ 650\ \text{m}.$$

Deduce the value of angle *BCA* and thence the triangular error.

Answer: Error = $3''$.

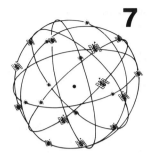

7 Orientation and position

In Chapter 6 we defined the bearing of a line at a station as the clockwise angle at the station between the direction to 'north' and the direction of that line. Furthermore we implied that 'north' could be true north, magnetic north or national grid north as applicable.

Orientation is synonymous with bearing from the surveyor's point of view, although the term azimuth rather than bearing can be encountered when referring to true north. In Fig. 7.1 the observer's terrestrial meridian is indicated as the line connecting the north and south poles and his station O. The true bearing of line OY at O is thus the horizontal clockwise angle between the direction to true north along the meridian at O and line OY.

Fig. 7.1

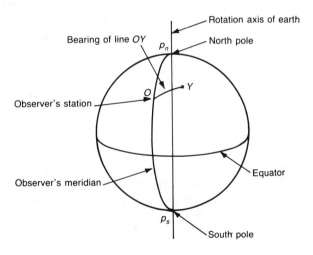

Two methods of determination of the true bearing of a line are described later in this chapter, but initially attention will be paid to the measurement of magnetic bearings with respect to the magnetic meridian which connects the magnetic north and south poles and the station.

Determination of magnetic bearing by compass

The fact that a magnetized needle, pivoted so that it can rotate freely in the horizontal plane, will come to rest in the magnetic meridian, has been used for centuries by navigators. Compasses are still used in surveying, the main instruments being (*a*) the prismatic compass, of which the best-known type is perhaps the military prismatic, widely used in rough and exploratory surveys, and (*b*) the compass theodolite, used also for such low-order surveys. Before dealing briefly with the compass and its use, however, it is necessary to consider one or two aspects of terrestrial magnetism in so far as they affect the accuracy of compass work.

The *magnetic* meridian, which is the line taken by the compass needle outside the range of any uncompensated masses of magnetized material, lies approximately N–S, and the angle between it and the *true* meridian is known as the *declination*. This varies from one locality to another and in any given locality it also varies with time.

Variations in the strength of the magnetic field at any one place produce the following variations in the declination on a time basis:

(*a*) *Secular variation*, the full cycle of which takes several centuries. The annual amount of this change is 5′–9′ at Greenwich. When magnetic bearings are given, the date, the declination, and the annual rate of change for the locality should also be given.

(*b*) *Diurnal variations*, which are more or less regular changes in the needle about its mean position during the course of the day. The maximum change on the daily basis is about 10′.

(ic) *Annual variations*, in which the period of variation is a year; these are so small that they may be ignored.

In addition, *magnetic storms* can cause sudden variations of as much as 1°. Of these various effects, the diurnal variation and magnetic storms are the most serious in reducing the accuracy of compass bearings. In addition there can be interference with the magnetic field caused by electric cables, small masses of iron, or iron ore and this is known as *local attraction*.

The prismatic compass

Figure 7.2 shows a 'card' type of prismatic compass, the needle being attached below the card upon which the arrow is painted. The needle and card are suspended on a pointed steel pivot working in a jewelled boss. Two sets of graduations, increasing clockwise, are provided on the card, the finer set being 'zeroed' at 180° from the arrowhead. The object whose bearing is required is sighted by means of a hairline engraved on the window placed in a vertical position. This is observed through the sighting slot of the prism casing when it has been swung upwards from the position shown into the reading position. The hairline is seen superimposed on the 'finer' graduations and thereby forms an index against which to read. The instrument must be held in a horizontal position when observing and bearings can be measured to above a quarter of a degree. When the needle is in the meridian the observer effectively looks at the south pole of the compass via the prism but a reading 0° is presented to him by those graduations. Thus when a bearing is being taken, a direct reading is obtained.

More precise tripod-mounted compasses are available, these are generally supplied with telescopes and in certain cases a vertical circle.

Fig. 7.2 Prismatic compasses

An example is the Wild B3 in which the telescope had a 2×
magnification and the needle is replaced by a graduated compass circle
which allowed bearings to be estimated to one tenth of a degree. In
addition, many modern theodolites and total station instruments come
supplied with compass attachments that allow the instrument to be
zeroed onto a northerly pointing.

Traversing with the compass

The compass can be used for rapid and exploratory surveys. Prismatic
compasses are generally held in the hand so that the length of the legs
in the traverse must be sufficiently long to reduce the effects of centring.
With such instruments, the *free-needle method* of surveying is used
in which the needle is floated before each reading so as to establish
the magnetic meridian relative to which the actual reading is taken.

Fig. 7.3

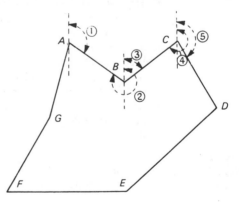

Referring to the traverse *ABCDEFGA* in Fig. 7.3, and starting at
A, the instrument is held horizontally and a sight (1) taken on *B*, the
forward bearing of *AB*. At *B*, angle (2), which is the *back bearing*
of *AB* (or the forward bearing *BA*), is read, and also angle (3), the
forward bearing of *BC*. This procedure is repeated round the traverse
and the readings should be booked as whole-circle bearings. The lengths
of the traverse legs must, of course, be measured.

The back bearing of each line should differ from the forward bearing by 180°, so that by comparison of the two, errors can be detected including, in particular, the presence of local attraction. The adjustment for this, and typical calculations, are shown in the following example. Bowditch's method is suitable for use in correcting a compass traverse; it gives greater angular distortion of the framework than the Transit rule, and this is required owing to the relatively large errors in angle measurement which occur in compass work.

Example 7.1 The following bookings were taken during a rough compass traverse round station *ABCDEA*. Since magnetic interference was suspected, forward and back bearings were taken at each station. If the magnetic declination was 12°W, plot the traverse to a scale of 1 in 5000, making any adjustments required and using co-ordinates relative to the true meridian.

Line	Length (m)	Forward bearing	Back bearing
AB	150	32°00′	213°30′
BC	54	78°30′	257°00′
CD	158	107°00′	285°00′
DE	106	120°00′	302°00′
EA	375	265°00′	85°00′

Considering each line in turn the forward and backward bearings should differ by 180°. This is true for line *EA* and we can assume that stations *A* and *E* are free from local attraction. Assuming the back bearing *DE* and forward bearing *AB* now to be free from error, corrections can be applied to the other bearings taken from *B* and *D*, which are affected by interference. At the same time, a study of the amounts of error at *B* and *D* shows that station *C* is free from interference. The corrected forward bearings are now tabulated and, after correction for declination, the easting and northing differences calculated. (Note that where the forward and backward bearings differ by only a few minutes, this probably only indicates diurnal variations and errors in reading: in such cases the mean of the two bearings is used.)

Line	Corrected forward bearing	True bearing	Length (m)	ΔE +(m)−	ΔN +(m)−	Adj ΔE +(m)−	Adj ΔN +(m)−	Point	Co-ordinates (m) E	N
AB	32°00′	20°00′	150	51	141	51	140	A	0	0
BC	77°00′	65°00′	54	49	23	49	23	B	51	140
CD	107°00′	95°00′	158	157	14	158	15	C	100	163
DE	122°00′	110°00′	106	100	36	100	36	D	258	148
EA	265°00′	253°00′	375	359	110	358	112	E	358	112
			843	357 359	164 160	358 358	163 163			

The easting and northing misclosures are -2 m and $+4$ m respectively and these errors are apportioned using the Bowditch rule. At a scale of 1:5000 an error of 1 m is represented by 0.2 mm and accordingly the corrections have been broadly taken to 1 m.

Determination of true bearing by gyroscope

The gyroscope has been used in navigation as a north-seeking device for a considerable period of time and surveying instruments are now available which allow the direct establishment of the meridian by theodolite without the need for calculations based on astronomical observations.

The development of the gyroscope for the precise transfer of bearings underground in mining surveying dates from about the beginning of the First World War, but an instrument capable of registering bearings to within one minute of arc did not appear until after the end of the Second World War. Even then it was of considerable weight, but advances in the design of gyroscopes for airborne inertial navigations sytems have allowed much more compact and portable units. For instance the Wild GAK1, described later, has a total mass of about 13.6 kg, and its accuracy is of the same order as that obtained in sun observations.

In this instrument a rotor or spinner is driven at speeds in excess of 20 000 rev/min, the axis of spin being in a horizontal-plane. Figure 7.4 shows the earth rotating about its axis with an angular velocity

Fig. 7.4

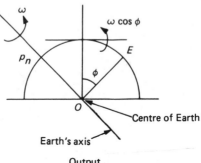

Fig. 7.5

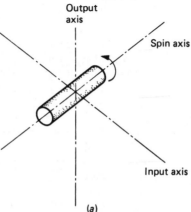

(a)

of ω; at a place in latitude ϕ an angular velocity of $\omega \cos \phi$ obtains about the meridian. The spinner of a gyroscope set up at this point will try to maintain its initial spatial position provided that its angular momentum, I_ω, is large, but the rotation of the earth itself pulls the spinner out of this plane. There is a consequent reaction in the form of a rotation, or precession, about the vertical (or output) axis of the spinner which holds until the spin axis lies in the meridian at the place. Figure 7.5 shows the three axes of the spinner and Fig. 7.6 shows the spin axis at an angle of α to the meridian. The earth's rotation causes an interference torque (M_E) equal to $I_\omega \omega \cos \phi$, and a consequent precessional couple of $M_E \sin \alpha$ is induced which forces the spinner into the meridian with an angular rate of $\omega \cos \phi \sin \alpha$. When the spinner is suspended (as it is in the GAK1 attachment) it will be seen that the maximum precessional couple occurs when ϕ is zero, i.e. at the equator, and that the gyro will float freely at the poles where $\cos \phi$ is zero. The precessing gyro does not align immediately on the meridian but oscillates about it.

Fig. 7.6

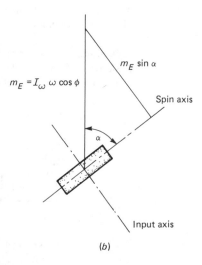

$$m_E = I_\omega \, \omega \cos \phi$$

$$m_E \sin \alpha$$

Spin axis

α

Input axis

(b)

GAK1 Attachment

The GAK1 attachment (Fig. 7.7) can be placed on instruments such as Wild T1A, T16 or T2A theodolites which must be suitably modified, and it is then connected to a control unit or converter. The oscillating system (Fig. 7.8(a)), containing the spinner and optical system with an index mark is carried by a tape suspended vertically within the 'chimney' of the supporting case and fastened at the top. When setting up, the theodolite telescope should be roughly pointed towards north by means of, say, prismatic compass or tubular compass observations, and the gyro attachment can then be fastened on the theodolite with the telescope *in the face-left position*. The gyro is clamped by means of a device underneath the supporting case, and then set to 'run'. A 'wait' indicator on the control box remains red for a short period of

Fig. 7.7 Wild GAK1 Gyroscope attachment (*Courtesy*: Leica UK Ltd)

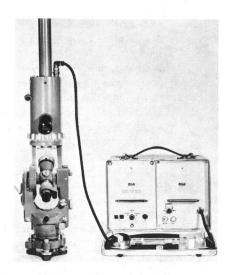

Fig. 7.8 (*a*) Part cross-section through GAK1 (*b*) Image of the index and auxiliary scale in the observation tube (*Courtesy*: Leica UK Ltd)

time before the green colour shows at 'measure', indicating that the spinner is now running at its correct operational speed and that it can be unclamped. As the spinner oscillates about the meridian the gyro index mark can be seen moving across an auxiliary scale in an observation tube, and when the spin axis and the line of sight of the theodolite are in the same vertical plane that mark should be centred in the middle of the scale, which is also marked by a V-shaped index (Fig. 7.8(*b*)).

Thus if the mid-position of the oscillations is established, the telescope line of sight at the station is oriented towards the north, and the purpose of all observations is to determine that position. A calibration constant

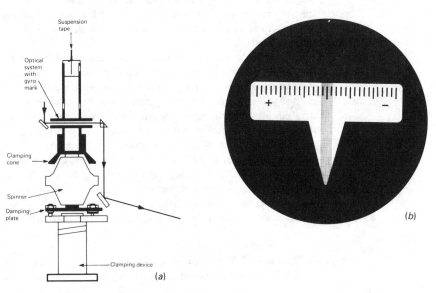

(E), which is the horizontal angle between the direction of the line of sight in the mid-position of oscillation and the meridian at a station, must however be determined for each instrument at frequent intervals by comparing the bearing, A_Z, determined by astronomical observations with that, A_G, determined by the gyrotheodolite, i.e. $E = A_Z - A_G$. The calibration constants of gyrotheodolites can show a systematic drift caused by unrequired torque on the gimbal system which, in the case of the suspended spinner, is the tape and the bearing of the spin axis. Chrzanowski has suggested in an article 'New techniques in mine orientation surveys', *Canadian Surveyor*, 24(I) 1970, that checks be made immediately before and after the underground work and on the same day.

The form of oscillation is sinusoidal and in the middle of an oscillation the index mark is moving at maximum speed in the relevant field of view but it slows down quite markedly as the turning points are approached and for a very short period is stationary at such points. If the gyro index mark has been kept centred within the V-shaped index by using the horizontal motion of the theodolite alidade, the corresponding horizontal circle reading of the theodolite can be obtained for this turning point.

Various methods of orientation are available including quick methods for pre-orientation before the adoption of a more precise method. In one of the quick methods the line of sight should be set within 15° of the estimated meridian and by following up the gyro index mark, with the V-shaped index in coincidence, horizontal circle readings are obtained for two successive turning points on opposite sides of the meridian. The mean horizontal circle reading gives the approximate meridian and the line of sight of the telescope can therefore be so established. It is essential that the two index marks be kept in coincidence so that the suspension tape is held free from torsion. Some initial practice is required for this and to facilitate the work some instruments are provided with a wider-range horizontal slow-motion screw. An accuracy of the order of ±3′ is quoted by Messrs Wild for this method.

The methods devised for the precise location of the centre of oscillation can be classified into tracking and non-tracking methods. One of the former group is an extension of the above quick method in that several successive turning points are observed, ensuring that the coincidence previously mentioned is maintained. The corresponding horizontal circle readings have to be obtained very quickly since the gyro index mark, relating to the spin axis, is only stationary for a very short period when it also has to be centred accurately on the zero division of the scale. Knowing the circle readings for pointings on the referring object (M, mean of readings on two faces) and the meridian (N), the two are related to give the geographical azimuth as $(M - N + E)$. Four to six turning points are observed for highest accuracy.

Example 7.2 Horizontal circle readings of 34°20.8′ and 214°20.4′ were measured face-left and face-right respectively by pointings on a reference mark from a station. Turning-point readings were given by gyro observation as follows:

Turning point left	Turning point right
353°44.0′	357°52.8′
353°46.6′	357°51.6′
353°47.0′	357°50.8′

If the calibration constant of the instrument is +2.1′ determine the azimuth of the reference mark from the station.

The horizontal circle value for the centre of oscillations can be determined by calculating the Schuler mean. Taking three successive turning point readings r_1, r_2 and r_3, as in Fig. 7.9, this mean (r_0) is given by the expression

$$r_0 = \frac{r_1 + 2r_2 + r_3}{4}$$

Hence the following obtain:

TP left	TP right	Schuler mean
353°44.0′		
	357°52.8′	355°49.0′
353°46.6′		355°49.4′
	357°51.6′	355°49.2′
353°47.0′		355°49.1′
	357°50.8′	
	Mean	355°49.2′

Fig. 7.9

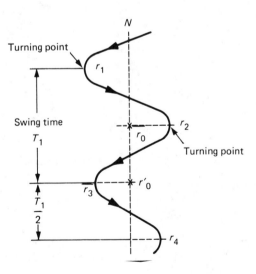

Azimuth of RM = 34°20.6′ + (360°00.0′) − 355°49.2′ + 2.1′
= **38°33.5′ clockwise from north.**

Thomas (*Chartered Surveyor*, March 1965) suggests the use of the four-point mean of

$$r_0 = \frac{r_1 + 3r_2 + 3r_3 + r_4}{8}$$

which represents the best fit to four observations using the method of least squares.

Another method, which falls into the second group, is that of timing the transits of the gyro index mark through the zero reading of the auxiliary scale by means of a stopwatch. In this method the approximate orientation (N') of the meridian is first obtained by, say, a pair of turning-point observations, to within $\pm 15'$ of the true meridian (N), and the telescope is clamped to maintain this direction, the horizontal circle reading being noted.

The gyro oscillations are now damped so that amplitudes ($a \pm \Delta a$) do not exceed the range of the auxiliary scale (15 divisions). To obtain an accuracy of $\pm 15''$, four or five transits are timed, but the gyro mark does not have to be continuously centred within the V-shaped index, the oscillation rhythm is not disturbed and the method is less tiring than the first method since the telescope remains fixed in position. It is essential, however, that a factor, known as the proportionality factor, be established.

Fig. 7.10

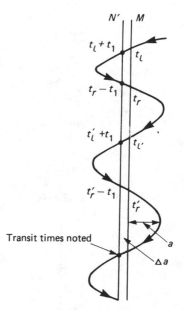

Figure 7.10 shows transit times of $t_l + t_1$, $t_r - t_1$, $t_l' + t_1$ etc., which imply swing times, one side or the other of the approximate meridian, of

$$(t_r - t_1) - (t_l + t_1) = t_r - t_l - 2t_1$$

and
$$(t_l' + t_1) - (t_r - t_1) = t_l' - t_r + 2t_1$$

when the three successive transits are timed. If the differences between these times be taken as Δt, then

$$\Delta t = (t'_l - t_r) - (t_r - t_l) + 4t_1$$

It will be apparent that $(t'_l - t_r) = (t_r - t_l)$ and so t_1, which equals the interval of time between transits across the approximate meridian and the centre-line direction of the clamped oscillations, has a value of $\Delta t/4$. Since the telescope remains position-fixed during the observations, so does the upper clamp of the suspension tape of the gyro. The swing time is therefore influenced by tape torsion and the centre-line position of the oscillations is identified with the resultant of the earth's rotation torque and the tape torsion torque.

In Fig. 7.11 the torques due to the earth's rotation and the tape torsion are represented by M_E and M_T respectively, the latter causing the centre line direction to be at γ from the true meridian. Since γ and Δa are small the relationship

$$\frac{M_E}{\sin \Delta a} = \frac{M_T}{\sin \gamma}$$

can be expressed as $\gamma = \dfrac{M_T}{M_E} \cdot \Delta a$ where $\dfrac{M_T}{M_E}$ is the torque ratio.

Therefore $\qquad \Delta N = \Delta a + \gamma = \Delta a \left(1 + \dfrac{M_T}{M_E} \right)$

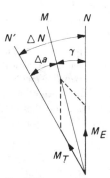

Fig. 7.11

When tape torsion is not present (as in the Turning Point Method) the swing time (T_1) is controlled by M_E alone, and for a particular gyro $T_1\sqrt{M_E} = K$.

Similarly, with the telescope line-of-sight position fixed, the swing time (T_2) of a whole period is given by the equation $T_2\sqrt{M_T + M_E} = K$.

Thus $\qquad\qquad \dfrac{M_T}{M_E} = \dfrac{T_1^2}{T_2^2} - 1$

and so, $\qquad\qquad \Delta N = \Delta a\, \dfrac{T_1^2}{T_2^2}$

Now $\Delta a = a \sin \omega' t_1$, and if observations are taken within the linear range of the sine curve (which applies to about one-sixth of the amplitude) then, neglecting damping effects since the GAK1 in fact is very lightly damped,

$$\Delta a = a\, \frac{\pi}{2}\, \frac{\Delta t}{T_2}, \quad \text{since} \quad \frac{\omega' T_2}{4} = \frac{\pi}{2} \quad \text{and} \quad t_1 = \frac{\Delta t}{4}$$

and so $\quad \Delta N = \dfrac{T_1^2}{T_2^3}\, \dfrac{\pi}{2}\, a\, \Delta t$

$$= ca'\, \Delta t$$

c, the proportionality factor, is equal to $(T_1^2/T_2^3)\,(\pi/2)m$, m being the value of one scale unit and a' the amplitude measured in those units, i.e. $a = a' \cdot m$. Its value can be determined by taking two different sets of observations related to approximate meridians, one east and one west of the true meridian. The reader can check that $c = (N_E' - N_W')/(a_E' \, \Delta t_E - a_W' \, \Delta t_W)$ where suffix E refers to the first and suffix W refers to the second of the two sets, with N' the relevant horizontal circle reading.

Once established, c can be considered an instrumental constant within a range of 1000 km even though it actually depends upon latitude.

Thus Δt, which is measured, and ΔN are related by the expression $\Delta N = c \cdot a' \cdot \Delta t$. Δt must not be greater than about 40 seconds so that the observations are taken in the linear range of the sine curve: a' is the mean of the amplitudes measured, left and right.

In the foregoing the centre of oscillations of the non-spinning freely-hanging gyro has been assumed to coincide with the scale index. Any residual error might well be regarded as a calibration constant within E. Reference can be made to the paper by H.S. Williams, 'A critical review of the practical use of the gyrotheodolite', *Survey Review*, No. 190, in this respect. In that paper he stresses the importance of good adjustment and levelling of the theodolite, and in his comprehensive review he also discusses the important point of possible deterioration of performance as the instrument ages.

Example 7.3 Horizontal circle readings of a reference mark were determined at a station as 326°34′00″ FL and 146°34′00″ FR and an approximate gyro orientation of 184°12.5′ was then established. Successive transit times and amplitudes were observed as shown in the table. Find the azimuth of the reference mark, taking the proportionality factor (c) as 0.047′/sec and the instrument calibration value (E) as −2.1′.

Time of transit (min) (secs)		Swing time Left + Right − (secs)	Time difference Δt (secs)	Amplitude reading	Amplitude a'
00	00.0				
		+229.3		+10.6	
03	49.3		+10.9		9.1
		−218.4		−7.6	
07	27.7		+11.1		
		+229.5		+10.6	
11	17.2		+11.0		9.1
		−218.5		−7.6	
14	55.7				

$$\Delta N = ca' \, \Delta t = +0.47 \times 9.1 \times 11.0$$
$$= +4.7'$$

Corrected azimuth of RM is now given as $M - N + E$

$$= 326°34'00'' - (184°12.5' + 4.7') - 2.1'$$
$$= \mathbf{142°14.7'}$$

It will be noted that the signs of Δt are related to those of the swing times, and that ΔN takes the sign of Δt.

It is good practice to allow the gyro to make two or three complete swings before starting the observation programme in this and other methods. In addition the set of readings could be repeated after positioning N' symmetrically at the other side of N.

The accuracy with which the auxiliary scale can be read is of paramount importance in so far as ΔN is concerned. R.C.H. Smith in his paper 'A modified GAK1 gyro attachment', *Survey Review* 183, has described the fitting of a micrometer to improve the reading accuracy to one-hundredth of a division. The micrometer is fitted immediately above the observation tube and it actuates a parallel plate placed in the downward part of the light path between the optical system with its measuring mark above the gyro, and the lower reflecting mirror shown in Fig. 7.8(a): the light passes through the graduated scale after reflection there.

The Sokkisha Gyro Station AP1-2 consists of a suspended gyro mounted on a total station instrument (SET 3) with external keyboard. Observations can be made using either the turning-point method or the transit method to establish the true meridian at a station. When this has been effected the SET 3 switches to azimuth display mode, revealing the true azimuth reading and its current horizontal circle reading in separate windows. The latter value can be converted to the former by actuating a key on the external keyboard and all further pointings can be so referenced. If the surveyor wishes, the gyro attachment can then be removed and the standard functions of SET 3 utilized, for example, to give co-ordinates of other stations or targets.

Determination of true bearing by observation to the sun

It was pointed out in the last section that the calibration factor E for a gyrotheodolite can be determined by relating the bearings of a line measured by astronomical observations and by the gyrotheodolite itself.

Traditionally, until recent times, the true bearing of a line was established by star, or sun, observations and, in fact, the latter approach still offers a very quick and simple method.

The earth rotates about its polar axis approximately once every twenty-four hours mean time (i.e. civil clock time). Accordingly the sun appears to move across the sky in a clockwise direction, its path being determinable and principally dependent upon two factors:

(a) the *latitude* of the observer (ϕ), measured north or south from the equator
(b) the variable, but predictable, parameter of the sun termed *declination* (δ). Values of δ are given in tables included in *The Star Almanac for Land Surveyors* and are interpolated for the date and time of observation. They are listed for each 6 hours of Greenwich Mean Time, now known as Universal Time (UT) and

The Star Almanac, which is published annually, records δ and other parameters at those times of UT throughout the year. δ is measured north or south from the celestial equator.

The sun pursues a curved path from sunrise in the east in the morning, to a maximum altitude in the middle of the day, to sunset in the west in the evening. However it must be remembered that the earth actually traces out an elliptical path around the stationary sun once every year. Thus since its polar axis is inclined at an angle of approximately $23\frac{1}{2}°$ to the normal to the plane of orbit round the sun, the sun's observed path varies throughout the year. This is reflected by the values of δ in *The Star Almanac*.

Fig. 7.12

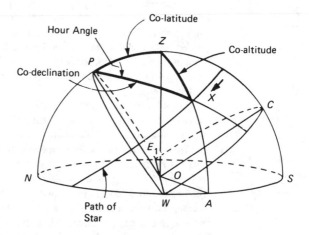

Figure 7.12 shows the celestial hemisphere for an observer in the northern hemisphere, the centre of the earth, O, being taken as the centre of that hemisphere. The sky can be considered to be the surface of the celestial hemisphere; the sun, and the stars as well, can be assumed to move over this surface. Z is the zenith of the observer, and is a point on the celestial sphere directly above the observer. P is the celestial north pole and WCE_1 is the celestial equator: these correspond to the terrestrial north pole and the terrestrial equator respectively. NE_1SW is the *true horizon* of the observer, N being the north point of the horizon. In the northern hemisphere the celestial pole is north of the zenith, while in the southern hemisphere the pole is south of the zenith.

The sensible horizon (Fig. 7.13) may be taken as the plane in which the level tube of a theodolite lies when it has been accurately set up.

The altitude and bearing of the sun are linked so that if the true altitude, $X\hat{O}A$ (Fig. 7.12), is determined, the bearing to the sun can be calculated by solving spherical triangle ZPX. Of course, there are two positions in which the sun reaches a particular altitude: one as it rises in the morning and the other as it sets in the afternoon, crossing meridian $NPZS$ at upper transit.

Fig. 7.13

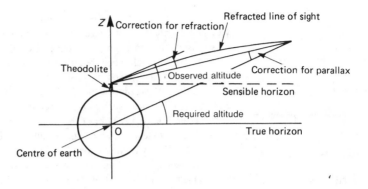

If ϕ and δ are known and the altitude (h) is measured, the bearing to the sun can be calculated using the formulae

$$\cos Z = \frac{\cos PX - \cos PZ \cos XZ}{\sin PZ \sin XZ}$$

or

$$\tan \frac{Z}{2} = \sqrt{\frac{\sin (S-XZ) \sin (S-PZ)}{\sin S \sin (S-PX)}}$$

where

$$2S = XZ + PX + PZ$$

If we denote altitude, latitude and declination as h, ϕ and δ respectively, $ZX = (90-h)$, $ZP = (90-\phi)$ and $PX = (90-\delta)$. This term co-altitude, etc. indicated on Fig. 7.12 can be encountered in this respect.

The first of the above two expressions can be written as

$$\cos A = (\sin \delta - \sin \phi \sin h) \sec \phi \sec h$$

where A, known as the azimuth angle, is the horizontal angle between the direction of true north and the direction to the sun, clockwise for a morning observation and anticlockwise for one taken in the afternoon (Fig. 7.14). Thus we can establish the bearing to the sun, and if the horizontal angle between a line and the sun is observed then the bearing of that line can be determined.

Fig. 7.14

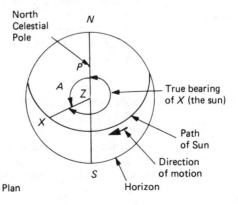

Plan

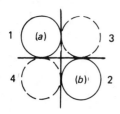

Fig. 7.15

Taking observations on the sun is difficult because in the field of view it presents a bright, large and moving target. Thus the observer must ensure that an appropriate dark filter glass is placed over the eyepiece of the telescope before pointing at the sun; otherwise permanent damage to the retina can occur. Due to its size it is difficult to place the intersection of the diaphragm lines on the sun's centre, and observations are usually made when the sun is touching both the vertical and horizontal hairs of the diaphragm, i.e. it is observed in one quadrant of the diaphragm with its limbs (or edges) in simultaneous contact. By observing in diagonally opposite quadrants on Face Left (*a*) and Face Right (*b*) (Fig. 7.15) the mean observations are equivalent to observing to the centre of the sun provided that they are carried out in quick succession.

Since the sun is moving across the field of view it is best to use just one slow-motion screw, thereby causing one crosshair only to be in motion. Say the sun is travelling downwards, i.e. its altitude is decreasing, then the observer should point at the sun on one face so that it appears in the top right-hand quadrant, well clear of both the horizontal and vertical hairs but moving towards them. Using the horizontal or vertical slow-motion screw the appropriate hair is brought into contact with the leading edge of the sun and by constant and careful manipulation of the slow-motion screw it is kept in contact until the sun touches the other hair as at position 3 in Fig. 7.15. At this point the circle readings are taken (and the altitude bubble position noted if the vertical circle is not compensated).

On changing face, a pointing is made such that the sun cuts both hairs in the bottom left quadrant of the graticule and in this position it is moving away from both hairs. Using the same slow-motion screw, the appropriate hair is brought to the trailing edge of the sun and so maintained until the sun is just breaking clear of the other hair. Again both circles, etc. are read. Either slow-motion device can be used, the choice possibly being dictated by the sun's motion, i.e. if it is moving faster vertically the vertical hair should be kept in contact. The basic principle is that on one face the sun is moving towards both hairs of the selected quadrant whilst on the opposite face it is moving away from both hairs of the diagonally opposite quadrant.

If the observations are taken in one quadrant only, a correction for the sun's semi-diameter would have to be applied to reduce the circle readings from the edges of the sun to its centre. Two corrections (Fig. 7.13) have to be made to the observed altitude to obtain the true altitude, *h*, of the sun.

(*a*) Correction for refraction

A ray of light from the sun is refracted as it passes through the atmosphere such that the body appears at a greater altitude than it actually is. An approximate value for this correction is $-58''$ cot h, where h the apparent altitude should be over 20° at least for this correction to be used. Accurate values call for readings of atmospheric pressure and air temperature and recourse to the *Refraction Tables of the Star Almanac*.

A measurable angle is subtended at the sun by the earth's radius, and the observed altitude of the sun from the earth's surface will be slightly smaller than the altitude which would be observed from the earth's centre.

The correction required is $+8.80''\cos h$, where h is the altitude corrected for refraction.

(1) Set up the theodolite at one end of the line whose direction is required and level it very carefully. Place a target at the other end to use as referring object (RO).

(2) On Face Left (say) observe the RO and record the horizontal circle reading.

(3) Place a dark filter over the eyepiece and point on the sun using the procedure given above. Record the horizontal and vertical circle readings and the time (the nearest minute will suffice).

(4) Change face and repoint on the sun using the appropriate procedure. Record the horizontal and vertical circle readings and the time.

(5) Remove the filter, point back to the RO and record the horizontal circle reading.

This is one observation, and for each determination two or three such observations should be taken, as implied in Fig. 7.15.

Example 7.4 Observations taken on the sun at a place in N 53°29′19″ gave the tabulated data. (Dec. 3°25′6″S at 12 h UT decreasing at 59″ per hour.) Determine the true bearing of RO.

Readings

Object	Face	Horizontal circle	Altitude	UT
RO	R	60°00′00″		
Sun ⨁	R	191°38′00″	22°56′30″	15 hr 12 min
Sun ⨁	L	12°39′30″	22°13′00″	15 hr 14 min
RO	L	240°00′00″		

$$\text{Mean altitude} = \frac{22°56′30'' + 22°13′00''}{2}$$

$$= 22°34′45''$$

$$\text{Refraction correction} = -58'' \cot 22°34′45''$$
$$= -2′19''$$

$$\text{Parallax} = +8.8'' \cos h$$
$$= +8.8'' \cos 22°32′26''$$
$$= +8.0''$$

Hence corrected altitude (h_1) = 22°32'34"

$$\text{Mean UT of observation} = \tfrac{1}{2}[15 \text{ hr } 12 \text{ min} + 15 \text{ hr } 14 \text{ min}]$$
$$= 15 \text{ hr } 13 \text{ min}$$
$$= 15.217 \text{ hr}$$

$$\text{Hence declination of sun } (\delta) = 3°25'6" - 3.217 \times 59"$$
$$= 3°21'56" \text{ S}$$

$$\text{Latitude of place} = 53°29'19" \text{ N}$$

$PZ = 90 - \phi = 36°30'41"$	$S - PZ = 62°09'20"$
$ZX = 90 - h_1 = 67°27'26"$	$S - ZX = 31°12'35"$
$PX = 90 + \delta = 93°21'56"$	$S - PX = 5°18'05"$

Therefore
$$2S = 197°20'03"$$
$$S = 98°40'01"$$

Now
$$\tan \frac{Z}{2} = \sqrt{\frac{\sin (S-ZX) \sin (S-PZ)}{\sin S \sin (S-PX)}}$$

whence
$$Z = 131°52'46"$$

Now mean horizontal angle between RO and sun
$$= \tfrac{1}{2}[131°38'00" + 132°39'30"]$$
$$= 132°08'45"$$

Therefore true bearing of RO $= 360° - 132°08'45" - 131°52'46"$
$$= \textbf{95°58'29" clockwise from north}$$

Note: (i) Sun's southerly declination gives a co-declination of $(90 + 3°21'56") = 93°21'56"$; (ii) the observations, on a clock-wise reading instrument, place the sun 'ahead' of RO; then the RO is placed relative to the sun as in Fig. 7.16; (iii) since observations are taken

Fig. 7.16

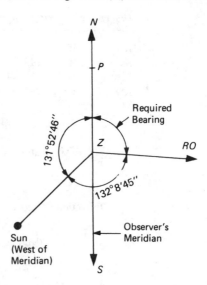

after the sun's upper transit, then the sun will be west of the observer's meridian; (iv) in general it is best to observe when the sun is at a reasonable altitude, say $20°$ to $30°$; (v) the azimuth angle could have been calculated using the cosine relationship.

Bearings of long lines

In Chapter 6 the statement was made that $\alpha_{BA} = \alpha_{AB} \pm 180°$ in respect of the bearings of line AB at its terminals. Over long distances the bearing of a great-circle line AB will change from α to $\alpha + \delta\alpha$ if A and B are at different latitudes, since the meridians at A and B are not parallel. Providing that $(\phi_A - \phi_B)$ and $\delta\theta$ are small it can be shown that $\delta\alpha = \delta\theta \sin[(\phi_A + \phi_B)/2]$ where $\delta\theta$ is the difference in longitude between A and B, since from Fig. 7.17

Fig. 7.17

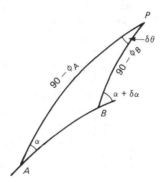

$$\tan \frac{\alpha + (180 - \alpha - \delta\alpha)}{2} = \cot \frac{\delta\theta}{2} \left[\frac{\cos \dfrac{(90 - \phi_B) - (90 - \phi_A)}{2}}{\cos \dfrac{(90 - \phi_B) + (90 - \phi_A)}{2}} \right]$$

whence
$$\cot \frac{\delta\alpha}{2} = \left(\frac{\cot \dfrac{\delta\theta}{2} \cos \dfrac{\phi_A - \phi_B}{2}}{\sin \dfrac{\phi_A + \phi_B}{2}} \right)$$

In so far as relatively short lines (up to 40 km, say) are concerned, lengths and bearings in respect of points of known latitude and longitude can be estimated by considering the meridians and parallels of latitude through A and B to give rectangular axes. The separation distances between the axes are then denoted by $\lambda(\phi_B - \phi_A)$ and $\mu\delta\theta$ respectively, where λ and μ are the lengths of $1''$ of latitude and $1''$ of longitude at the mean latitude of $(\phi_A + \phi_B)/2$.

Thus in Fig. 7.18 we have

$$\lambda(\phi_B - \phi_A) = l \cos \left(\alpha + \frac{\delta\alpha}{2} \right) \quad \text{and} \quad \mu\delta\theta = l \sin \left(\alpha + \frac{\delta\alpha}{2} \right)$$

Fig. 7.18

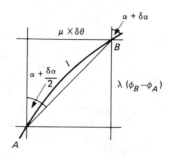

These formulae are reasonably valid for latitudes not exceeding 60°. If the latitudes and longitudes of the terminal stations are given then $\delta\alpha$, α and l can be determined directly. In other cases an indirect approach may be needed as shown in the following example.

Example 7.5 A line AB of length 10 854 m has a bearing of 46°21′20″ at A in latitude 52°20′15″N and longitude 24°28′50″E. Estimate the reverse bearing at B.

Latitude	Length of 1″ of latitude	Length of 1″ of longitude
52°20′	30.9022 m	18.9364 m
52°25′	30.9107 m	18.9008 m

It will be noted that $(\phi_A + \phi_B)/2$ is not known and therefore some approximations must be made.

In the first instance we can assume that the latitude of A is nearly equal to the mid-latitude of AB, i.e. 52°20′15″N.

Thus $\qquad \lambda = 30.9022 + \dfrac{15}{300} \times 0.0085 = 30.9026$ m.

Then the approximate latitude difference

$$= \frac{l \cos \alpha}{\lambda} = \frac{10\ 854 \cos 46°21′20″}{30.9026}$$

$$= 242.4″, \text{ neglecting } \delta\alpha/2$$

The approximate mid-latitude is therefore 52°22′16″ to the nearest second.

For this value $\mu = 18.9364 - \dfrac{136}{300} \times 0.0356 = 18.9203$ m

and $\qquad \delta\theta = \dfrac{10\ 854 \sin 46°21′20″}{18.9023} = 415.1″, \text{ neglecting } \delta\alpha$

This allows an assessment of $\delta\alpha$ to now be made since

$$\delta\alpha = \delta\theta \, \sin \frac{\phi_A + \phi_B}{2} = 415.1'' \times \sin 52°22'16''$$

$$= 328.8''$$
$$= 05'28.8''$$

Now
$$\delta\theta = \frac{10\,854 \sin (46°21'20'' + 02'44'')}{18.9203}$$

$$= 415.4'', \text{ as a further estimation,}$$

with
$$\delta\alpha = 415.4'' \times \sin 52°22'16'' = 329''$$

A check can be made at this stage to revise the first approximation of latitude difference, and the reader can check that $\delta\phi$ is now about 242.2'', which still implies a mid-latitude of $52°22'16''$ to the nearest second. Therefore the value of $\delta\alpha = 05'29''$ can be accepted, to give the bearing of AB at B equal to $46°26'49''$ and the reverse bearing at B as **226°26'49''**.

The Transverse Mercator Projection of Great Britain: grid bearing

The grid or projection bearing of a line cannot be directly observed since it depends upon the projection co-ordinate system in use. In this section the Transverse Mercator Projection of Great Britain is described briefly, and the relationship between grid bearing and true bearing explained.

The position of a point can be defined by its latitude and longitude, the 'axes' being the equator and the Greenwich meridian respectively.

Fig. 7.19

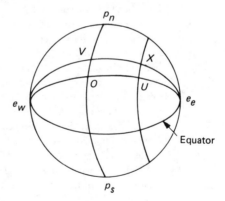

However, another reference meridian p_sOp_n (Fig. 7.19) could be adopted and an origin O could be established on that meridian as required. Point X could be located by its perpendicular distance VX along the great circle e_wVe_e from p_sOp_n and its distance OV from O along that meridian. VX can be defined as the easting of X, and OV as the northing of X in this system. All points having the same easting as X would lie on the small circle XU and all points having the same northing would lie on e_wVe_e, but it will be apparent that the latitudes of points on e_wVe_e will vary. Thus northing co-ordinates measured from e_wOe_e along the relevant small circles, e.g. UX, would be smaller than distances measured along p_sOp_n. It will be seen, therefore, that from this aspect alone, if p_sOp_n and e_wOe_e were laid down as straight lines, mutually at right angles and in a plane, X would have to be displaced so that OV was its actual northing co-ordinate, and not UX.

Plane rectangular axes are, however, very convenient for recording the positions of points by co-ordinates, and the National Grid reference system of the Ordnance Survey consists of a series of lines parallel to the adopted central meridian with a series of lines at right angles thereto, to give a square grid. The positions of the triangulation stations in Great Britain are defined by their National Grid co-ordinates to 0.001 m, and are available from the Ordnance Survey. In addition, the National Grid lines are superimposed on maps published by the Ordnance Survey and detail points can be uniquely referenced thereto; for instance the 1:2500 maps show the grid at 100 m intervals. This grid was obtained by a Transverse Mercator Projection, the origin being at 49°N on the central meridian through 2°W. The earth is not a true sphere as implied in Fig. 7.1 since it actually approximates to a spheroid, the solid produced when an ellipse is rotated about its minor axis. The ocean surface when considered motionless gives an equipotential surface of gravitation, a plumb-line being perpendicular to it at any point. As mentioned in Chapter 3, heights can be determined relative to the geoid, or mean sea-level surface (assumed continued through land masses) which is very nearly a spheroid. However, since it is not exactly a spheroid, the geoid is not taken to be suitable for geodetic computation, and a spheroid of reference is adopted.

Observations have been made at various times in various parts on the surface of the earth, and the form of the spheroid of reference indicated by these observations varies somewhat, though not greatly. As a result, a spheroid of reference may be chosen to suit the area under consideration. Spheroidal or geodetic co-ordinates of latitude and longitude may be related to an origin as on the celestial sphere.

The Airy spheroid of reference was taken to represent the earth in the Transverse Mercator Projection of Great Britain, its major and minor axes being 6377542 m and 6356236 m respectively. In the basic projection the spheroid would be represented on a cylinder touching it along the central meridian, and the scale would be constant along that meridian but would increase to either side. However, in the actual projection some adjustments were made, resulting in the scale being correct on lines nearly parallel to the central meridian, and about two-thirds towards the edges of the projection.

The north–south lines on the grid are parallel to the central meridian throughout, and therefore grid north at any point off the meridian will not be the same as true north at that point. Also, since the grid is square (being based on 10 km sides) the scales of distance along the lines perpendicular to the central meridian had to be increased to agree with the scales parallel to that meridian. The scale is therefore constant in all directions at any one point, although it changes slightly with distance from the central meridian, and the projection is said to be orthomorphic. To convert true distances (S) measured along the spheroid to grid distances (s) derived from the grid co-ordinates, a scale factor (F) must be applied in accordance with the expression $s = S \times F$. Values of F are listed in *Constants, Formulae and Methods used in Transverse Mercator Projection* published by HMSO, F being 1.00000 at easting distances of 180 km from the central meridian and 0.99960 on the meridian. It must be pointed out that the National Grid has in fact been based for convenience on a false origin 400 km to the west and 100 km to the north of the 49°N, 2°W origin so that all eastings will in fact be positive (at 2°W they are 400 km) and no northing will exceed 1000 km.

Furthermore, to convert grid bearings to true bearings a convergence factor C must be introduced to allow for the angle between the north−south grid line and the meridian at a point. C_1 and C_2 are not equal since the meridians at A and B are not parallel (Fig. 7.20). In addition, for very long sighting distances and in very accurate work, an extra correction known as the $(t - T)$ correction is to be applied, to allow for the fact that line of sight between two points will be effectively curved in the projection as in Fig. 7.20. Data for these corrections can be found in *Projection Tables for the Transverse Mercator Projection of Great Britain* published by HMSO.

Fig. 7.20

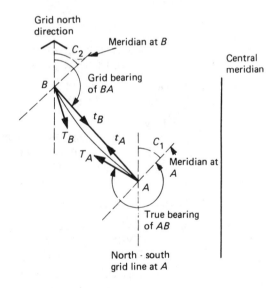

Example 7.6 In Fig. 7.20 the grid co-ordinates of A and B are 356724.280 mE, 420483.740 mN and 340526.570 mE, 455316.140 mN respectively. Determine the true bearings of AB and BA together with the spheroid distance AB.

$$\text{Grid bearing of } BA = \tan^{-1} \frac{356724.280 - 340526.570}{420483.740 - 455316.140}$$

$$= 155°03'38.5''$$

$$\text{Grid length } BA = \frac{356724.280 - 340526.570}{\sin(180° - 155°03'38.5'')}$$

$$= 38414.35 \text{ m}$$

To determine the value of C_1 and C_2 we use the general expression

$$c'' = Q(\text{XVI}) - Q^3(\text{XVII}) + Q^5(\text{XVIII})$$

in which $Q = (E - 400000) \times 10^{-6}$ and (XVI) (XVII) (XVIII) are values tabulated in the columns of the Projection Tables. These values are listed for corresponding latitudes which in turn may be found from Table I against northing values.

Thus for A

$$Q = (356724.280 - 400000) \times 10^{-6}$$
$$= -0.04327572$$

and so
$$Q^3 = -0.00008105$$
$$Q^5 = -0.00000015$$

Since $N_A = 420483.74$ m, Table I gives $\phi_A = 53°40'50.11''$

Whence (XVI) $= 43919.541$
(XVII) $= 1021.561$
(XVIII) $= 37.832$

$$C_1'' = (-0.04327572 \times 43919.541) - (0.00008105 \times 1021.561)$$
$$- (0.00000015 \times 37.832)$$

Hence $C_1 = -1900.73''$
$$= -31'40.7''$$

Similarly for B

$$Q = (340526.570 - 400000) \times 10^{-6}$$
$$= -0.05947343$$
$$Q^3 = -0.00021036$$
$$Q^5 = -0.00000074$$

Since $N_B = 455316.140$ m Table I gives $\phi_B = 53°59'37.30''$

Whence (XVI) $= 44425.473$
(XVII) $= 1048.887$
(XVIII) $= 39.504$

$$C_2'' = (-0.05947343 \times 44425.473) - (0.00021036 \times 1048.887)$$
$$- (0.00000074 \times 39.504)$$

Hence $C_2 = -2642.36''$
$$= -44'2.4''$$

$(t - T)$ corrections are given by the general expression,

$$(t_1 - T_1)'' = (2y_1 + y_2)(N_1 - N_2)(\text{XXIII}) \times 10^{-9} \text{ m}$$

in which $y = (E - 400000)$

and so $\quad y_A = - 43275.720$

$\qquad y_B = - 59473.430$

Thus at A

$\qquad 2y_A = - 86551.440$

$\qquad y_B = - 59473.430$

$\quad 2y_A + y_B = - 146024.870$ m

and $N_A - N_B = - 34832.400$ m

Value $(XXIII)_m$ is taken against the mean latitude of A and B. Now $N_m = (N_A + N_B)/2 = 437899.940$ m and so from Table I, $\phi_m = 53°50'13.71''$. Thus for line AB $(XXIII)_m = 0.84417$.

Whence

$(t_A - T_A)'' = - 146024.87 \times - 34832.40 \times 0.84417 \times 10^{-9}$

$\qquad\qquad = +4.3''$

Similarly at B

$\quad N_B - N_A = + 34832.400$ m

$\qquad 2y_B = - 118946.860$

$\qquad y_A = - 43275.720$

$\quad 2y_B - y_A = - 162222.580$ m

and

$(t_B - T_A) = - 162222.580 \times 34832.400 \times 0.84417 \times 10^{-9}$

$\qquad\qquad = -4.8''$

The calculated values are linked together as,

True bearing = Grid Bearing + $C - (t - T)$

Whence

	AB	BA
Grid bearing	335°3'38.5''	155°03'38.5''
C	−31'40.7''	−44'2.4''
−(t−T)	−4.3''	+4.7''
True bearing	**334°31'53.5''**	**154°19'40.9''**

Note that in the projection the line of sight is concave to the central meridian and this is true for lines either side of that meridian.

Local scale factor (F) can also be derived from the data in the Projection Tables based upon the mean easting and northing using Tables XXI and XX2. The general expression is,

$$F = F_0 [1 + Q^2(XXI) + Q^4(XXII)]$$

in which $\qquad F_0 = 0.99960127$

Table XXV however gives F values directly to five places and the relative value therefrom will be adopted in this example.

For E_m = 348625.425 m
F = 0.99963
S = $\dfrac{\text{grid length}}{F}$ = $\dfrac{38414.35}{0.99963}$

Spheroid distance = **38428.57 m**

National Grid co-ordinates mentioned above are referred to the Ordnance Survey Great Britain 1936 datum derived from the adjustment of the re-triangulation (see Chapter 6) carried out between 1935 and 1952. Further checking of the re-triangulation revealed some variations over the country and, accordingly, further observations, both angular and linear, were made, together with azimuth evaluations. The consequent re-adjustment of the triangulation is termed the Ordnance Survey Great Britain 1970 (Scientific Network) and co-ordinates are available for the primary stations: the 1936 datum co-ordinates will, however, continue to be used for the National Grid mapping.

Position determination

The traditional method for the direct determination of position has been by astronomical observation. The relative positions and movements of the stars are well catalogued and so with a combination of altitude, direction and time observations to the stars, the position in terms of latitude and longitude of a ground station can be calculated. There is less call for direct position determination nowadays since most countries are covered by a primary horizontal control scheme, and the absolute position of any new local survey work can be established by including a national reference point in the local survey.

In very remote areas, or where the terrain is totally unsuitable for the classic survey methods of triangulation and traverse, there is, however, a need for direct position determination. For the majority of this work, visual observation to the stars has been replaced by electromagnetic measurements to or from artificial earth satellites. In addition, the relative positions of survey stations can be determined directly by inertial techniques originally developed for aircraft navigation. Reference can be made to a report in *New Civil Engineer*, 17 January 1980 on their value to the highway engineer.

Global positioning systems

In previous editions we have discussed various systems for the continuous observation of targets travelling at speed above the Earth, e.g. the Askania Potentiometer Theodolite and the Wild BC4 Ballistic Camera. The former was capable of tracking satellites to determine their orbits whilst the latter could track missiles, three of these instruments being needed for position fix by resection.

During the last two decades the US Navy's Transit Satellite System has been available to the surveyor and this too was discussed in the last edition. It is now being superseded by the Global Positioning System (GPS) which is to be operational in the early 1990s and which is scheduled to give enhanced accuracy. A total of 21 satellites will be

Fig. 7.21 Geotracer
100 GPS receiver
(*Courtesy*: Geotronics
(UK) Ltd)

in orbit in the system, placed within six near-circular orbital planes
inclined at 55° to the equator as shown at the chapter heading. The
satellites are at altitudes in excess of 20 000 km (those of Transit were
at about 1000 km) and they will orbit the Earth every twelve hours.
This disposition should virtually ensure that at least four satellites can
be observed from any point on Earth.

By virtue of an oscillator, having a frequency of 10.23 MHz, each
satellite continuously transmits two carrier signals at frequencies of
1575.42 MHz (L1) and 1227.60 MHz (L2). These carriers are phase
modulated by binary code signals which contain time information, one
being termed the coarse acquisition (C/A) code and the other the precise
(P) code. The L1 carrier is modulated with both codes, C/A at a
frequency of 1.023 MHz and P at a frequency of 10.23 MHz: the L2
carrier is modulated with the P-code only. Each satellite has a unique
C/A and P-code for its identification. Access to the P-code is not
guaranteed, and in certain eventualities will only be available to military
users, thus the C/A code is the one used in general practice by the
surveyor. In addition to the codes, parameters relating to the satellite
position are transmitted: these are determined from data collected at
four tracking stations and updated from a control station in Colorado,
USA.

At the survey station an antenna and receiver are required and due
to the restriction mentioned above for the P-code some manufacturers
arrange for L1 tracking only by their receivers. The instrument shown
in Fig. 7.21 is so designed. The transmitted signals are received by
the antenna, mounted on a tripod over the survey station, processed

electronically, and passed by cable to the receiver/data logger where a microprocessor reduces the data. The Geotronics Geotracer 100 receiver, has eight L1 channels, i.e. transmission circuits, for the signals from the satellite, but there is an option available which provides for dual frequency tracking by eight or twelve channels of L1 and L2.

GPS is a uni-directional method of distance measurement which depends upon accurate time measurement, facilitated by precise synchronization of clocks in both satellite and receiver. Identification of the number of wavelengths transmitted by the satellite is determined by combining carrier phase and coarse acquisition code measurements. The frequency of the carrier wave transmitted by the satellite is subjected to the Doppler effect. Recourse may be made to a physics textbook in this respect but, essentially, it can be said that as a source transmitting waves of constant frequency (f_S) directly approaches a receiver, the frequency (f) received is enhanced by the Doppler shift such that if the source has a velocity, v_S, the frequency will have a constant value of $f_S v/(v - v_S)$ in which v is the transmission velocity. Since generally the receiver will lie out of the orbital plane of the satellite frequency f, will not be constant. The signal being received is compared with a constant reference signal generated in the receiver and phase difference is determined similar to the logic of EDM. Since the theoretical L1 wavelength is 19 cm, distance can be established to very fine tolerances. Also the reference C/A code generated in the receiver can be related to that received from the satellite to give the time shift, which effectively is the difference between transmission time and reception time, both measured in the time frame of the satellite. This is transformed into distance by application of the velocity of the electromagnetic wave. The distance established at this stage is known as pseudo-range since any clock error has not been corrected.

Fig. 7.22

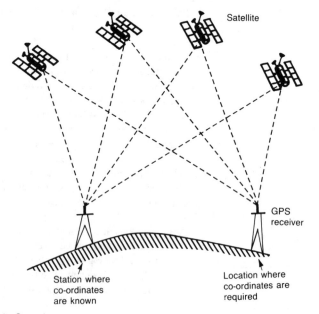

Satellite

GPS receiver

Station where co-ordinates are known

Location where co-ordinates are required

When using the pseudo-range method a minimum of four satellites should be visible so that X, Y and Z co-ordinates can be determined together with a correction for any timing errors. If two receivers are involved, one being established at a known point whilst the other is at a point to be fixed, and both receivers track the same satellites as in Fig. 7.22 the technique of relative point positioning is being invoked. The four satellites should be reasonably spread and in at least two orbital paths. The accuracy claimed for base line measurement by Geotracer is 10 mm $+ 2D \times 10^{-6}$, D being the measured length.

Co-ordinates resulting from GPS measurements are in the system of the satellite datum which is the World Geodetic System of 1984 (WGS 84); this is an Earth-centred cartesian co-ordinate system (X, Y, Z). For use in the United Kingdom these have to be transformed to heights above the reference ellipsoid and the National Grid system.

Further information on applications of GPS can be found in *New Civil Engineer*, 4 June 1987.

Exercises 7

1 Describe the projection and the National Reference system used by the Ordnance Survey in the production of maps and plans of Great Britain, and give an example of a National Grid reference on the 1/2500 scale.

With the aid of neat sketches explain the meaning of the following.

(i) Local scale factor
(ii) Convergence of meridians
(iii) The $(t - T)$ direction correction
(iv) The local mean radius of the earth. [*Eng. Council*]

2 Horizontal circle readings of a reference mark were determined at a station as 82°36.0′ (FL) and 262°36.2′ (FR) and were followed by an approximate gyro orientation of 343°22.6′. The following transit times and amplitudes were then booked in sequence: 00 min 00.0 sec, amplitude -10.7, 03 min 52.5 sec, amplitude $+7.5$, 07 min 29.8 sec, amplitude -10.7, 11 min 22.5 sec, amplitude $+7.5$, 14 min 59.9 sec. Determine the azimuth of the reference mark, given that the proportionality factor was 0.047′/s and that the calibration factor was $+2.0′$.

Answer: 99°22.0′

3 In order to determine the bearing of a survey line XY a theodolite was set up at X and observations, face left and face right, made on the sun during the afternoon.

Using the following information, determine the bearing of the line:
Corrected mean alttitude of the sun 17°38′11″
UT of observation 15 h 18 m 14 s.

Extract from the *Star Almanac* for the day of observation

UT	δ
0	9°46′.8 S
6	9°52′.2 S
12	9°57′.6 S
18	10°03′.0 S

Latitude of station X 51°32′25″N

Horizontal angle from sun to station Y (measured clockwise) 55°26′28″

Answer: 281°31′58″

4 Two stations A and B have latitudes of 52°20′10″N and 52°24′20″N respectively and longitudes of 02°14′10″W and 02°23′40″W respectively. Determine the distance between the stations and the bearings of AB and BA.

	Length of 1″	*Length of 1″*
Latitude	*of longitude*	*of latitude*
52°20′	18.9364 m	30.9022 m
52°25′	18.9008 m	30.9107 m

Answer: 13 266.8 m, 305°40′50″, 125°33′28″.

5 A line XY of length 10 560.45 m has an aximuth of 54°43′40″ at X in latitude 53°22′00″N and longitude 02°16′30″W. Determine, to the nearest second, the reverse azimuth of the line at Y.

	Length of 1″	*Length of 1″*
Latitude	*of latitude*	*of longitude*
53°20′	30.9155 m	18.5065 m
53°25′	30.9160 m	18.4704 m

Answer: 234°49′55″   (*Salford*)

6 The following observations were taken by a gyrotheodolite established at station P:

Horizontal circle readings onto reference object Q

Face right	59°31.5′
Face left	239°31.7′
Turning point left	*Turning point right*
174°16.4′	179°47.0″
174°17.9′	179°45.6′
174°19.2′	179°44.4′

Given that the calibration constant for the instrument was $-2.3″$ calculate the bearing of line PQ.

Answer: 242°27.2′

7 In order to determine the National Grid co-ordinates and level of a survey station D, slope distances from D to two survey stations, E and S, of known co-ordinate and level values, have been measured with a precise electronic distance measuring instrument and the reciprocal vertical angles have been observed with a first-order theodolite. From the data listed below determine the mean National Grid co-ordinates and level of station D, which is to the West generally of station E.

Observed data
Mean observed slope distances
DE 1875.9644 m
DS 3825.3140 m

Mean observed verical angles
DE + 04°38′25″   DS + 02°45′19″
ED − 04°39′15″   SD − 02°47′02″

Known data
NG co-ordinates and levels above Ordnance Datum of stations *E* and *S*.

Station *E*: 323 048.355 (E) 343 362.105 (N) 352.220 AOD.
Station *S*: 323 679.139 (E) 340 431.408 (N) 385.115 AOD.

Local scale factors
Station *E* 0.99967396
Station *S* 0.99967278
Station *D* 0.99967754

Assume the local mean radius of the earth to be 6380.851 km.

[*Eng. Council*]

Answer: 321327.876 mE, 342631.677 mN, 200.253 m

8 Derive from first principles the following equation used for the reduction of measured slope distances:

$$K = \left[\frac{D^2 - \Delta h^2}{\left(1 + \dfrac{h_1}{R}\right)\left(1 + \dfrac{h_2}{R}\right)} \right]^{1/2}$$

where K is the ellipsoidal chord distance, D the measured slope distance, Δh the difference in the heights, h_1 and h_2, of the stations above mean sea level, R the radius of curvature of a line traced out on the reference spheroid by a vertical plane passing through the two survey stations.

Explain why additional adjustments are necessary in the chord to spheroid correction when measuring distances in excess of 10 km with electromagnetic distance measuring equipment.

The measured slope distance between two survey stations is 2999.137 m. The heights of the two stations above mean sea level are 352.220 m and 385.115 m. Assuming the radius of curvature of a line traced out on the reference spheroid by a vertical plane passing through the two stations to be 6380.851 km, and the local scale factor to be 0.9996734, compute the National Grid distance between the two stations.

[*Eng. Council*]

Answer: 2998.021 m

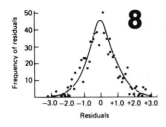

8 Analysis and adjustment of measurements

In Chapters 1 and 2, the differences between mistakes, systematic errors and random errors were discussed in general terms. It was pointed out that surveying work should be free from mistakes, and that systematic errors, where known, should be corrected; this chapter concerns itself, therefore, with small random accidental errors. These are equally likely to be positive or negative, considered with respect to the 'true' value of the quantity being measured. In precise surveying, knowledge of such errors and of their adjustment becomes important.

Distributions

If a large number of measurements is made of an angle or line all with equal precision, a range of results will be obtained. Consider as an example the following measurements of an angle which were taken in a test on a Watts Theodolite (Hilger & Watts Technical Report BS 7).

Mean angle	Mean angle	Mean angle	Mean angle
94°4′19.00″	94°4′20.00″	94°4′22.25″	94°4′21.00″
94°4′21.75″	94°4′20.00″	94°4′19.50″	94°4′20.50″
94°4′22.50″	94°4′21.25″	94°4′19.75″	94°4′20.50″
94°4′21.25″	94°4′21.25″	94°4′20.50″	94°4′20.50″
94°4′20.25″	94°4′22.00″	94°4′18.75″	94°4′20.25″
94°4′21.50″	94°4′21.50″	94°4′18.75″	94°4′20.50″
			94°4′19.00″

If we divide the range of variation into a number of groups and count the number of observations falling in each group, the results can be plotted as a *histogram*, as in Fig. 8.1. The horizontal axis is divided into segments corresponding to the ranges of the groups, and on each segment is constructed a rectangle whose area is proportional to the group frequency. In the example, since the range is the same for all the groups, the heights of the rectangles are proportional to the frequencies. It is a reasonable supposition, and found to be so in practice, that if the number of observations was made very much larger and the groups much narrower, the irregular shape would become a smooth curve as shown dotted in Fig. 8.1. Such a curve is referred to as a *frequency curve*.

If the results are expressed as proportional frequencies as in Fig. 8.2, the curve can then be referred to as the probability distribution.

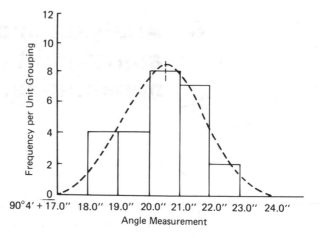

Fig. 8.1 Histogram with smooth curve superimposed

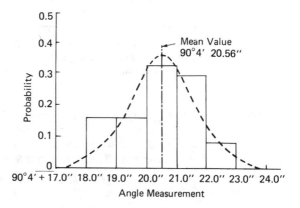

Fig. 8.2 Probability curve

Mean Value
90°4′ 20.56″

The probability of the occurrence of a given observation is defined as the *proportional frequency* with which it occurs in a large number of observations.

In most cases, the observed results are the only direct information available, and the problem is to deduce the probability distribution. This is only possible if a very large number of observations have been made, whereas in most cases relatively few are available. It is often possible, however, to make an assumption as to the general form of the probability distribution with a fair degree of confidence, using as a basis either past experience or knowledge of similar situations. In surveying where we are concerned with direct measurement, the distribution can be assumed to be of the normal (or Gaussian) type.

This is in fact the most commonly used of all the theoretical distributions; it gives a bell-shaped probability curve, and is expressed mathematically as

$$dp = \frac{1}{\sigma\sqrt{2\pi}} \exp - (x_1 - \mu)^2 / 2\sigma^2 \cdot dx \qquad (8.1)$$

If the data are normally distributed, the probability that x will assume a value between x_1 and $(x_1 + dx)$ is then given by the equation (8.1) in which μ is the true mean of the population and σ is the population standard deviation. The population is the complete set of values of a given variate and the number of individual values is generally very large or even infinite.

The distribution is thus defined by μ and σ, where μ is the central value and σ is a measure of the spread. A change in σ alters the shape of the probability curve, an increase causing the curve to flatten and spread outwards.

Arithmetic mean

In practice, μ and σ are not known and have to be estimated from the limited observations taken. Instead of the population mean μ, we use the arithmetic mean of the observations:

$$\bar{x} = \frac{\Sigma x}{n} \tag{8.2}$$

where n is the sample size and where $\bar{x} \to \mu$ as $n \to \infty$.

Equation (8.2) can be written in the form

$$\bar{x} = x_0 + \frac{\Sigma(x - x_0)}{n} \tag{8.3}$$

where x_0 is a datum quantity lower in value than the smallest observation.

This method of calculating mean values is particularly useful when dealing with angle measurement. Thus, the mean of the angles given earlier in this chapter may be obtained, by putting $x_0 = 90°04'18''$, as

$$90°04'18'' + [(1 + 3.75 + 4.5 + 3.25 + 2.25 + 3.5$$
$$+ 2 + 2 + 3.25 + 3.25 + 4 + 3.5 + 4.25$$
$$+ 1.5 + 1.75 + 2.5 + 0.75 + 0.75 + 3$$
$$+ 2.5 + 2.5 + 2.5 + 2.25 + 2.5 + 1) \div 25]$$
$$= 90°04'18'' + [64.00 \div 25] = 90°04'18'' + 2.56''$$
$$= 90°04'20.56''$$

Standard deviation

The standard deviation of the population is given by

$$\sigma = \sqrt{\frac{\Sigma(x - \mu)^2}{n}} \tag{8.4}$$

Since μ is not known, the arithmetic mean, $\bar{x}$, of the sample must be used. It can be shown that $\Sigma(x - \bar{x})^2$ is less than $\Sigma(x - \mu)^2$ since over n observations

$$\Sigma(x - \mu)^2 - \Sigma(x - \bar{x})^2 = n\mu^2 - 2\mu\Sigma x + 2\Sigma x \left(\frac{\Sigma x}{n}\right) - n\left(\frac{\Sigma x}{n}\right)^2$$

$$= n\left(\mu - \frac{\Sigma x}{n}\right)^2$$

which is always positive.

There is thus an underestimation of the standard deviation when $\bar{x}$ is used in place of μ and $(n-1)$ is therefore used in place of n in the divisor to correct for this. The estimate of the population standard deviation derived from the sample is denoted by s, where

$$s = \sqrt{\frac{\Sigma(x-\bar{x})^2}{n-1}} \qquad (8.5)$$

in which $(x-\bar{x})$ is termed a *residual* to distinguish it from the error of a reading, which is its deviation from the true (but unknown) mean, μ. Thus as $n \to \infty$, $s \to \sigma$.

The square of the standard deviation is known as the variance, V, and is also used as a measure of dispersion or spread; but in surveying, the standard deviation is most commonly used. It has the same dimensions as the variable and is thus easier to comprehend as a measure of the spread of the distribution.

The computation of the term $\Sigma(x-\bar{x})^2$ may give rise to awkward numbers and use should be made of the identity

$$\Sigma(x-\bar{x})^2 = \Sigma x^2 - \frac{(\Sigma x)^2}{n} \qquad (8.6)$$

Put $S = \Sigma(x)$; then

$$s = \sqrt{\frac{\Sigma x^2 - (S^2/n)}{n-1}} \qquad (8.7)$$
$$= \sqrt{\frac{n\Sigma x^2 - S^2}{n(n-1)}}$$

Since S has already been determined in the computation of $\bar{x}$, the arithmetic mean is reduced, and the device given in equation (8.3) can be used to reduce it further. As an example, the standard deviation of the observations given earlier is calculated in Table 8.1, a datum value, x_0, of $90°04'18''$, being used.

Table 8.1

x_0 $90°04'18''$	$(x - x_0) = X$	$(x - x_0)^2 = X^2$
	1.00	1.00
	3.75	14.06
	4.50	20.25
	3.25	10.56
	2.25	5.06
	3.50	12.25
	2.00	4.00
	2.00	4.00
	3.25	10.56
	3.25	10.56
	4.00	16.00
	3.50	12.25
	4.25	18.06

1.50	2.25
1.75	3.06
2.50	6.25
0.75	0.56
0.75	0.56
3.00	9.00
2.50	6.25
2.50	6.25
2.50	6.25
2.25	5.06
2.50	6.25
1.00	1.00
$S = 64.00$	$\Sigma X^2 = 191.35$

Therefore
$$s = \sqrt{\frac{n\Sigma X^2 - S^2}{n(n-1)}}$$

$$= \sqrt{\frac{687.75}{600}} = 1.07''$$

(Note that the mean can be calculated from the same computation.)

Standard deviation and standard error

If as equation (8.1) the substitution

$$u = \frac{(x-\mu)}{\sigma}$$

is made, then
$$dp = \frac{1}{\sqrt{2\pi}} e^{-u^2/2}\, du \qquad (8.8)$$

This is known as the standardized form of the normal equation; if the value of μ is known, dp/du is determined uniquely. Equation (8.8) represents the probability of occurrence of observations for which u lies between u and $u+du$, and by integration gives the total probability of observations having u less than a given value. The definite integral has been evaluated for a wide range of values of u. The normal curve (Fig. 8.3) is symmetrical about $u = 0$, the areas on each side being equal to $\frac{1}{2}$ (the total area under the curve must equal 1). It is thus only necessary to consider positive values of u, and in Fig. 8.3 the shaded area α to the right of $+u_1$ is the probability that u will be greater than or equal to u_1. It is also the probability that u will be less than or equal to $-u_1$ and the total probability that u will lie outside the range $+u_1$ to $-u_1$ is 2α. For $u = 3\sigma$, the probability that a value will lie outside the range 3σ to -3σ is about 0.003 or 1 chance in 400, and for $u = 2\sigma$ the probability is about 0.045 or about 1 chance in 20. The extent to which a single observation may depart from its true value, therefore, is measured by the standard deviation.

In addition, it can be shown that if the population is normal, with

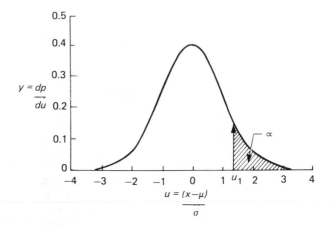

Fig. 8.3 Normal probability curve

$y = \dfrac{dp}{du}$

$u = \dfrac{(x-\mu)}{\sigma}$

mean μ and the standard deviation σ, then the mean of a random sample (e.g. a set of independent observations made under uniform conditions) is also normally distributed with mean μ and standard deviation $\sigma/\sqrt{n}$. This is an indication of how the reliability of the mean is related to sample size.

The standard deviation of the mean is usually referred to as its *standard error* and this is inversely proportional to the square root of n. If the sample is increased four times, the standard error of the mean is halved and this in turn is a measure of the precision of the estimate. Thus, from the equations already given,

$$\text{SE}_x = \text{standard error of a single observation} = s$$

$$= \sqrt{\frac{\Sigma(x-\bar{x})^2}{n-1}} = s_{\bar{x}} \times \sqrt{n} \qquad (8.9)$$

$$\text{SE}_{\bar{x}} = \text{standard error of the mean} = s_{\bar{x}}$$

$$= \sqrt{\frac{\Sigma(x-\bar{x})^2}{n(n-1)}} = \frac{\text{SE}_x}{\sqrt{n}} \qquad (8.10)$$

It will be observed that standard deviation and standard error of a single observation are similarly defined by equations (8.5) and (8.10), and in fact the two terms seem to be used synonymously by surveyors.

For estimates of a population mean derived from samples of the same size, the standard error of the estimates is that error which is numerically exceeded with a relative frequency of about one in three in a large number of trials. In surveying it was the convention for many years to estimate the value of the *probable error, E,* such that the probability of an error greater than E is $\frac{1}{2}$. The term probable error is now virtually obsolete in surveying practice, but if required it can be calculated as $0.6745\ \text{SE}_x$.

Rejection of observations

The assumptions made in the preceding paragraphs are that we are considering small random errors which are normally distributed, and from a consideration of the normal distribution it can be said that the chance of a result deviating from the mean by more than three times the standard deviation is about 1 in 400, and by more than four times the standard deviation, 1 in 10 000. Thus, when the result deviates by a large amount from the mean, there is a possibility that unusual circumstances have operated and that the result does not belong to the same normal distribution. It is reasonable in such a case to reject the doubtful observation, and this is the usual practice if the deviation from the mean is more than three times the standard deviation.

In the previous example, the reader can confirm that the largest residual is 1.94″. The standard deviation was calculated to be 1.07″, so that any observation giving a residual greater than 3.21″ would be considered unreliable and consequently would be rejected.

Combinations of errors

Many observations in surveying are of a quantity which must be calculated from quantities whose observed or estimated values are liable to error, and it is necessary to estimate the precision of the required quantity. In theodolite readings, the error in angle measurement can be considered as a function of the error in angle reading and the error in signal bisection. In the precise spirit level, precision is influenced by error in registered reading error and bubble displacement.

If $z = f(x, y)$, where x and y are independent variables (i.e. the probability that either of them will assume a prescribed value does not depend on the value assumed by the other), then, to the first order,

$$\delta z = \frac{\partial f}{\partial x} \delta x + \frac{\partial f}{\partial y} \delta y$$

Since δx and δy are deviations from their mean values, the partial derivatives are given their values at $(\bar{x}, \bar{y})$, and it can be shown that, for any number of independent variables,

$$\sigma_z^2 = \left(\frac{\partial f}{\partial x}\right)^2 \sigma_x^2 + \left(\frac{\partial f}{\partial y}\right)^2 \sigma_y^2 + \left(\frac{\partial f}{\partial a}\right)^2 \sigma_a^2 \qquad (8.11)$$

and so, if $z = xy$ and x is a constant free from error

$$\sigma_z = x \cdot \sigma_y \qquad (8.12)$$

Also, if $z = x+y+a+ \ldots,$

$$\sigma_z^2 = \sigma_x^2 + \sigma_y^2 + \ldots$$

and the standard deviation is

$$\sigma_z = \sqrt{\sigma_x^2 + \sigma_y^2 + \ldots} \qquad (8.13)$$

For a fuller treatment, see, for example, *Statistical Methods for Technologists*, by Paradine and Rivett (EUP).

Example 8.1 A measuring tape of length 20 m has been calibrated to have a standard deviation of 1 mm. If it is used to measure a length of 180 m what is the standard deviation of this measurement? Assuming an accidental error of magnitude

2 mm occurred when measuring each bay, find the standard deviation of the whole measurement.

It will be noted that nine 20 m bays have been measured, and so in equation (8.12) $x = 9$. Hence, standard deviation due to standardizing

$$= 1 \times 9$$
$$= 9 \text{ mm}$$

Also from (8.13) standard deviation due to measurement

$$= \sqrt{9 \times 2^2} = 2\sqrt{9}$$
$$= 6 \text{ mm}$$

Standard deviation of whole

$$= \sqrt{(9^2 + 6^2)}$$
$$= \mathbf{11 \text{ mm}}$$

If x and y are not independent then equation (8.11) becomes

$$\sigma_z^2 = \left(\frac{\partial f}{\partial x}\right)^2 \sigma_x^2 + \left(\frac{\partial f}{\partial y}\right)^2 \sigma_y^2 + 2\sigma_{xy} \left(\frac{\partial f}{\partial x}\right)\left(\frac{\partial f}{\partial y}\right) \qquad (8.14)$$

Similarly if $u = g(x, y)$

$$\sigma_u^2 = \left(\frac{\partial g}{\partial x}\right)^2 \sigma_x^2 + \left(\frac{\partial g}{\partial y}\right)^2 \sigma_y^2 + 2\sigma_{xy} \left(\frac{\partial g}{\partial x}\right)\left(\frac{\partial g}{\partial y}\right)$$

In each case σ_{xy} is the covariance of x and y, and the coefficient of correlation is given by $\sigma_{xy}/\sigma_x \cdot \sigma_y$. The covariance of u and z is

$$\sigma_{uz} = \left(\frac{\partial f}{\partial x}\right)\left(\frac{\partial g}{\partial x}\right)\sigma_x^2 + \left(\frac{\partial f}{\partial y}\right)\left(\frac{\partial g}{\partial y}\right)\sigma_y^2$$
$$+ \sigma_{xy}\left[\left(\frac{\partial f}{\partial x}\right)\left(\frac{\partial g}{\partial x}\right) + \left(\frac{\partial f}{\partial y}\right)\left(\frac{\partial g}{\partial y}\right)\right] \qquad (8.15)$$

The term covariance implies a measure of mutual dependence, two different observed quantities being correlated if they have common variables. Values of the coefficient of correlation lie between 0 and 1, the former implying that x and y are uncorrelated and the latter that they are linear functions exactly.

Example 8.2 The bearing and length of a traverse line have been observed to be $38°45'20''$ and 169.08 m respectively. Assuming that standard deviations of 20 seconds and 50 mm apply to the observations, calculate the standard deviations and covariance of the coordinate differences of the line.

$$\Delta E = f(l, \alpha) = l \sin \alpha = 105.85 \text{ m}$$

$$\Delta N = g(l, \alpha) = l \cos \alpha = 131.85 \text{ m}$$

$$\delta_{\Delta E}^2 = \left(\frac{\partial f}{\partial l}\right)^2 \sigma_l^2 + \left(\frac{\partial f}{\partial \alpha}\right)^2 \sigma_\alpha^2$$

$$= (\sin \alpha)^2 \sigma_l^2 + (l \cos \alpha)^2 \sigma_\alpha^2$$

In practice σ will be replaced by s with

$$s_l = 0.05 \text{ m}$$

and $s_\alpha = \dfrac{20}{206\ 265} = 9.7 \times 10^{-5} \text{ rad}$

$$s_{\Delta E}^2 = (0.6260)^2(0.05)^2 + (169.08 \times 0.7798)^2(9.7 \times 10^{-5})^2$$

$$s_{\Delta E} = \mathbf{34 \ mm}$$

Similarly

$$S_{\Delta N}^2 = \left(\frac{\partial g}{\partial l}\right)^2 s_l^2 + \left(\frac{\partial g}{\partial \alpha}\right)^2 s_\alpha^2$$

$$= (\cos \alpha)^2 s_l^2 + (-l \sin \alpha)^2 s_\alpha^2$$

$$S_{\Delta N} = \mathbf{40 \ mm}$$

also

$$s_{\Delta E \Delta N} = \left(\frac{\partial f}{\partial l}\right)\left(\frac{\partial g}{\partial l}\right) s_l^2 + \left(\frac{\partial f}{\delta \alpha}\right)\left(\frac{\partial g}{\delta \alpha}\right) s_\alpha^2$$

$$s_{\Delta E \Delta N} = \mathbf{1089 \ mm^2}$$

Note that $s_{l\alpha}$ is zero since the bearing and length measurements are independent. Coefficient of correlation

$$= \frac{s_{\Delta E \Delta N}}{s_{\Delta E} \cdot s_{\Delta N}} = 0.80$$

Weighted observations

It may be necessary to assess the best value of a quantity for which the observed values are of different reliability. As examples, readings might be made by different observers, or with different instruments; in lines of levels between two points, different distances might be covered.

If the true value of a quantity is μ_x and the observed values are x_1, x_2, ..., x_r, corresponding standard deviation being σ_1, σ_2, ..., σ_r, then, if the errors are distributed normally, the probabilities of the observed values are proportional to

$$\exp - (x_1 - \mu_x)^2/2\sigma_1^2, \quad \exp - (x_2 - \mu_x)^2/2\sigma_2^2 \text{ etc. (from eqn. (8.1))}$$

The total probability of the set of observations is thus proportional to

$$\exp - \Sigma (x_r - \mu_x)^2/2\sigma_r^2$$

Choose the value of x which makes this a maximum, i.e.

$$\sum \frac{(x_r - \mu_x)^2}{2\sigma_r^2} \text{ is a minimum} \tag{8.16}$$

Differentiating with respect to x and equating to zero,

$$\sum \frac{(x_r - \mu_x)}{\sigma_r^2} = 0$$

$$\mu_x = \frac{\sum \dfrac{x_r}{\sigma_r^2}}{\sum \dfrac{1}{\sigma_r^2}} \tag{8.17}$$

Writing

$$w_r\, \sigma_r^2 = \sigma^2$$

where σ^2 is the variance of an observation of unit weight (which may be termed a standard observation)

$$\mu_x = \frac{\sum x_r \dfrac{w_r}{\sigma^2}}{\sum \dfrac{w_r}{\sigma^2}} = \frac{\sum w_r\, x_r}{\sum w_r} \tag{8.18}$$

i.e. the best estimate of μ_x is given by a weighted mean (the arithmetic mean of the observations reduced to the same standard) in which the weights are inversely proportional to the squares of the standard deviations of the several observations (i.e. inversely proportional to the variances).

It will be realized by the reader that weights can also be assigned inversely to the squares of either the probable errors or the standard errors of the several observations.

The best value thus derived by equation (8.18) can also be assessed as to its precision. If the variance of the weighted mean is denoted by σ_m^2, it can be shown that

$$\sigma_m^2 = \frac{1}{(\Sigma w_r)^2} \{w_1^2\, \sigma_1^2 + w_2^2\, \sigma_2^2 + \dots\} = \frac{1}{(\Sigma w_r)^2} \Sigma w_r^2\, \sigma_r^2$$

$$= \frac{1}{(\Sigma w_r)^2} \sum \left(\frac{\sigma^2}{\sigma_r^2}\right)^2 \sigma_r^2 = \frac{1}{(\Sigma w_r)^2} \sigma^2\, \Sigma w_r$$

$$= \frac{\sigma^2}{\Sigma w_r}$$

Hence

$$w_m = \frac{\sigma^2}{\sigma_m^2} = \Sigma w_r \tag{8.19}$$

i.e. the weight of the value of μ_x given by equation (8.18) is the sum of the weights of the several observations.

Note also that equations (8.9) and (8.10) now need modification.
Standard deviation of a single observation of weight w_1

$$= \sqrt{\frac{\Sigma w_r (x - \bar{x}_w)^2}{(n-1) w_1}}$$

Standard error of the weighted mean $\bar{x}_w$

$$= \sqrt{\frac{\Sigma w_r (x - \bar{x}_w)^2}{(n-1)\Sigma w_r}}$$

> **Example 8.3** Let us say that the data given at the beginning of the chapter were provided by observer A, i.e. a mean of $90°04'20.56''$ and a standard deviation of $\sqrt{(687.75/600)}$ whilst the same number of observations by B resulted in a mean of $90°04'23.45''$ and a standard deviation of $\sqrt{(876.25/600)}$. What is the best value of the angle and its standard error?

From equation (8.17)

$$\frac{w_A}{w_B} = \frac{s_B^2}{s_A^2} = \frac{876.25}{687.75}$$

i.e. the ratio of the weights is

$$w_A : w_B = 1.27:1$$

which is a measure of the relative reliability of the two sets of readings.

From equation (8.18), the best value of the angle is given by

$$x = \frac{Aw_A + Bw_B}{w_A + w_B} = 90°04' + \frac{(20.56)(876.25) + (23.45)(687.75)}{876.25 + 687.75}$$

$$= \mathbf{90°04'21.83''}$$

From equation (8.19)

$$w_x = w_A + w_B = 2.27 w_B$$

$$s_x^2 = \frac{w_B}{2.27 w_B} \cdot \left(\frac{876.25}{600 \times 25}\right) \quad \text{since } s_{\bar{x}B} = \sqrt{\frac{876.25}{600 \times 25}}$$

$$s_x = \mathbf{0.16''}$$

This is the standard error of the derived best value.

Confidence limits

Another, and often convenient, manner of expressing the precision of an estimate is to give limits which, with a stated probability, include the true value. Such limits are termed *confidence limits*, that is to say they are limits within which it can be said, with a given degree of confidence, that the true value lies. Provided the distribution is known — and, as stated previously, in surveying the assumption is made that the original data are distributed normally — such limits can be calculated for any statistic such as the mean or the standard deviation.

For a normal population with mean μ, and standard deviation σ, the probability that the mean of n observations will lie within the range $\mu - (3\sigma/\sqrt{n})$ to $\mu + (3\sigma/\sqrt{n})$ is 0.997 and, conversely, if the true mean μ is not known, the probability that μ lies within the range $-3\sigma/\sqrt{n}$ to $+3\sigma/\sqrt{n}$ about the estimated mean $\bar{x}$ is also 0.997. This implies that in repeated measurements of the same sort, the assertion that the true value of μ lies within the range $\bar{x} \pm (3\sigma/\sqrt{n})$ would be right in 99.7 per cent of cases. These limits are thus the 99.7 per cent confidence limits for μ. It can be stated with 99.7 per cent confidence that μ lies somewhere in this range, its most probable value being $\bar{x}$. The degree of confidence is called the *confidence coefficient* and the interval between the limits the *confidence interval*. Common intervals are the 95 per cent interval, which is roughly $\bar{x} - (2\sigma/\sqrt{n})$ to $\bar{x} + (2\sigma/\sqrt{n})$, and the 50 per cent interval, which is roughly $\bar{x} - (\frac{2}{3}\sigma/\sqrt{n})$ to $\bar{x} + (\frac{2}{3}\sigma/\sqrt{n})$. In general,

$$\text{Lower confidence limit} = \bar{x} - (u_\alpha \sigma/\sqrt{n})$$

and $\qquad \text{Upper confidence limit} = \bar{x} + (u_\alpha \sigma/\sqrt{n})$

where u_α corresponds to a stated value of α as shown in Fig. 8.3. The confidence coefficient is then, for both limits, $100(1 - 2\alpha)$ per cent.

In the preceding paragraphs the assumption was made that σ was known exactly, whereas in most cases an estimate s has to be made using equation (8.5). This estimated standard deviation is also subject to some uncertainty, and the confidence limits for μ the true mean will be farther apart than if σ were known. W. S. Gossett (who wrote under the name of Student) allows for this uncertainty by using t in place of u, and the distribution of t was tabulated by R. A. Fisher (Table 8.2). Where the sample is very large, the uncertainty in s is very small and t becomes almost identical with u. Thus, for values of $\alpha = 0.25, 0.10, 0.05, 0.025, 0.001$ and 0.0005 (Table 8.2 gives values of confidence coefficients of 50 per cent, 80 per cent, 90 per cent, 95 per cent, 99.8 per cent and 99.9 per cent respectively),

Table 8.2

ϕ	t_{50}	t_{80}	t_{90}	t_{95}	$t_{99.8}$	$t_{99.9}$
∞	0.674	1.282	1.645	1.960	3.090	$3.291 = u$
25	0.684	1.316	1.708	2.060	3.450	3.725
10	0.700	1.372	1.812	2.228	4.144	4.587
5	0.727	1.476	2.015	2.571	5.893	6.869
1	1.000	3.078	6.314	12.706	318.31	636.62

The symbol ϕ is used to denote the number of degrees of freedom and equals $n - 1$. Then

$$\text{Lower confidence limit} = \bar{x} - (t_\alpha s/\sqrt{n})$$

and $\qquad \text{Upper confidence limit} = \bar{x} + (t_\alpha s/\sqrt{n})$

where t_α is read off from the table using the appropriate number of degrees of freedom. As an example, the 95 per cent confidence intervals will be evaluated for the mean from the tests on the Watts theodolite.

$$n = 25, \quad \text{therefore} \quad \phi = 24.$$

From statistics tables,

$$t = 2.06(\alpha = 0.025)$$
$$\bar{x} = 90°04'20.56''$$

$$s = 1.07'' \quad \frac{s}{\sqrt{n}} = 0.21''$$

Therefore the 95-per-cent confidence limits are

$$90°04'20.56'' \pm 0.43''$$

As a further example, if an engineer accepts that 90 per cent of all plotted co-ordinates should lie within 1 mm of their true position when plotted to scale on a plan, then

$$1.645 \frac{s}{\sqrt{n}} = 1$$

i.e.

$$\frac{s}{\sqrt{n}} = 0.6 \text{ mm, say}$$

This in fact implies that 68 per cent of the plotted co-ordinates should lie within 0.6 mm of their true plotted position. Using the table it will be seen that 50 per cent should lie within 0.674×0.6 mm = 0.4 mm of their true position whilst 99.8 per cent should lie within 1.9 mm of their true position.

The concept of confidence limits can be found in BS 5606: 1990 (*Code of Practice for Accuracy in Building*) in which it is pointed out that dimensional deviations of some kind are inevitable during the construction process, so that only by chance are intended sizes likely to be achieved. The Code is based upon the results of a survey of building accuracy which revealed that each combination of construction methods and materials demonstrated a unique pattern of accuracy, below or above the intended size, which has been termed characteristic accuracy. This is defined in terms of (i) displacement of the mean and (ii) standard deviation. These parameters have been used to establish permissible deviations largely, but not wholly, based on three times the standard deviation.

Adjustment of errors

In precise surveying, errors of the nature just discussed give rise to small discordances. In plane surveying involving angle measurement, such discrepancies can be detected by the normal rules of geometry: for example, (i) the angles closing the horizon, or a complete round at a point, should sum to 360°; (ii) the angles of a small triangle must sum to 180°; (iii) the sum of the angles of a polygon is fixed by the number of its sides. In triangulation, more difficult problems of adjustment can arise since, besides the exact angle sums, there are other exact relations to be satisfied due to the fact that a particular side may occur in more than one triangle; calculation of the spherical excess may also be required if the triangles are large (*see* Chapter 6). In levelling, the reduced levels of a number of points may be determined by direct measurement. If further observations are made of the differences in level between selected pairs of points, it is almost inevitable that some small inconsistency will be found between the two sets of observations.

These inconsistencies and discordances require to be adjusted and allocated, and the method of least squares allows a solution. Indirect mention has been made of this earlier in the chapter, where from statement (8.16) it will be seen that we have expressed the criterion of the best value in the form

$$\Sigma w_r(x_r - \mu_x)^2 \text{ is a minimum} \tag{8.20}$$

The true error is now known, and the statement (8.20) can be expressed in the form: the best value of an unknown quantity, for which discordant values of different weights have been observed, is that which makes the weighted sum of the squares of the residuals a minimum. This is the *principle of least squares*, and in surveying it can be used to incorporate exact relations by extension of the method used in the section dealing with weighted means. In the detailed working out, two approaches are possible:

(i) Reduction to a minimum number of unknowns.
(ii) Lagrange's method of undetermined multipliers, more usually known in surveying as the *method of correlates*.

These are perhaps best studied by application to particular problems, and examples of both are given.

Example 8.4 Some levellings were carried out with the following results:

	Rise or fall	Weight
P to Q	+4.32 m	1
Q to R	+3.17 m	1
R to S	+2.59 m	1
S to P	−10.04 m	1
Q to S	+5.68 m	2

The reduced level of P is known to be 134.31 m above datum. Determine the levels of the other points.

Let the best values of the heights of Q, R and S above P be q, r and s respectively. The residuals are then

$$r_1 = 4.32 - q \qquad r_4 = 10.04 - s$$
$$r_2 = 3.17 - (r - q) \qquad r_5 = 5.68 - (s - q)$$
$$r_3 = 2.59 - (s - r)$$

The sum (say E) of the weighted squares of the residuals is to be a minimum and thus

$$\frac{\partial E}{\partial q} = 0; \qquad \frac{\partial E}{\partial r} = 0; \qquad \frac{\partial E}{\partial s} = 0.$$

$$E = 1[4.32 - q]^2 + 1[3.17 - (r-q)]^2 + 1[2.59 - (s-r)]^2$$
$$+ 1[10.4 - s]^2 + 2[5.68 - (s-q)]^2$$
$$= 200.74 + 20.42q + 4q^2 + 2r^2 - 2rq - 1.16r - 47.98s$$
$$+ 4s^2 - 2rs - 4sq$$

Differentiating with respect to q, r and s in turn and equating to zero, three equations, known as *normal equations*, are produced from which the values of q, r and s may be found:

$$+8q - 2r - 4s + 20.42 = 0$$
$$-2q + 4r - 2s - 1.16 = 0$$
$$-4q - 2r + 8s - 47.98 = 0$$

whence $q = 4.33$ m, $r = 7.47$ m, and $s = 10.03$ m.

The required values are now obtained as

Q 138.64 m

R 141.78 m

S 144.34 m

The residuals r_1, etc. for the five observations can be evaluated and used to obtain the standard deviation of an unadjusted observation of unit weight from the general expression

$$s^2 = \frac{\Sigma w_1 r_1^2}{n-m} \tag{8.21}$$

where n is the number of observation equations and m is the number of unknowns. In this example specific weights were designated to the various legs and the reader would derive the value of 0.037 m for s on writing $m = 3$. It has been mentioned in Chapter 3 that different lengths can affect the precision, and in this respect it is usually taken that the standard deviation is proportional to the square root of the length. Thus, for a single levelling along one leg the weight can be taken to be inversely proportional to its length. Unit weight is often given to height differences so determined over 1 km distances and (8.17) allows the standard deviations of levellings carried out over different lengths and numbers of runs to be then determined.

Rather than obtain the expression for E by expanding the quadratics and then differentiating to produce the normal equations, it is possible to draw up a table of coefficients of q, r and s from the expressions for the observation errors. The separate values obtained upon squaring and differentiating with respect to q give, on addition

$$\frac{\partial E}{\partial q} = (2q - 8.64) + (2q - 2r + 6.34) + (4q - 4s + 22.72) = 0$$

and on dividing by 2

$$\frac{\partial E}{\partial q} = (q - 4.32) + (q - r + 3.17) + (2q - 2s + 11.36) = 0$$

It will be seen that the terms in brackets equal the weighted errors multiplied by the coefficients of q for those errors. Similarly

$$\frac{\partial E}{\partial r} = (r-q-3.17)+(r-s+2.59) = 0$$

and $\frac{\partial E}{\partial s} = (s-r-2.59)+(s-10.04)+(2s-2q-11.36) = 0$

Again the separate terms are equal to the weighted errors multiplied by the relevant coefficient. The three equations which now follow are in each case half those previously produced but give the same values of q, r and s. The work can, however, be conveniently tabulated as follows from the expression for the residuals

Wt.	q	r	s	N
1	-1	0	0	$+4.32$
1	$+1$	-1	0	$+3.17$
1	0	$+1$	-1	$+2.59$
1	0	0	-1	$+10.04$
2	$+1$	0	-1	$+5.68$

the coefficients are entered in turn as they apply: in the last column the numerical value is entered. Now multiply each line by the coefficient of the unknown under consideration on that line, including the value of the weight.

For q which appears on three lines

$1(-1)(-1)q+1(-1)(+4.32)$ from the 1st line

$1(+1)(+1)q+1(+1)(-1)r+1(+1)(+3.17)$ from the 2nd line

$2(+1)(+1)q+2(+1)(-1)s+2(+1)(+5.68)$ from the 5th line

which give

$$(q-4.32)+(q-r+3.17)+(2q-2s+11.36)$$

respectively.

Hence $\qquad\qquad 4q-r-2s+10.21 = 0$

For r which appears on two lines

$$1(-1)(+1)q+1(-1)(-1)r+1(-1)(+3.17)$$
$$1(+1)(+1)r+1(+1)(-1)s+1(+1)(+2.59)$$

whence $\qquad\qquad -q+2r-s-0.58 = 0$

and for s which appears on three lines

$$1(-1)(+1)r+1(-1)(-1)s+1(-1)(+2.59)$$
$$1(-1)(-1)s+1(-1)(+10.04)$$
$$2(-1)(+1)q+2(-1)(-1)s+2(-1)(+5.68)$$

whence $\qquad\qquad -2q-r+4s-23.99 = 0$

Three normal equations for the solutions of q, r and s are thus obtained.

Note that our criterion can also be the minimization of

$$\Sigma\,[w_r^{1/2}(x_r-\mu_x)]^2$$

The residuals or observation equations can be multiplied by $\sqrt{w_r}$ as applicable, the weighting column not being included now. In our example the last line would be entered as $+\sqrt{2}$ for q, 0 for r, $-\sqrt{2}$ for s and $+5.68\sqrt{2}$ for N, the three normal equations being obtained in the same manner.

The sum of the weighted squares of n residuals,

$$w_1 r_1^2+w_2 r_2^2+\ \dots\ w_n r_n^2$$

can be written in matrix notation as

$$(r_1\ r_2\ \dots\ r_n)\begin{vmatrix} w_1 & 0 & \dots & 0 \\ 0 & w_2 & \dots & 0 \\ \cdot & \cdot & & \cdot \\ \cdot & \cdot & & \cdot \\ \cdot & \cdot & & \cdot \\ 0 & 0 & & w_n \end{vmatrix}\begin{vmatrix} r_1 \\ r_2 \\ \cdot \\ \cdot \\ \cdot \\ r_n \end{vmatrix}$$

or $r^T\,Wr$, in which r is a column vector and this is the format normally used by computer software.

If observations $k_1, k_2\ \dots\ k_n$ are related to unknowns $x_1, x_2, \dots x_m$ by observation equations of the form

$$\begin{aligned} a_{11}\,x_1+a_{12}\,x_2+\ \dots\ a_{1m}\,x_m &= k_1 \\ a_{21}\,x_1+a_{22}\,x_2+\ \dots\ a_{2m}\,x_m &= k_2 \end{aligned}$$

$$\begin{aligned} \cdot \\ \cdot \\ \cdot \end{aligned}$$

$$a_{n1}\,x_1+a_{n2}\,x_2+\ \dots\ a_{nm}\,x_m = k_n$$

we can express these as $Ax = k$.

Moreover, if x_1, x_2, etc. are to be the best values of the unknowns then the residuals can be expressed as $r = -Ax+k$. Thus we minimize $(-Ax+k)^T\,W(-Ax+k)$ which leads to

$$A^T W\,Ax = A^T Wk$$

or
$$x = (A^T WA)^{-1}\,A^T Wk \qquad (8.22)$$

If, in the previous example, we write the observation equations as

$$\begin{aligned} x_1 &= 4.32 \\ x_2-x_1 &= 3.17 \\ x_3-x_2 &= 2.59 \\ x_3 &= 10.04 \\ x_3-x_1 &= 5.68 \end{aligned}$$

Then
$$A = \begin{vmatrix} +1 & 0 & 0 \\ -1 & +1 & 0 \\ 0 & -1 & +1 \\ 0 & 0 & +1 \\ -1 & 0 & +1 \end{vmatrix}$$

and

$$W = \begin{vmatrix} 1 & 0 & 0 & 0 & 0 \\ 0 & 1 & 0 & 0 & 0 \\ 0 & 0 & 1 & 0 & 0 \\ 0 & 0 & 0 & 1 & 0 \\ 0 & 0 & 0 & 0 & 2 \end{vmatrix}$$

$$x = \begin{vmatrix} x_1 \\ x_2 \\ x_3 \end{vmatrix} \quad \text{and} \quad k = \begin{vmatrix} 4.32 \\ 3.17 \\ 2.59 \\ 10.04 \\ 5.68 \end{vmatrix}$$

Method of correlates

The method of correlates may be applied here to give a solution. In this method either the individual errors or their corrections are taken as the unknowns, and when they have been determined the observed values can be adjusted accordingly. In Fig. 8.4 if corrections of ϵ_1 ... ϵ_5 are given to the level differences P to Q ... Q to S respectively, the reader will find that the condition equations

$$\epsilon_1 + \epsilon_2 + \epsilon_3 + \epsilon_4 = -0.04$$

and

$$\epsilon_1 + \epsilon_5 + \epsilon_4 = +0.04$$

obtain for closed circuits $PQRSP$ and $PQSP$, starting from P to Q.

Note the way in which consistency in direction has been maintained between the two circuits in Fig. 8.4. Had circuits $PQSP$ and $QSRQ$ been adopted for compatibility in QS, levellings S to R and R to Q would have been written as -2.59 m and -3.17 m respectively.

Fig. 8.4

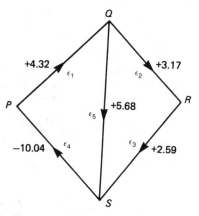

For the least squares condition $\Sigma w_i \epsilon_i^2$ is to be a minimum, and on differentiation with respect to the unknown corrections in turn we obtain $w_1 \epsilon_1 \delta \epsilon_1 = 0$, $w_2 \epsilon_2 \delta \epsilon_2 = 0$, etc.
Hence

$$w_1 \epsilon_1 \delta \epsilon_1 + w_2 \epsilon_2 \delta \epsilon_2 + w_3 \epsilon_3 \delta \epsilon_3 + w_4 \epsilon_4 \delta \epsilon_4 + w_5 \epsilon_5 \delta \epsilon_5 = 0 \quad (1)$$

also so that the condition equations are not invalidated

$$\delta \epsilon_1 + \delta \epsilon_2 + \delta \epsilon_3 + \delta \epsilon_4 = 0 \quad (2)$$

and $$\delta\epsilon_1 + \delta\epsilon_5 + \delta\epsilon_4 \qquad = 0 \qquad\qquad (3)$$

If we multiply equations (2) and (3) by $-x$ and $-y$ respectively and add to equation (1) to obtain:

$$\delta\epsilon_1(w_1\epsilon_1-x-y) + \delta\epsilon_2(w_2\epsilon_2-x) + \delta\epsilon_3(w_3\epsilon_3-x)$$
$$+ \delta\epsilon_4(w_4\epsilon_4-x-y) + \delta\epsilon_5(w_5\epsilon_5-y) = 0$$

Now $\delta\epsilon_1$, $\delta\epsilon_2$ etc. are independent quantities and so

$$w_1\epsilon_1 = w_4\epsilon_4 = x+y$$
$$w_2\epsilon_2 = w_3\epsilon_3 = x$$
$$w_5\epsilon_5 = y$$
$$w_1 = w_2 = w_3 = w_4 = 1 \qquad w_5 = 2$$

These values may be substituted into the correction equations to give:

$$(x+y) + (x) + (x) + (x+y) = -0.04$$
$$(x+y) + (\tfrac{y}{2}) + (x+y) \qquad = 0.04$$

Whence $\qquad x = -0.03 \qquad\qquad y = +0.04$

Thus
$$\epsilon_1 = \epsilon_4 = \quad 0.01$$
$$\epsilon_2 = \epsilon_3 = -0.03$$
$$\epsilon_5 \quad = \quad 0.02$$

Whence rise from P to Q is 4.33 m whilst the fall from S to P is $-10.04 + 0.01 = -10.03$ m.

The standard deviation of an adjusted observation of unit weight can be derived from the general expression

$$s^2 = \frac{\Sigma w_1 r_1^{\,2}}{n_c} \qquad\qquad (8.23)$$

where n_c is the number of condition equations. In the above example $\Sigma w_1 r_1^{\,2} = 0.0028$ m^2, $n_c = 2$ and so, as previously, $s = 0.037$ m (with $s_5 = 0.026$ m). It has already been mentioned that the standard deviation of a single run of levelling over distance l may be taken to be proportional to $\sqrt{l}$, so that the weight of the mean observation from x runs over that leg can be considered to be x/l in its relation with other legs. Accordingly, if l is expressed in kilometres and r in metres, s will be expressed in m/$\sqrt{}$km units. Thus for a single run over K kilometres distance, the corresponding standard deviation implied by equation (8.14) is $s\sqrt{K}$ m.

Other considerations apply in respect of the weights (w_r) of adjusted values since, when deriving corrections, weights (w_r) were assumed. Equation (8.18) still applies in the form

$$w_r(s_r')^2 = (s')^2$$

say, and the standard deviation, s', of an adjusted observation of unit weight may be expressed as

$$s' = s\sqrt{\frac{n-n_c}{n}} \qquad\qquad (8.24)$$

in which n is the number of observation equations.

Example 8.5   *A, B, C* and *D* form a round of angles at a station such that

$$(A + B + C + D) = 360°$$

Their observed values are

$$A = 110°37'45'', \qquad B = 82°15'35''$$
$$C = 66°24'40'', \qquad D = 100°42'10''$$

$(A+B)$ was measured separately twice and found to average $192°53'25''$. Find the best value of each angle if all six measurements were of equal reliability.

The expression $A+B+C+D = 360°$ forms a condition equation and allows the elimination of one unknown since it is a requirement which all four best values must satisfy. Since $(A+B)$ has been measured twice, the weight of the mean may be taken as twice that of a single observation.

If a, b and c be the most probable values of angles A, B and C respectively, then the errors of observation are

$$110°37'45'' - a, \qquad 82°15'35'' - b, \qquad 66°24'40'' - c,$$
$$259°17'50'' - (a+b+c), \qquad 192°53'25'' - (a+b)$$

And so, on tabulating, we get

Wt.	a	b	c	N
1	−1	0	0	+110°37'45''
1	0	−1	0	+82°15'35''
1	0	0	−1	+66°24'40''
1	−1	−1	−1	+259°17'50''
2	−1	−1	0	+192°53'25''

Thus

$$4a+3b+c-755°42'25'' = 0$$
$$3a+4b+c-727°20'15'' = 0$$
$$a+b+2c-325°42'30'' = 0$$

whence

$$a = \mathbf{110°37'46''}$$
$$b = \mathbf{82°15'36''}$$
$$c = \mathbf{66°24'34''}$$

and so

$$d = 360° - (a+b+c) = \mathbf{100°42'04''}$$

Mention has been made in Chapter 6 of the use of least squares in the adjustment of triangulation figures, but before deriving typical solutions some basic principles need to be indicated. In Fig. 8.5 *ABCD* shows a quadrilateral with diagonals in which eight angles can be measured. When adjusting the measured values certain condition equations have to be satisfied. In this particular case two sets of three angle equations can be immediately established, although four independent equations are ultimately involved. Thus we have

Fig. 8.5

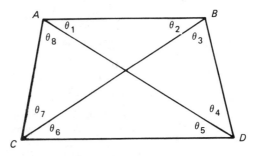

either
$$\theta_1+\theta_2+\theta_3+\theta_4 = 180°$$
$$\theta_3+\theta_4+\theta_5+\theta_6 = 180°$$
$$\theta_5+\theta_6+\theta_7+\theta_8 = 180°$$

or
$$\theta_1+\theta_2+\theta_3+\theta_4+\theta_5+_6+\theta_7+\theta_8 = 360°$$
$$\theta_1+\theta_2 = \theta_5+\theta_6$$
$$\theta_3+\theta_4 = \theta_7+\theta_8$$

The fourth equation is established using the relationships between the sides

$$AB = BD\ \frac{\sin \theta_4}{\sin \theta_1} = DC\ \frac{\sin \theta_6}{\sin \theta_3}\ \frac{\sin \theta_4}{\sin \theta_1}$$

$$= AC\ \frac{\sin \theta_8}{\sin \theta_5}\ \frac{\sin \theta_6}{\sin \theta_3}\ \frac{\sin \theta_4}{\sin \theta_1}$$

$$= AB\ \frac{\sin \theta_2}{\sin \theta_1}\ \frac{\sin \theta_8}{\sin \theta_5}\ \frac{\sin \theta_6}{\sin \theta_3}\ \frac{\sin \theta_4}{\sin \theta_1}$$

Thus $\sin \theta_1 \cdot \sin \theta_3 \cdot \sin \theta_5 \cdot \sin \theta_7 = \sin \theta_2 \cdot \sin \theta_4 \cdot \sin \theta_6 \cdot \sin \theta_8$. This relationship has been traditionally evaluated with books of log tables in the form:

$$\Sigma \log \sin \theta_1 = \Sigma \log \sin \theta_2$$

In the above, the assumption has been made that the triangles are small: spherical excess has to be considered otherwise.

The relationship between log sines of angles is used in the 'method of equal shifts' to balance out closing errors in triangulation figures. Whilst the method gives a quick hand calculation it does not use the least square approach and as such has very little mathematical justification for the adjustments. With the introduction of computers, speed of calculation is no longer important and the method is largely redundant. The reader is directed to *Solving Problems in Surveying* by Bannister and Baker (Longman) if examples and more detail are required.

Example 8.6 A tunnel was required to be set out between two points A and B on a line running approximately West to East.

The points were not intervisible so a line *CD* was set out with *C* and *D* approximately south of *A* and *B* respectively.

The length of *CD* and the following angles were measured:

$$A\hat{C}B = 45°24'10'', \qquad B\hat{C}D = 37°14'12''$$
$$C\hat{D}A = 29°38'52'', \qquad A\hat{D}B = 63°19'35''$$
$$C\hat{A}D = 67°43'04'', \qquad C\hat{B}D = 49°47'08''$$

Adjust the angles to the nearest second assuming that those taken at *A* and *B* are of twice the weight of the other four. Explain, quoting the relevant formulae, how you would compute the length *A* to *B* and the alignment angles $C\hat{A}B$ and $A\hat{B}D$.

(London)

Let the angles be designated θ_1 etc. as shown in Fig. 8.5.

Since θ_1 and θ_2 have not been measured, at this stage only two conditions can be satisfied, and these are

$$\theta_5+\theta_6+\theta_7+\theta_8 = 180°00'00''$$
$$\theta_3+\theta_4+\theta_5+\theta_6 = 180°00'00''$$

The angular errors in those two triangles are $+18''$ and $-13''$ respectively and these require equal and opposite total corrections.

Let $\epsilon_3, \epsilon_4, \ldots, \epsilon_8$ be the respective individual corrections to the measured values, $\theta_3, \theta_4, \ldots, \theta_8$.

Then
$$\epsilon_3+\epsilon_4+\epsilon_5+\epsilon_6 = +13''$$
$$\epsilon_5+\epsilon_6+\epsilon_7+\epsilon_8 = -18''$$

For the least squares condition $\Sigma w_3\epsilon_3^2$ is to be a minimum.

We obtain
$$w_3\epsilon_3\delta\epsilon_3 + w_4\epsilon_4\delta\epsilon_4 + w_5\epsilon_5\delta\epsilon_5 + w_6\epsilon_6\delta\epsilon_6$$
$$+ w_7\epsilon_7\delta\epsilon_7 + w_8\epsilon_8\delta\epsilon_8 = 0 \qquad (1)$$

also
$$\delta\epsilon_3 + \delta\epsilon_4 + \delta\epsilon_5 + \delta\epsilon_6 = 0 \qquad (2)$$

and
$$\delta\epsilon_5 + \delta\epsilon_6 + \delta\epsilon_7 + \delta\epsilon_8 = 0 \qquad (3)$$

since the equations for the corrections must not be invalidated by the small increments $\delta\epsilon_3$, etc. Multiply equations (2) and (3) by $-x$ and $-y$ respectively and add to equation (1) to obtain:

$$\delta\epsilon_3(w_3\epsilon_3-x) + \delta\epsilon_4(w_4\epsilon_4-x) + \delta\epsilon_5(w_5\epsilon_5-x-y)$$
$$+ \delta\epsilon_6(w_6\epsilon_6-x-y) + \delta\epsilon_7(w_7\epsilon_7-y) + \delta\epsilon_8(w_8\epsilon_8-y) = 0$$

Thus
$$w_3\epsilon_3 = w_4\epsilon_4 = x$$
$$w_5\epsilon_5 = w_6\epsilon_6 = x+y$$
$$w_7\epsilon_7 = w_8\epsilon_8 = y$$

But $w_3 = w_8 = 2w$, and $w_4 = w_5 = w_6 = w_7 = w$

therefore
$$\epsilon_3 = \frac{x}{2w}, \qquad \epsilon_4 = \frac{x}{w}, \qquad \epsilon_5 = \frac{x+y}{w},$$

etc., and these values may be substitued in the correction equations to be satisfied, to give:

$$\frac{x}{2w} + \frac{x}{w} + \frac{x+y}{w} + \frac{x+y}{w} = +13''$$

and
$$\frac{x+y}{w} + \frac{x+y}{w} + \frac{y}{w} + \frac{y}{2w} = -18''$$

i.e.
$$3.5x + 2y = +13w$$
$$2x + 3.5y = -18w$$

Whence $\qquad x = 9.9w \qquad$ and $\qquad y = -10.8w$

The values of ϵ_3, ϵ_4, ϵ_5, etc. (correct to the nearest second), are now deduced to be:

$$\epsilon_3 = +5'', \qquad \epsilon_4 = +10'', \qquad \epsilon_5 = -1''$$
$$\epsilon_6 = -1'', \qquad \epsilon_7 = -11'', \qquad \epsilon_8 = -5''$$

These corrections are applied to the measured values to give the best values of the angles as:

$$\theta_3 = 49°47'08'' + 05'' = 49°47'13''$$
$$\theta_4 = 63°19'35'' + 10'' = 63°19'45''$$
$$\theta_5 = 29°38'52'' - 01'' = 29°38'51''$$
$$\theta_6 = 37°14'12'' - 01'' = 37°14'11''$$
$$\theta_7 = 45°24'10'' - 11'' = 45°23'59''$$
$$\theta_8 = 67°43'04'' - 05'' = 67°42'59''$$

In this example it will be seen that the values determined satisfy the condition that

$$\theta_3 + \theta_4 = \theta_7 + \theta_8 = 113°06'58''$$

and it also follows that

$$\theta_1 + \theta_2 = 63°53'02''$$

whilst
$$\frac{\sin \theta_1}{\sin \theta_2} = \frac{\sin \theta_4 \sin \theta_6 \sin \theta_8}{\sin \theta_3 \sin \theta_5 \sin \theta_7} = \alpha \text{ say}$$

Then
$$\frac{\sin \theta_1 - \sin \theta_2}{\sin \theta_1 + \sin \theta_2} = \frac{\alpha - 1}{\alpha + 1}$$

$$= \frac{2 \sin \dfrac{\theta_1 - \theta_2}{2} \cos \dfrac{\theta_1 + \theta_2}{2}}{2 \sin \dfrac{\theta_1 + \theta_2}{2} \cos \dfrac{\theta_1 - \theta_2}{2}} = \frac{\tan \dfrac{\theta_1 - \theta_2}{2}}{\tan \dfrac{\theta_1 + \theta_2}{2}}$$

$(\theta_1 - \theta_2)$ may now be determined, since $(\theta_1 + \theta_2)$ is known and α can be calculated using the adjusted values of θ_3, etc. The individual values of θ_1 and θ_2 are then found from the values of $(\theta_1 + \theta_2)$ and $(\theta_1 - \theta_2)$, as $44°40'35''$ and $22°12'27''$ respectively.

Two-point resection

The log sin side relationship can be used to solve a resection problem. For example in Fig. 8.5, let A and B be control points, whilst C and D are stations to be fixed in position without occupying A or B. Initially it is necessary to deduce the values of θ_3 and θ_8 before setting up the 'side equations', accepting four measured values. Thus

$$\theta_3 = 180° - (\theta_4 + \theta_5 + \theta_6) = 49°47'26''$$
$$\theta_8 = 180° - (\theta_5 + \theta_6 + \theta_7) = 67°42'46''$$

Whence

$$\frac{\sin \theta_1}{\sin \theta_2} = \frac{\sin \theta_4 \sin \theta_6 \sin \theta_8}{\sin \theta_3 \sin \theta_5 \sin \theta_7} = 1.8599812$$

Therefore

$$\frac{\tan \dfrac{\theta_1 - \theta_2}{2}}{\tan \dfrac{\theta_1 + \theta_2}{2}} = \frac{0.8599812}{2.8599812}$$

Since $\theta_1 + \theta_2 = \theta_5 + \theta_6$
$$= 66°53'56''$$

It follows that

$$\frac{\theta_1 - \theta_2}{2} = 11°13'56''$$

and $\theta_1 = 44°40'28''$
$\theta_2 = 22°12'36''$

These values of course are not so reliable as those calculated previously when θ_3 and θ_8 had been measured. The length and bearing of AB will be known from the co-ordinate data and accordingly there is sufficient information to calculate the co-ordinates of C and D.

Example 8.7 An engineering triangulation in the form of a quadilateral $ABCD$, with a central station O, is interlocked with a primary triangulation so that angle ABC is common to both. The established primary value of this angle is $83°03'06''$. Adjust the given observed values, which are of equal weight, ensuring that the final value of angle ABC is $83°03'06''$.

Triangle	Central angle	Left-hand angle	Right-hand angle
AOB	123°35'30''	39°00'20''	17°24'00''
BOC	63°40'10''	65°39'10''	50°41'10''
COD	88°22'50''	28°39'30''	62°57'55''
DOA	84°21'30''	38°49'40''	56°48'35''

Angle	Log sine	Difference for 1''	Angle	Log sine	Difference for 1''
1	$\bar{1}.7989238$	26.0×10^{-7}	2	$\bar{1}.4757304$	67.1×10^{-7}
3	$\bar{1}.9595489$	9.5×10^{-7}	4	$\bar{1}.8885651$	17.2×10^{-7}
5	$\bar{1}.6808660$	38.6×10^{-7}	6	$\bar{1}.9497468$	10.7×10^{-7}
7	$\bar{1}.7972547$	26.1×10^{-7}	8	$\bar{1}.9226514$	13.8×10^{-7}
	$\bar{1}.2365934$			$\bar{1}.2366937$	

In Fig. 8.6 the left-hand angles are odd numbered: it will be noted that in any triangle they are to the left of the observer looking outwards from O.

When the central point has been occupied our condition equations now demand that the sum of the angles at this point be $360°$ and that the sum of the angles in each of the constituent triangles be $180°$ when small: the log sine relationship still obtains. Normally six condition equations are required for the adjustment of a figure such as this but, due to the extra constraint in respect of the fixed value of ABC, a seventh equation is involved in this example.

Fig. 8.6

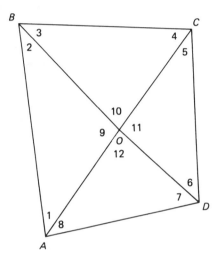

On summation of the various groups of angles, errors of $-10''$, $+30''$, $+15''$ and $-15''$ are present in triangles AOB, BOC, COD and DOA respectively. In addition, the observed central angles sum to $360°00'00''$ whilst an error of $+4''$ obtains at station B.

Let $e_1, e_2, \ldots, e_{12}$ be the respective individual corrections (in seconds of arc), to the measured values.

Then
$$e_9 + e_{10} + e_{11} + e_{12} = 0$$
$$e_1 + e_2 + e_9 = +10$$
$$e_3 + e_4 + e_{10} = -30$$
$$e_5 + e_6 + e_{11} = -15$$
$$e_7 + e_8 + e_{12} = +15$$
$$e_2 + e_3 = -4$$

The sum of the log sines of the left-hand angles should equal that of the right-hand angles, and it will be noted that after the observations the latter sum exceeds the former by 0.0001003. Since the corrections e_1, e_2 etc. are expressed in seconds of arc, they must be multiplied by the respective values of log sine $1''$ to conform with the logarithmic values. Denoting the log sine $1''$ values as $f_1, f_2, \ldots, f_3$, we obtain the further equation

$$e_1 f_1 + e_3 f_3 + e_5 f_5 + e_7 f_7 + \bar{1}.2365934$$
$$= e_2 f_2 + e_4 f_4 + e_6 f_6 + e_8 f_8 + \bar{1}.2366937$$

i.e. $e_1 f_1 + e_3 f_3 + e_5 f_5 + e_7 f_7 - e_2 f_2 - e_4 f_4 - e_6 f_6 - e_8 f_8$
$$= 0.0001003$$

Multiplying throughout by 10^7 and writing $v_1 = f_1 \times 10^7$, $v_2 = f_2 \times 10^7$, etc., we obtain the log sine relationship

$$e_1 v_1 + e_3 v_3 + e_5 v_5 + e_7 v_7 - e_2 v_2 - e_4 v_4 - e_6 v_6 - e_8 v_8 = 1003$$

where $v_1 = 26.0$, $v_2 = 67.1$ etc.

For the least squares condition

$$\Sigma e_1{}^2 \text{ is to be a minimum, and on differentiation}$$

$$e_1 \delta e_1 + e_2 \delta e_2 + \dots e_{12} \delta e_{12} = 0 \quad (1)$$
$$\delta e_9 + \delta e_{10} + \delta e_{11} + \delta e_{12} = 0 \quad (2)$$
$$\delta e_1 + \delta e_2 + \delta e_9 = 0 \quad (3)$$
$$\delta e_3 + \delta e_4 + \delta e_{10} = 0 \quad (4)$$
$$\delta e_5 + \delta e_6 + \delta e_{11} = 0 \quad (5)$$
$$\delta e_7 + \delta e_8 + \delta e_{12} = 0 \quad (6)$$
$$\delta e_2 + \delta e_3 = 0 \quad (7)$$

and $v_1 \delta e_1 + v_3 \delta e_3 + v_5 \delta e_5 + v_7 \delta e_7 - v_2 \delta e_2 - v_4 \delta e_4$
$$- v_6 \delta e_6 - v_8 \delta e_8 = 0 \quad (8)$$

Multiply equations (2) to (8) by $-\lambda_1$ to $-\lambda_7$ respectively and add to equation (1) to obtain

$$\delta e_1 (e_1 - \lambda_2 - v_1 \lambda_7) + \delta e_2 (e_2 - \lambda_2 - \lambda_6 + v_2 \lambda_7)$$
$$+ \delta e_3 (e_3 - \lambda_3 - \lambda_6 - v_3 \lambda_7)$$
$$+ \delta e_4 (e_4 - \lambda_3 + v_4 \lambda_7)$$
$$+ \delta e_5 (e_5 - \lambda_4 - v_5 \lambda_7) + \delta e_6 (e_6 - \lambda_4 + v_6 \lambda_7)$$
$$+ \delta e_7 (e_7 - \lambda_5 - v_7 \lambda_7) + \delta e_8 (e_8 - \lambda_5 + v_8 \lambda_7)$$
$$+ \delta e_9 (e_9 - \lambda_1 - \lambda_2) + \delta e_{10} (e_{10} - \lambda_1 - \lambda_3)$$
$$+ \delta e_{11} (e_{11} - \lambda_1 - \lambda_4) + \delta e_{12} (e_{12} - \lambda_1 - \lambda_5)$$
$$= 0$$

Thus, since δe_1, δe_2 etc. are independent quantities,

$$e_1 = \lambda_2 + v_1 \lambda_7 \qquad e_2 = \lambda_2 + \lambda_6 - v_2 \lambda_7$$

etc., and on substitution in the original correction conditions, the following equations result:

$$4\lambda_1 + \lambda_2 + \lambda_3 + \lambda_5 = 0$$
$$\lambda_1 + 3\lambda_2 + \lambda_6 + v_1 \lambda_7 - v_2 \lambda_7 = +10$$
$$\lambda_1 + 3\lambda_3 + \lambda_6 + v_3 \lambda_7 - v_4 \lambda_7 = -30$$
$$\lambda_1 + 3\lambda_4 + v_5 \lambda_7 - v_6 \lambda_7 = -15$$
$$\lambda_1 + 3\lambda_5 + v_7 \lambda_7 - v_8 \lambda_7 = +15$$
$$\lambda_2 + \lambda_3 + 2\lambda_6 + v_3 \lambda_7 - v_2 \lambda_7 = -4$$

and

$$\lambda_2 (v_1 - v_2) + \lambda_3 (v_3 - v_4) + \lambda_4 (v_5 - v_6) + \lambda_5 (v_7 - v_8)$$
$$+ \lambda_7 (\Sigma v_1{}^2) + \lambda_6 (v_3 - v_2) = 1003$$

These equations can now be expressed in matrix form as:

$$\begin{pmatrix} 4 & 1 & 1 & 1 & 1 & 0 & 0 \\ 1 & 3 & 0 & 0 & 0 & 1 & -41.1 \\ 1 & 0 & 3 & 0 & 0 & 1 & -7.7 \\ 1 & 0 & 0 & 3 & 0 & 0 & +27.9 \\ 1 & 0 & 0 & 0 & 3 & 0 & +12.3 \\ 0 & 1 & 1 & 0 & 0 & 2 & -57.6 \\ 0 & -41.1 & -7.7 & +27.9 & +12.3 & -57.6 & 8039 \end{pmatrix} \begin{pmatrix} \lambda_1 \\ \lambda_2 \\ \lambda_3 \\ \lambda_4 \\ \lambda_5 \\ \lambda_6 \\ \lambda_7 \end{pmatrix} = \begin{pmatrix} 0 \\ +10 \\ -30 \\ -15 \\ +15 \\ -4 \\ 1003 \end{pmatrix}$$

and solved for

$$\lambda_1 = 5.139, \quad \lambda_2 = 0.852, \quad \lambda_3 = -14.978$$
$$\lambda_4 = -8.799, \quad \lambda_5 = 2.368, \quad \lambda_6 = 11.521$$
and $\quad \lambda_7 = 0.224$

The individual values of $e_1, e_2, \ldots, e_{12}$ are now deduced as follows,

$$e_1 = \lambda_2 + v_1\lambda_7 = 6.67''$$
$$e_2 = \lambda_2 + \lambda_6 - v_2\lambda_7 = -2.66'' \text{ etc.}$$

Correction	e_1	e_2	e_3	e_4
Value	+6.67″	−2.66″	−1.33″	−18.83″

Correction	e_5	e_6	e_7	e_8
Value	−0.15″	−11.20″	+8.21″	−0.72″

Correction	e_9	e_{10}	e_{11}	e_{12}
Value	+6.00″	−9.84″	−3.66″	+7.51″

These values can be rounded off to the nearest second.

Adjustment of a traverse

Semi-rigorous methods of adjustment of traverse data were described in Chapter 6. Angular errors were apportioned equally amongst the angles: this is a convenient but entirely arbitrary technique. Positional misclosures were distributed between all traverse legs proportional to the length of the leg: this had some theoretical background but assumed equal reliability in the length and the bearing of the traverse lines. It is possible to include the angles and distances in a comprehensive solution based on least squares by initially evaluating easting and northing differences using bearings derived from the unadjusted angles and then forming three condition equations. An outline of the method as applied to a loop traverse is given below.

The three condition equations are:

$$d\theta_1 + d\theta_2 + \ldots d\theta_n = \text{angular closure}$$
$$\text{correction (seconds)} = P_1, \text{ say}$$

$$(l_1 + dl_1)\sin(\gamma_1 + d\gamma_1) + \ldots = \text{easting correction} = P_2, \text{ say}$$
$$(l_1 + dl_1)\cos(\gamma_1 + d\gamma_1) + \ldots = \text{northing correction} = P_3, \text{ say}$$

l_1 and dl_1 refer to measured lengths and their corrections respectively, whilst γ_1 and $d\gamma_1$ refer to the bearings of those lines and their corrections.

Allowing for the fact that dl and $d\gamma$ are small, we can expand the equations relating to the misclosures of latitude and departure as follows:

$$l_1 \sin \gamma_1 + dl_1 \sin \gamma_1 + l_1 \cos \gamma_1 \cdot d\gamma \sin 1'' + \ldots = P_2$$
$$l_1 \cos \gamma_1 + dl_1 \cos \gamma_1 - l_1 \sin \gamma_1 \cdot d\gamma \sin 1'' + \ldots = P_3$$

in which $d\gamma$ is expressed in seconds.

Since the bearings of the lines are calculated successively using the angular measurements,

$$d\gamma_n = \sum_{i=1}^{n} d\theta_1$$

(i.e. $d\theta_1$ appears in all $d\gamma$ values, $d\theta_2$ appears in all but $d\gamma_1$, etc.) and the three condition equations can be written to contain the unknowns $d\theta_1$ etc. and dl_1 etc., as follows:

$$d\theta_1 + d\theta_2 + \ldots d\theta_n = P_1$$

$$d\theta_1 \sum_{1}^{n} l \cos \gamma \cdot \sin 1'' + d\theta_2 \sum_{2}^{n} l \cos \gamma \cdot \sin 1'' \ldots$$

$$+ d\theta_n l_n \cos \gamma_n \cdot \sin 1'' + \sin \gamma_1 \, dl_1$$

$$+ \sin \gamma_2 \, dl_2 + \ldots \sin \gamma_n \, dl_n$$

$$= \left(P_2 - \sum_{1}^{n} l \sin \gamma \right)$$

and

$$-d\theta_1 \sum_{1}^{n} l \sin \gamma \cdot \sin 1'' - d\theta_2 \sum_{2}^{n} l \sin \gamma \cdot \sin 1'' \ldots$$

$$-d\theta_n l_n \sin \gamma_n \cdot \sin 1'' + \cos \gamma_1 \, dl_1$$

$$+ \cos \gamma_2 \, dl_2 + \ldots \cos \gamma_n \, dl_n$$

$$= \left(P_3 - \sum_{1}^{n} l \cos \gamma \right)$$

Weights need to be applied to the corrections $d\theta_1$ etc. and dl_1 etc. and these may be derived in the usual way as the inverse of the squares of the estimated standard errors.

Correct estimation of these values is not easy and Ashkenazi in 'Adjustment of control networks for precise engineering surveys', *Chartered Surveyor*, January 1970 has given a trial-and-error approach to attain this, the standard deviation of an observation of unit weight derived by the expression

$$s = \sqrt{\frac{\Sigma w_1 r_1^2}{n-m}}$$

mentioned previously, then having a value of unity. It will be appreciated that angles and lengths are physically dissimilar quantities and this aspect is also covered by Ashkenazi.

Only the most precise traverses are likely to be adjusted by this approach and the attention of the reader is drawn to comments by P. Berthon-Jones 'A comparison of the precision of traverses adjusted by Bowditch Rule and by least squares.' *Survey Review*, No. 164.

Adjustment by variation of co-ordinates

The principle of least squares is used in this method of adjustment, which is applicable to triangulation, trilateration or traverse. Primarily the method requires the determination of approximate co-ordinates for stations to be fixed, and thence the adjustment of bearings of lines computed therefrom: distances can be introduced also.

Fig. 8.7

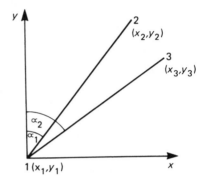

Figure 8.7 shows two lines 12 and 13 connecting stations 1, 2 and 3, whose approximate co-ordinates (x_1, y_1), (x_2, y_2) and (x_3, y_3) have been derived (directly from unadjusted measured lengths and angles, say), so that computed bearings α_1 and α_2 can be determined,

i.e.
$$\tan \alpha_1 = \frac{x_2 - x_1}{y_2 - y_1} \text{ and } \tan \alpha_2 = \frac{x_3 - x_1}{y_3 - y_1}$$

If small changes such as dx_1, dy_1 etc. be applied to those values the corresponding changes in bearing can be evaluated by differentiation as follows:

$$\sec^2 \alpha_1 \, d\alpha_1 = \frac{(y_2 - y_1)}{(y_2 - y_1)^2} \, dx_2$$

$$\therefore d\alpha_1 = \frac{\cos \alpha_1}{l_1} \, dx_2 = \frac{(y_2 - y_1)}{l_1^2} \, dx_2$$

$$\sec^2 \alpha_1 \, d\alpha_1 = \frac{(y_2 - y_1)}{(y_2 - y_1)^2} \, dx_1$$

$$\therefore d\alpha_1 = -\frac{\cos \alpha_1}{l_1} \, dx_1 = -\frac{(y_2 - y_1)}{l_1^2} \, dx_1$$

$$\sec^2 \alpha_1 \, d\alpha_1 = -\frac{(x_2 - x_1)}{(y_2 - y_1)^2} \, dy_2$$

$$\therefore d\alpha_1 = -\frac{\sin \alpha_1}{l_1} \, dy_2 = -\frac{(x_2 - x_1)}{l_1^2} \, dy_2$$

$$\sec^2 \alpha_1 \, d\alpha_1 = \frac{(x_2 - x_1)}{(y_2 - y_1)^2} \, dy_1$$

$$\therefore \, d\alpha_1 = \frac{\sin \alpha_1}{l_1} \, dy_1 = \frac{(x_2 - x_1)}{l_1^2} \, dy_1$$

where l_1 is the length of line 12. Therefore either

$$d\alpha_1 = \frac{\sin \alpha_1}{l_1} \, dy_1 - \frac{\cos \alpha_1}{l_1} \, dx_1 - \frac{\sin \alpha_1}{l_1} \, dy_2 + \frac{\cos \alpha_1}{l_1} \, dx_2$$

or

$$d\alpha_1 = \frac{(x_2 - x_1)}{l_1^2} \, dy_1 - \frac{(y_2 - y_1)}{l_1^2} \, dx_1$$

$$- \frac{(x_2 - x_1)}{l_1^2} \, dy_2 + \frac{(y_2 - y_1)}{l_1^2} \, dx_2$$

Similarly from

$$l_1^2 = (y_2 - y_1)^2 + (x_2 - x_1)^2$$

either

$$dl_1 = \frac{-(y_2 - y_1)}{l_1} \, dy_1 - \frac{(x_2 - x_1)}{l_1} \, dx_1$$

$$+ \frac{(y_2 - y_1)}{l_1} \, dy_2 + \frac{(x_2 - x_1)}{l_1} \, dx_2$$

or

$$dl_1 = -\cos \alpha_1 \, dy_1 - \sin \alpha_1 \, dx_1 + \cos \alpha_1 \, dy_2 + \sin \alpha_1 \, dx_2$$

Corresponding relationships can be established for line 13 involving $d\alpha_2$ or dl_2.

The computed value of C of angle 213 is $(\alpha_2 - \alpha_1)$, measurement being clockwise, and if this be adjusted by the small increments $d\alpha_1$ and $d\alpha_2$ we can write its best value as

$$(\alpha_2 + d\alpha_2 - \alpha_1 - d\alpha_1) = (C + d\alpha_2 - d\alpha_1)$$

If O is the observed value of that angle we have the relationship for the residual r as

$$r = (O - C) - (d\alpha_2 - d\alpha_1) \tag{8.25}$$

from the observation equation for that angle.

A similar expression arises for line 12, in the form $r = (O - C) - dl_1$, O being the measured length and C the computed length. Geographic co-ordinates can be used but change in azimuth over long lengths can now be involved. It is more convenient to adjust in the context of the projection plane having introduced the $(t - T)$ correction (see Chapter 7). One further general point is that theodolite observations obtained by the method of rounds give directions, i.e. angles between the stations and the reference station as well as individual angles between the various stations. These directions can be used to form observation equations if required.

Fig. 8.8

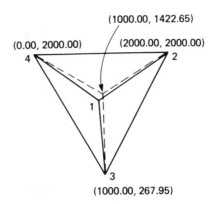

(1000.00, 1422.65)

(0.00, 2000.00)   (2000.00, 2000.00)

(1000.00, 267.95)

Example 8.8 Stations 2, 3 and 4 (Fig. 8.8) have the following respective co-ordinates (2000.00 mE, 2000.00 mN) (1000.00 mE, 267.95 mN) and (0.00 mE, 2000.00 mN). Station 1 was established very near to the centre of the circle which circumscribes Stations 2, 3 and 4, and the following readings were recorded by a theodolite set up at Station 1.

Pointing on	Horizontal circle reading
2	00°00′00″
3	119°59′51.0″
4	240°00′23.5″

Determine the co-ordinates of Station 1.

It will be noted that triangle 2 3 4 is equilateral. The centre of the circumscribing circle will be at the centroid of the triangle, whose co-ordinates are 1000.00 mE and 1422.65 mN. These will be taken as the approximate co-ordinates of Station 1, with

$$l_{12} = l_{13} = l_{14} = l = 1154.70 \text{ m}$$

Also since 2, 3 and 4 are fixed points they are not subject to correction, and so dx_2, dy_2, etc., have zero value.

From the data

$x_2 - x_1 =$	1000.00 m	$y_2 - y_1 =$	577.35 m
$x_3 - x_1 =$	0.00 m	$y_3 - y_1 =$	-1154.70 m
$x_4 - x_1 =$	-1000.00 m	$y_4 - y_1 =$	577.35 m

Angle 2 1 3

$$O = 119°59'51.0''$$

$$C = 120°00'00''$$

$$d\alpha_{13} = \frac{1154.70}{l^2} dx_1$$

$$d\alpha_{12} = \frac{1000.00}{l^2} dy_1 - \frac{577.35}{l^2} dx_1$$

Residual

$$r_1 = \frac{-9.0}{206\ 265} - \left[\frac{1154.70}{l^2} dx_1 - \frac{1000.00}{l^2} dy_1 + \frac{577.35}{l^2} dx_1 \right]$$

$$= \frac{-9.0}{206\ 265} - \frac{1732.05}{l^2} dx_1 + \frac{1000.00}{l^2} dy_1$$

Angle 3 1 4

$$O = 120°00'32.5''$$

$$C = 120°00'00''$$

$$d\alpha_{14} = -\frac{1000.00}{l^2} dy_1 - \frac{577.35}{l^2} dx_1$$

$$d\alpha_{13} = \frac{1154.70}{l^2} dx_1$$

Residual

$$r_2 = \frac{32.5}{206\ 265} - \left[-\frac{1000.00}{l^2} dy_1 - \frac{577.35}{l^2} dx_1 - \frac{1154.70}{l^2} dx_1 \right]$$

$$= \frac{32.5}{206\ 265} + \frac{1732.05}{l^2} dx_1 + \frac{1000.00}{l^2} dy_1$$

On tabulating from the expressions for the residuals to determine the best values for dx_1 and dy_1, as explained earlier in the chapter, we have

dx_1	dy_1	N
$- \dfrac{1732.05}{1154.70^2}$	$+ \dfrac{1000.00}{1154.70^2}$	$- \dfrac{9.0}{206\ 265}$
$+ \dfrac{1732.05}{1154.70^2}$	$+ \dfrac{1000.00}{1154.70^2}$	$+ \dfrac{32.5}{206\ 265}$

Whence
$$dx_1 = -0.08 \text{ m}$$
$$dy_1 = -0.08 \text{ m}$$

Hence the co-ordinates of Station 1 are

999.92 mE

1422.57 mN

Note:

(1) In the calculations $(O - C)$ has been expressed in radian measure to conform with $d\alpha$.

(2) Angles only were present in the observation equations since no distances were measured. The angles were taken to be of equal weight but had distances also been measured the weighting of observations would have been essential. The weights not only allow for the different reliability of the observations but also, since the angles and lengths are pysically dissimilar quantities, they

compensate for the difference in units. (Radian measure can, however, be adopted for the bearings and angles if no other sensible weighting is possible.)

(3) Station 1 was known to be very near to the centroid, so that its approximate co-ordinates needed relatively small corrections. In practice if x_1 and y_1 had been 'coarsely' estimated then a number of iterations could well have been needed.

(4) In Fig. 8.8 the four stations form a triangle with a central point. In the general case of nine angles measured, a solution can be obtained as for the quadrilateral of Fig. 8.6. However, had all lengths been measured also, then fifteen observation equations could be obtained for a solution by variation of co-ordinates, one equation arising for each side and one for each angle. Holding only station 3 as fixed, six co-ordinate corrections (i.e. two per station) can be derived using statement 8.22.

Exercises 8

1 Two EDM were used to measure line XY and the following readings recorded:

Instrument A (m)	Instrument B (m)
156.276	156.285
156.279	156.275
156.280	156.280
156.277	156.278
156.280	156.279
156.279	
156.281	
156.279	

What is the relative precision of the two instruments?

(*Salford*)

Answer: 1:7.89

2 In a topographical survey, the difference in level between two points A and B is found by three routes, via C, D, E and F, via G and H, and via J, the distances being:

Route (1)	Route (2)	Route (3)
$AC = 120$ m	$AG = 240$ m	$AJ = 294$ m
$CD = 162$ m	$GH = 306$ m	$JB = 462$ m
$DE = 300$ m	$HB = 384$ m	
$EF = 258$ m		
$FB = 132$ m		

The sections on route (1) are each levelled four times, those on route (2) eight times, and on route (3) twice; the established differences in level thus obtained being 30.81 m, 30.57 m and 31.08 m, respectively.

If the error in any section at each levelling is proportional to its length, and the usual laws for combinations of readings hold, find the most probable value for the difference in level between A and B.

(*London*)

Answer: 30.72 m

3 In a round of levels the following data were collected:

	Length (m)	Level difference
A to B	1200	Rise 5.263 m
B to C	1500	Rise 2.144 m
C to D	2100	Fall 1.278 m
D to A	1000	Fall 6.138 m
B to D	3200	Rise 0.863 m
		(measured twice)

Determine the most probable values of the levels of B, C and D above A. *(Salford)*

Answer: 5.267 m, 7.412 m, 6.135 m

4 In the course of the precise levelling of a certain area the following results were obtained. Determine the most probable level of B, C and D above A.

Level line	Level difference	Weight
A to B	Rise 4.727 m	1
B to C	Rise 1.580 m	1
C to D	Rise 3.540 m	3
D to A	Fall 9.846 m	1
B to D	Rise 5.125 m	2
A to C	Rise 6.315 m	2

(Salford)

Answer: B, 4.728 m; C, 6.312 m; D, 9.851 m.

5 During a round of levels *ABCDA* it was found that B was 3.71 m above A, C was 4.69 m above B, and D was 6.12 m above B, 1.58 m above C and 9.75 m above A. If C was known to be 8.35 m above A, determine the most probable heights of B and D above A, all levellings being of equal weight. *(Salford)*

Answer: 3.69 m, 9.83 m. (Note that C, being known, must not be corrected.)

6 The magnitude of angle A was obtained as follows

(i) by measurement six times to give an average value of $90°24'28''$, the error of a single measurement being $\pm 4''$, and

(ii) by addition of angle B $(55°14'15'')$ and angle C $(35°10'20'')$, each of which had been measured once only by methods having errors of $\pm 3''$ and $\pm 5''$ respectively.

Determine the most probable value for angle A.

Answer: $90°24'29''$.

7 In quadrilateral *ABCD* angles were measured as follows

$BAC = 55°02'15''$	$DCA = 42°16'42''$
$CAD = 59°29'09''$	$ACB = 34°48'42''$
$CBD = 54°00'58''$	$ADB = 29°20'36''$
$DBA = 36°08'22''$	$BDC = 48°53'35''$

Adjust the angles using the method of least squares assuming equal weights apply.

Answer: $BAC = 55°02'01''$, $CAD = 59°29'13''$.

8 A precise traverse was run to fix the co-ordinates of stations P11 to P14 (Fig. 8.9). An abstract of the data is given below. Use a least squares adjustment procedure to determine the best co-ordinates for the new stations.

Fig. 8.9

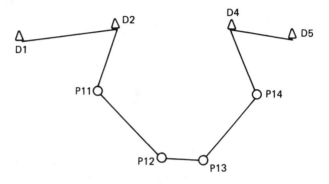

Known co-ordinates

	E (m)	N (m)
D1	1975.328	8847.314
D2	2345.083	8901.159
D4	2780.360	8911.363
D5	3003.060	8877.048

Observed horizontal clockwise angles

D1	D2	P11	290°04'41''	estimated s.e.
D2	P11	P12	123°39'21''	in angle ±4''
P13	P12	P11	133°31'17''	
P14	P13	P12	129°39'32''	
P13	P14	D4	116°54'55''	
P14	D4	D5	303°13'03''	

Observed distances (m)

D2	P11	233.998	estimated s.e. in distance
P11	P12	327.334	±0.01 m
P12	P13	179.258	
P13	P14	293.638	
P14	D4	263.604	

Answer:

P11	2297.262 mE	8672.093 mN
P12	2526.919	8438.813
P13	2706.167	8442.034
P14	2889.483	8671.426

9 To fix a new control station *P* from two existing control stations *A* and *B*, the observations as listed below were taken. Perform

a least squares adjustment to derive the most probable co-ordinates for the new station.

Observed bearings
$$AP = 103°53'31'' \pm 3.0''$$
$$BP = 14°24'43'' \pm 3.0''$$

Observed distances
$$AP = 4662.63 \pm 0.07 \text{ m}$$
$$BP = 3821.68 \pm 0.07 \text{ m}$$

Observed angle
$$BPA = 89°28'54'' \pm 4.2''$$

Co-ordinates (metres)

	E	N
A	6 497.03	15 589.48
B	10 072.52	10 768.52
P	11 023.46	14 469.74 (provisional)

Answer: 11 023.55 mE, 14 469.98 mN.

10 By application of the principle of least squares find the best estimates of the variables x and y from the following observation equations

$$
\begin{aligned}
2x + y &= +1.0 \\
x + 2y &= -1.0 \\
x + y &= +0.1 \\
x - y &= +2.2 \\
2x &= +1.9
\end{aligned}
$$

You should assume the observations are of equal weights.

(London)

Answer: $x = 1.02$, $y = -1.05$

9 Areas and volumes

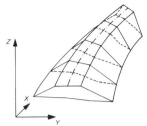

The calculations connected with the measurement of areas of land, etc., and of volumes and other quantities connected with engineering and building works, are dealt with in this chapter. Areas are considered first of all, since the computation of areas is involved in the calculation of volumes, and are dealt with under the following headings:

(1) Mechanical integration — the planimeter. (2) Areas enclosed by straight lines. (3) Irregular figures.

The planimeter

Mechanical integration by the planimeter can be applied to figures of all shapes, and although the construction and application of the instrument are simple, the accuracy achieved is of the highest degree, particularly when measuring irregular figures.

Planimeters (Fig. 9.1) consist essentially of:

(a) The pole block, which is fixed in position on the paper by a fine retaining needle.
(b) The pole arm, which is pivoted about the pole block at one end and the integrating unit at the other.
(c) The tracing arm (which may be either fixed or variable in length) attached at one end to the integrating unit and carrying at the other end the tracing point or optical tracer.

Fig. 9.1 Fixed bar planimeter

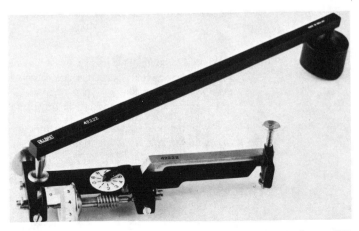

Fig. 9.2

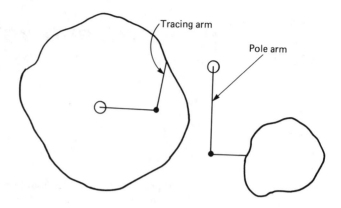

Tracing arm

Pole arm

(*d*) The measuring unit, consisting of a hardened steel integrating disc carried on pivots; directly connected to the disc spindle is a primary drum divided into 100 parts, and readings of 1/1000th of a revolution of the integrating disc are obtained, either by estimation using an index mark, or by vernier, on an opposite drum. Another indicator gives the number of complete revolutions of the disc.

Principles and operation

It can readily be shown that if the pole is suitably placed relative to the figure to be measured and the tracing point moved round the outline of the figure, then the integrating disc will register an amount proportional to the area of the figure. In the fixed-arm instrument, the drum is so graduated that the areas are given directly, and correction factors must be applied when working on plans.

Example 9.1 What is the area of a piece of land which has a plan area of 1613 mm^2 as measured by a fixed-arm planimeter if the scale of the plan is 1/2500?

On this scale, 1 mm^2 represents 2500 mm^2.

Hence $\qquad$ 1613 mm^2 $\equiv$ 1613 $\times$ 2500 $\times$ 2500 mm^2

$\qquad\qquad$ or **1.008 hectare.**

Where the length of the tracing arm is variable, the bar is graduated so that on being set at a particular graduation, direct readings of areas may be made on a plan of a particular scale.

A planimeter can be used in two ways (Fig. 9.2).

(*a*) with the pole *outside* the figure to be measured, or
(*b*) with the pole *inside* the figure to be measured.

(*a*) is the more convenient and should be used whenever possible, although (*b*) allows larger areas to be covered for one setting.

With the pole *outside* the figure the procedure for measuring any area, the plan being on a flat horizontal surface, is:

(1) Place the pole outside the area in such a position that the tracing point can reach any part of the outline.

(2) With the tracing point placed on a known point on the outline, read the vernier.

(3) Move the tracing point *clockwise* around the outline, back to the known point, and read the vernier again.

(4) The difference between the two readings, multiplied by the scale factor, gives the area.

(5) Repeat until three consistent values are obtained, and the mean of these is taken.

With the pole *inside* the figure, the procedure is as above, but a constant, engraved on the tracing arm, is applied to the difference in readings in each case. This constant represents the area of the *zero circle* of the planimeter, i.e. of that circle which will be swept out when the plane of the integrating disc lies exactly through the pole and the integrating disc does not revolve at all. It is added when the measured area is larger than the zero circle area, but when the measured area is smaller then it has to be subtracted from that constant.

The foregoing discussion refers to conventional mechanical planimeters. Figure 9.3 illustrates Tamaya's Digital Planimeter, Planix 7, which incorporates integrated circuit technology. It has a tracer arm with a tracer lens and tracer point, but the pole and pole arm have been eliminated and it will be noted that there are now rollers, with contact rings, on an axle.

Initially, the tracer arm is set on the approximate centre line of the area to be measured and after the power, supplied by a nickel-cadmium battery, is switched on, the unit of measurement, i.e. m^2, is selected, and the scale fed in via the relevant keys on the keyboard. A reference start point is selected as above, or marked, on the perimeter and the tracer point positioned thereon. The 'start' key is activated, causing the display to register zero, and the tracer point moved clockwise along the perimeter to return to the reference point. Motion of the system is sensed by an electro shaft-encoder generating pulses which are processed electronically so that the measured area is displayed digitally. This instrument can cater for dual scales, i.e. different horizontal and vertical scales, and has various other facilities also.

Roller planimeters are more versatile than polar planimeters and the

Fig. 9.3 Planix 7 Digital Planimeter (*Courtesy*: Hall & Watts Ltd)

most up-to-date version is the Tamaya Planex 5000 which is termed a 'digital area line meter'. It bears some resemblance to the Planex 7 although its keyboard is mounted over the rollers. It possesses two rotary encoders which facilitate the evaluation of co-ordinates and hence the computation of areas and line lengths. In the latter case the horizontal and vertical scales of the plan must be the same. There are two modes of operation, the point mode for length measurement of straight lines without tracing and the stream mode for measurement of curved lines. This instrument can be interfaced to a computer, in which eventuality the co-ordinates can be output in addition to the area and line data.

Areas enclosed by straight lines

Such areas include those enclosed inside the survey lines of a tape and offset survey or theodolite traverse, and fields enclosed by straight-line boundaries.

Simple triangles

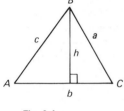

Fig. 9.4

Where the area is triangular in shape (Fig. 9.4), or is made up of a series of triangles the following formulae are used:

$$\text{Area} = \tfrac{1}{2}(\text{base} \times \text{perpendicular height})$$
$$= \tfrac{1}{2}\,bh$$

but
$$h = a \sin C$$

Therefore
$$\text{area} = \tfrac{1}{2}\,ba \sin C$$

Also
$$\text{area} = \sqrt{S(S-a)(S-b)(S-c)}$$

where
$$S = \frac{a+b+c}{2}$$

Any area bounded by straight lines can be divided completely into a series of triangles, and the total area then derived by summing the areas of the individual triangles.

Area by co-ordinates

In the particular case of traverse surveys which are plotted from co-ordinates, it is more convenient to calculate the area from the co-ordinates themselves.

Fig. 9.5

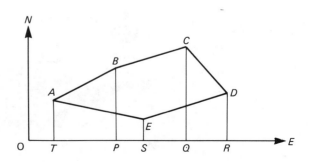

Consider closed traverse ABCDEA (Fig. 9.5), whose stations have co-ordinates E_A, N_A; E_B, N_B; etc., relative to two axes whose origin is O.

Area $ABCDEA$

$$= \text{areas } (ABPT + BCQP + CDRQ - DESR - EATS)$$

$$= \left(\frac{N_A+N_B}{2}\right) \times (E_B-E_A) + \left(\frac{N_B+N_C}{2}\right) \times (E_C-E_B)$$

$$+ \left(\frac{N_C+N_D}{2}\right) \times (E_D-E_C) - \left(\frac{N_D+N_E}{2}\right) \times (E_D-E_E)$$

$$- \left(\frac{N_E+N_A}{2}\right) \times (E_E-E_A)$$

$$= \tfrac{1}{2}[(N_A E_B + N_B E_C + N_C E_D + N_D E_E + N_E E_A) - (N_B E_A + N_C E_B + N_D E_C + N_E E_D + N_A E_E)]$$

$$= \tfrac{1}{2}[N_A(E_B-E_E) + N_B(E_C-E_A) + N_C(E_D-E_B) + N_D(E_E-E_C) + N_E(E_A-E_D)]$$

In general $\quad$ area $= \tfrac{1}{2}\left[\displaystyle\sum_{i=1}^{n} N_i(E_{i+1} - E_{i-1})\right]$

Note: (1) The formula could be developed by projecting the traverse legs onto the northing rather than the easting axis. In this case

$$\text{Area} = \tfrac{1}{2}[E_A(N_B-N_E) + E_B(N_C-N_A) + E_C(N_D-N_B) + E_D(N_E-N_C) + E_E(N_A-N_D)]$$

(2) the formula will give negative answers if the figure is lettered anticlockwise.

Example 9.2 Determine the area in hectares enclosed by the line of a closed traverse survey ABCDE from the following data.

Station	E(m)	N(m)
A	100.00	200.00
B	206.98	285.65
C	268.55	182.02
D	292.93	148.80
E	191.74	85.70

Station	N_i	E_{i+1}	E_{i-1}	$N_i(E_{i+1}-E_{i-1})$
A	200.00	206.98	191.74	3048.00
B	285.65	268.55	100.00	48146.31
C	182.02	292.93	206.98	15644.62
D	148.80	191.74	268.55	− 11429.33
E	85.70	100.00	292.93	− 16534.10

$$\frac{2)38875.50}{19437.75 \text{ m}^2}$$

Therefore $\qquad\qquad$ Area = **1.944 hectares**

Division of an area by a line of known bearing

If it is required to divide the area *ABCDEA* shown in Fig. 9.6 into two parts by a line *XY*, the procedure is as follows:

(1) Calculate the total area enclosed by *ABCDEA*.
(2) Draw a line at the known bearing through one of the stations which will divide the area approximately in the manner required, say BE_1.
(3) Determine the bearings of lines *EB*, EE_1 and *BA* and so deduce the angles α, β and γ.
(4) Calculate the length of *EB* from the known co-ordinates, and thus find EE_1 and BE_1 using the sine rule relationship

$$\frac{EE_1}{\sin \alpha} = \frac{EB}{\sin \beta} = \frac{BE_1}{\sin \gamma}$$

(5) Determine area ABE_1EA.
(6) The required area *AXYEA* = area ABE_1EA − area XBE_1YX. Hence, area XBE_1YX is found.
(7) To locate line *XY* it will be necessary to calculate the distance d separating the actual line *XY* and the trial line BE_1.

Area XBE_1YX
$$= \text{area } PBE_1QP + \text{area } QE_1Y - \text{area } PBX$$
$$= d \cdot BE_1 + \tfrac{1}{2}d[d \cdot \tan (\beta - 90)] - \tfrac{1}{2}d[d \cdot \tan(90 - \delta)]$$

All the terms in the above equation with the exception of d are known, so that the problem may be solved.

Fig. 9.6

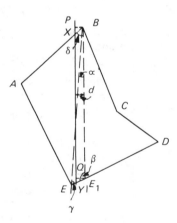

Irregular figures

The following methods may be used for determining the area of irregular curvilinear figures such as ponds, lakes and the areas enclosed between survey lines and natural boundaries.

1. 'Give and take' lines

The whole area is divided into triangles, or trapezoids, the irregular boundaries being replaced by straight lines so arranged that any small areas excluded from the survey by the straight line are balanced by

Fig. 9.7

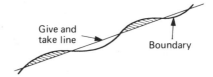

Give and take line

Boundary

other small areas outside the survey which are now included (*see* Fig. 9.7).

The positions of the lines are estimated by eye, using either a thin transparent straight-edge or a silk thread, and the lines are then drawn in faintly on the plan. The areas of the resulting triangles or trapezoids are calculated by the methods already described.

2. Counting squares

An overlay of squared tracing paper is laid on the drawing. The number of squares and parts of squares which are enclosed by the figure under consideration is now counted and, knowing the scale of the drawing and the size of the squares on the overlay, the total area of the figure is computed.

Fig. 9.8

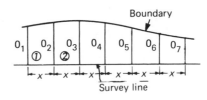

Boundary

O_1 O_2 O_3 O_4 O_5 O_6 O_7

① ②

Survey line

3. Trapezoidal rule

Figure 9.8 shows an area bounded by a survey line and a boundary. The survey line is divided into a number of small equal intercepts of length x, and the offsets O_1, O_2, etc., measured, either directly on the ground or by scaling from the plan. If x is short enough for the length of boundary between the offsets to be assumed straight, then the area is divided into a series of trapezoids.

$$\text{Area of trapezoid (1)} = \frac{O_1 + O_2}{2} \cdot x$$

$$\text{Area of trapezoid (2)} = \frac{O_2 + O_3}{2} \cdot x$$

$$\text{Area of trapezoid (6)} = \frac{O_6 + O_7}{2} \cdot x$$

Summing up, we get

$$\text{Area} = \frac{x}{2} \{O_1 + 2O_2 + 2O_3 + \dots + O_7\}$$

In the general case with n offsets, we get

Fig. 9.9

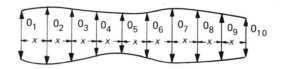

$$\text{Area} = x \left\{ \frac{O_1 + O_n}{2} + O_2 + O_3 + \ldots + O_{n-1} \right\}$$

If the area of a narrow strip of ground is required, this method may be used by running a straight line down the strip as shown in Fig. 9.9, and then measuring offsets at equal intercepts along this. By the same reasoning it will be seen that the area is given by

$$\text{Area} = x \left\{ \frac{O_1 + O_n}{2} + O_2 + O_3 + \ldots + O_{n-1} \right\}$$

Example 9.3 Calculate the area of the plot shown in Fig. 9.9 if the offsets, scaled from the plan at intervals of 10 m, are:

Offset	O_1	O_2	O_3	O_4	O_5
Length (m)	16.76	19.81	20.42	18.59	16.76
Offset	O_6	O_7	O_8	O_9	O_{10}
Length (m)	17.68	17.68	17.37	16.76	17.68

$$\text{Area} = 10 \left\{ \frac{16.76 + 17.68}{2} + 19.81 + 20.42 + 18.59 \right.$$
$$\left. + 16.76 + 17.68 + 17.68 + 17.37 + 16.76 \right\}$$
$$= 1622.9 \text{ m}^2$$
$$= \mathbf{0.162 \text{ hectare}}$$

---

4. Simpson's Rule This method, which gives greater accuracy than other methods, assumes that the irregular boundary is composed of a series of parabolic arcs. It is essential that the figure under consideration be divided into an *even* number of equal strips.

Referring to Fig. 9.8, consider the first three offsets, which are shown enlarged in Fig. 9.10.

The portion of the area contained between offsets O_1 and O_3

Fig. 9.10

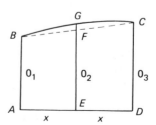

$$= ABGCDA$$

$$= \text{trapezoid } ABFCDA + \text{area } BGCFB$$

$$= \frac{O_1+O_3}{2} \cdot 2x + \frac{2}{3} \text{ (area of circumscribing parallelogram)}$$

$$= \frac{O_1+O_3}{2} \cdot 2x + \frac{2}{3} \cdot 2x \left(O_2 - \frac{O_1+O_3}{2} \right)$$

$$= \frac{x}{3} \{3O_1 + 3O_3 + 4O_2 - 2O_1 - 2O_3\}$$

$$= \frac{x}{3} \{O_1 + 4O_2 + O_3\}$$

For the next pair of intercepts, area contained between offsets O_3 and O_5

$$= \frac{x}{3} \{O_3 + 4O_4 + O_5\}$$

For the final pair of intercepts, area contained between offsets, O_5 and O_7

$$= \frac{x}{3} \{O_5 + 4O_6 + O_7\}$$

Summing up, we get

$$\text{Area} = \frac{x}{3} \{(O_1 + O_7) + 2(O_3 + O_5) + 4(O_2 + O_4 + O_6)\}$$

In the general case,

$$\text{Area} = \frac{x}{3} (X + 2O + 4E)$$

where X = sum of first and last offsets,

O = sum of the remaining odd offsets,

E = sum of the even offsets.

Simpson's Rule states, therefore, that the area enclosed by a curvilinear figure divided into an even number of strips of equal width is equal to one-third the width of a strip, multiplied by the sum of the two extreme offsets, twice the sum of the remaining odd offsets, and four times the sum of the even offsets.

Example 9.4 In a tape and offset survey the following offsets were taken to a fence from a survey line:

Chainage (m)	0	20	40	60	80
Offset (m)	0	5.49	9.14	8.53	10.67
Chainage (m)	100	120	140	160	180
Offset (m)	12.50	9.75	4.57	1.83	0

Find the area between the fence and the survey line.

The reader should note that the term 'chainage' refers to a cumulative increase in distance measured from a starting point on the line, i.e. zero chainage.

It will be seen that there are ten offsets, and since Simpson's Rule can be applied to an *odd* number of offsets only, it will be used here to calculate the area contained between the first and ninth offsets. The residual triangular area between the ninth and tenth offsets is calculated separately. It is often convenient to tabulate the working.

Offset no.	Offset	Simpson multiplier	Product
1	0	1	0
2	5.49	4	21.96
3	9.14	2	18.28
4	8.53	4	34.12
5	10.67	2	21.34
6	12.50	4	50.00
7	9.75	2	19.50
8	4.57	4	18.28
9	1.83	1	1.83

$$\Sigma = 185.31$$

$$\text{Area } (O_1 - O_9) = \tfrac{20}{3} \cdot 185.31 = 1235.40 \text{ m}^2$$

$$\text{Area } (O_9 - O_{10})$$
$$= \tfrac{20}{2} \cdot 1.83 = \underline{18.30 \text{ m}^2}$$
$$1253.70 \text{ m}^2$$

= 0.125 hectare

Volumes: earthwork calculations

The excavation, removal and dumping of earth is a frequent operation in building or civil engineering works. In the construction of a sewer, for example, a trench of sufficient width is excavated to given depths and gradients, the earth being stored in some convenient place (usually the side of the trench) and then returned to the trench after the laying of the pipe. Any material left over after re-instatement must be carted away and disposed of. In basement excavation, probably all the material dug out will require to be carted away, but for embankments the earth required will have to be brought from some other place.

In each case, however, payment will have to be made for labour, plant, etc., and this is done on the basis of the calculated volume of material handled. It is therefore essential that the engineer or surveyor should be able to make good estimates of volumes of earthwork.

There are three general methods for calculating earthworks: (1) by cross-sections, (2) by contours, (3) by spot heights.

(1) Volume from cross-sections

In this method, cross-sections are taken at right angles to some convenient line which runs longitudinally through the earth works and, although it is capable of general application, it is probably most used on long narrow works such as roads, railways, canals, embankments, pipe excavations, etc. The volumes of earthwork between successive cross-sections are calculated from a consideration of the cross-sectional areas, which in turn are measured or calculated by the general methods already given, i.e. by planimeter, division into triangles, counting squares, etc.

In long constructions which have constant formation width and side slopes it is possible to simplify the computation of cross-sectional areas by the use of formulae. These are especially useful for railways, long embankments, etc., and formulae will be given for the following types of cross-section: (*a*) sections level across, (*b*) sections with a cross-fall, (*c*) sections part in cut and part in fill, (*d*) sections of variable levels.

Fig. 9.11

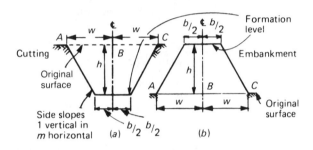

(*a*) Sections level across (Fig. 9.11)

Depth at centre line (or height in case of embankment)
$$= h \text{ units}$$
$$\text{Formation width} = b \text{ units}$$
$$\text{Side width} = w$$

The sloping side has to fall (or rise) a vertical height of h units from original level to final formation level. Since this side slopes in such a way that m units is the horizontal projection for every single unit vertical rise, then in h units the side gives a horizontal projection of mh units.

$$AB = BC = w = \frac{b}{2} + mh$$

$$\text{Area of cross section} = \frac{1}{2}\left[2\left(\frac{b}{2} + mh\right) + b\right]h$$

$$= h(b + mh) \tag{9.1}$$

Example 9.5 At a certain station an embankment formed on level ground has a height at its centre line of 3.10 metres. If the breadth of formation is 12.50 metres, find (*a*) the side widths,

> and (b) the area of cross-section, given that the side slope is
> 1 vertical to $2\frac{1}{2}$ horizontal.

$$b = 12.50 \text{ metres}, \quad m = 2.5$$

Therefore
$$\text{Area} = h(b+mh)$$
$$= 3.10(12.50+2.5 \times 3.10)$$
$$= \textbf{62.78 m}^2$$

Side widths are both equal to $\dfrac{b}{2} + mh$

$$= 6.25+2.5 \times 3.10$$
$$= \textbf{14.0 m}$$

(b) Sections with cross-fall (Fig. 9.12)

In this case the existing ground has a cross-fall or transverse gradient relative to the centre line, and the side widths are not equal since the section is not symmetrical about the centre line.

Fig. 9.12

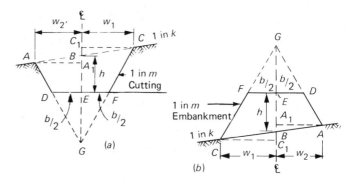

Now
$$C_1B = \frac{w_1}{k}$$

since this is the difference in level between C and B due to a gradient of 1 in k over a distance of w_1; similarly

$$A_1B = \frac{w_2}{k}$$

Also, if the side slopes intersect at G, then GE will be the vertical difference in level over a horizontal distance of $b/2$.

Hence
$$GE = \frac{b}{2m}$$

Since triangles C_1CG and EFG are similar,

$$\frac{CC_1}{EF} = \frac{GC_1}{GE}$$

$$\frac{w_1}{b/2} = \frac{\dfrac{b}{2m} + h + \dfrac{w_1}{k}}{b/2m}$$

Therefore
$$w_1 = \left(\frac{b}{2} + mh\right)\left(\frac{k}{k-m}\right) \tag{9.2}$$

Also
$$\frac{AA_1}{DE} = \frac{GA_1}{GE}$$

$$\frac{w_2}{b/2} = \frac{\dfrac{b}{2m} + h - \dfrac{w_2}{k}}{b/2m}$$

Hence
$$w_2 = \left(\frac{b}{2} + mh\right)\left(\frac{k}{k+m}\right) \tag{9.3}$$

The area of the cutting or the embankment is the area $ACFDA$,

$$= \text{area } BCG + \text{area } ABG - \text{area } DFG$$

$$= \tfrac{1}{2}w_1\left(\frac{b}{2m} + h\right) + \tfrac{1}{2}w_2\left(\frac{b}{2m} + h\right) - \tfrac{1}{2}b\,\frac{b}{2m}$$

$$= \frac{1}{2m}\left\{\left(\frac{b}{2} + mh\right)(w_1 + w_2) - \frac{b^2}{2}\right\} \tag{9.4}$$

Difference in level between C and F

$$= h + \frac{w_1}{k} \tag{9.5}$$

Difference in level between A and D

$$= h - \frac{w_2}{k} \tag{9.6}$$

This type of section is sometimes known as a 'two-level section', since two levels are required to establish the cross-fall of 1 in k.

Example 9.6 Calculate the side widths and cross-sectional area of an embankment to a road with formation width of 12.50 metres, and side slopes 1 vertical to 2 horizontal, when the centre height is 3.10 metres and the existing ground has a cross fall of 1 in 12 at right angles to the centre line of the embankment.

Referring to Fig. 9.12(b).

$$w_1 = \left(\frac{b}{2} + mh\right)\left(\frac{k}{k-m}\right)$$

where $b = 12.50$, $\quad k = 12$, $\quad m = 2$, $\quad h = 3.10$ m

Hence $w_1 = \left(\dfrac{12.50}{2} + 2 \times 3.10\right)\left(\dfrac{12}{10}\right)$

$= 14.94$ m

$w_2 = \left(\dfrac{b}{2} + mh\right)\left(\dfrac{k}{k+m}\right)$

$= \left(\dfrac{12.50}{2} + 2 \times 3.10\right)\left(\dfrac{12}{14}\right)$

$= 10.67$ m

From equation (9.4),

$\text{area} = \dfrac{1}{2m}\left\{\left(\dfrac{b}{2} + mh\right)(w_1 + w_2) - \dfrac{b^2}{2}\right\}$

$= \dfrac{1}{4}\left\{12.45(14.94 + 10.67) - \dfrac{12.50^2}{2}\right\}$

$= \mathbf{60.18\ m^2}$

Fig. 9.13

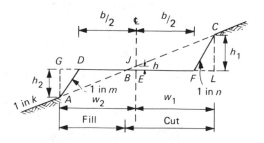

(c) Sections part in cut and part in fill (Fig. 9.13)

Inspection of Fig. 9.13 shows that the 'cut' position is similar to the 'cut' position right of the centre line of Fig. 9.12(a) and hence

$$w_1 = \left(\dfrac{b}{2} + nh\right)\left(\dfrac{k}{k-n}\right) \qquad (9.7)$$

when we directly substitute n for m.

The 'fill' position is similar to the 'fill' position left of the centre line of Fig. 9.12(b) except that h is now effectively negative, and so

$$w_2 = \left(\dfrac{b}{2} - mh\right)\left(\dfrac{k}{k-m}\right) \qquad (9.8)$$

Note that when $h = 0$, $w_2 = \dfrac{b}{2}\left(\dfrac{k}{k-m}\right)$

$\text{Area of fill} = \tfrac{1}{2}h_2 \cdot DB = \tfrac{1}{2}h_2\left(\dfrac{b}{2} - kh\right)$

$= \dfrac{1}{2}\left(\dfrac{2w_2 - b}{2m}\right)\left(\dfrac{b}{2} - kh\right)$

$$= \frac{1}{2} \frac{(b/2 - kh)^2}{(k - m)} \qquad (9.9)$$

$$\text{Area of cut} = \tfrac{1}{2}h_1 \cdot BF = \tfrac{1}{2}h_1 \left(\frac{b}{2} + kh \right)$$

$$= \frac{1}{2} \left(\frac{2w_1 - b}{2n} \right) \left(\frac{b}{2} + kh \right)$$

$$= \frac{1}{2} \frac{(b/2 + kh)^2}{(k - n)} \qquad (9.10)$$

When the cross-section is in fill at the centre line, instead of being in cut as in Fig. 9.13 then, replacing $+h$ by $-h$, the following modified formulae obtain:

$$\text{Area of fill} = \frac{1}{2} \frac{(b/2 + kh)^2}{(k - m)}$$

$$\text{Area of cut} = \frac{1}{2} \frac{(b/2 - kh)^2}{(k - n)}$$

Example 9.7 A road has a formation width of 9.50 metres and side slopes of 1 vertical to 1 horizontal in cut, and 1 vertical to 3 horizontal in fill. The original ground had a cross-fall of 1 vertical to 5 horizontal. If the depth of excavation at the centre line is 0.5 m, calculate the side widths and the areas of cut and fill.

Form formulae (9.7), (9.8), (9.9) and (9.10), we get

$$w_1 = \left(\frac{k}{k - n} \right) \left(\frac{b}{2} + nh \right)$$

$$= \left(\frac{5}{5 - 1} \right) (4.75 + 1 \times 0.5)$$

$$= \textbf{6.56 m}$$

$$w_2 = \left(\frac{k}{k - m} \right) \left(\frac{b}{2} - mh \right)$$

$$= \left(\frac{5}{5 - 3} \right) (4.75 - 3 \times 0.5)$$

$$= \textbf{7.88 m}$$

$$\text{Area of fill} = \frac{1}{2} \left(\frac{(b/2 - kh)^2}{(k - m)} \right) = \frac{1}{2} \left(\frac{(4.75 - 5 \times 0.5)^2}{(5 - 3)} \right)$$

$$= \textbf{1.27 m}^2$$

$$\text{Area of cut} = \frac{1}{2} \left(\frac{(b/2 + kh)^2}{(k - n)} \right) = \frac{1}{2} \left(\frac{(4.75 + 5 \times 0.5)^2}{(5 - 1)} \right)$$

$$= \textbf{6.57 m}^2$$

(d) Sections of variable level

The type of section shown in Fig. 9.14 is sometimes referred to as a 'three-level section' since at least three levels are required on each cross-section to enable the ground slopes to be calculated. The side-width formulae are the same as those developed in section (b), and are

$$w_1 = \left(\frac{b}{2} + mh\right)\left(\frac{k}{k-m}\right) \qquad (9.11)$$

$$w_2 = \left(\frac{b}{2} + mh\right)\left(\frac{l}{l+m}\right) \qquad (9.12)$$

Fig. 9.14

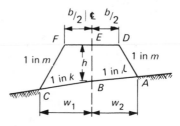

If BA were falling away from the centre line,

$$w_2 = \left(\frac{b}{2} + mh\right)\left(\frac{l}{l-m}\right)$$

$$\text{Area of cross-section} = \tfrac{1}{2}w_1\left(h + \frac{b}{2m}\right) + \tfrac{1}{2}w_2\left(h + \frac{b}{2m}\right)$$

$$-\tfrac{1}{2}b \cdot \frac{b}{2m}$$

$$= \frac{1}{2}\left\{(w_1+w_2)\left(h + \frac{b}{2m}\right) - \frac{b^2}{2m}\right\}$$

$$= \frac{1}{2m}\left\{(w_1+w_2)\left(mh + \frac{b}{2}\right) - \frac{b^2}{2}\right\} \qquad (9.13)$$

Note: The section shown in Fig. 9.14 is probably the most complex one to be worthy of analysis in the manner shown; it is worth using on long constructions, where many cross-sections are involved, because the use of a formula means (a) that levels are taken at predetermined points on each section, and (b) that computation of areas and volumes can be tabulated and computerized.

Computation of volumes

Having determined the various areas of cross-section, the volumes of earth involved in the construction are computed by one of the following methods: (1) mean areas, (2) end areas, (3) prismoidal formula.

(1) Volumes by mean areas

In this method the volume is determined by multiplying the mean of the cross-sectional areas by the distance between the end sections. If the areas are $A_1, A_2, A_3 \ldots A_{n-1}, A_n$ and the distance between the two extreme sections A_1 and A_n is L, then

$$\text{Volume} = V = \frac{A_1 + A_2 + A_3 \ldots A_{n-1} + A_n}{n} \cdot L$$

The method is not a very accurate one.

(2) Volumes by end areas

If A_1 and A_2 are the areas of two cross-sections distance D apart, then the volume V between the two is given by

$$V = D \cdot \frac{A_1 + A_2}{2} \qquad (9.14)$$

This expression is correct so long as the area of the section mid-way between A_1 and A_2 is the mean of the two, and such can be assumed to be the case so long as there is no wide variation between successive sections. If there is such a variation, then a correction must be applied, and this refinement is dealt with in the next method. In general, however, in view of the usual irregularities in ground surface which exist between successive cross-sections, and the problems of bulking and settlement normally associated with earthworks, it is reasonable to use the end-areas formula for normal estimating.

For a series of consecutive cross-sections, the total volume will be:

$$\text{Volume} = \Sigma V = \frac{D_1(A_1 + A_2)}{2} + \frac{D_2(A_2 + A_3)}{2}$$

$$+ \frac{D_3(A_3 + A_4)}{2} + \ldots$$

if $D_1 = D_2 = D_3$, etc. $= D$

$$\Sigma V = D \left\{ \frac{A_1 + A_n}{2} + A_2 + A_3 + \ldots + A_{n-1} \right\} \qquad (9.15)$$

which is sometimes referred to as the trapezoidal rule for volumes.

Example 9.8 An embankment is formed on ground which is level transverse to the embankment but falling at 1 in 20 longitudinally so that three sections 20 metres apart have centre-line heights of 6.00, 7.60 and 9.20 metres respectively above original ground level. If side slopes of 1 in 1 are used, determine the volume of fill between the outer sections when the formation width is 6.00 m, using the trapezoidal rule.

Using equation (9.1),

$$\begin{aligned}
A &= h(b + mh) \\
A_1 &= 6.00(6.00 + 6.00) = 72.00 \text{ m}^2 \\
A_2 &= 7.60(6.00 + 7.60) = 103.36 \text{ m}^2 \\
A_3 &= 9.20(6.00 + 9.20) = 139.84 \text{ m}^2
\end{aligned}$$

Note that the mid-area A_2 is not the mean of A_1 and A_3.

$$V = \frac{20.00}{2} \, [72.00 + 2 \times 103.36 + 139.84]$$

$$= \mathbf{4185.6 \ m^3}$$

In the above example, if the method of end areas is applied between the two extreme sections, we get

$$V = \frac{40.00}{2} \, (72.00 + 139.84)$$

$$= 4236.8 \ m^3$$

an over-estimation of over 1 per cent compared with the first method.

Fig. 9.15

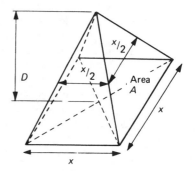

As an extreme case of such discrepancy, consider the volume of a pyramid (Fig. 9.15) of base area A and perpendicular height D.

$$\text{True volume} = \tfrac{1}{3} \, \text{base area} \times \text{height}$$
$$= \tfrac{1}{3} \cdot D \cdot A$$

But by end areas, $\qquad V = D \cdot \dfrac{A+0}{2}$

$$= \tfrac{1}{2} D \cdot A$$

(3) Volumes by prismoidal formula

If the volume of earth between two successive cross-sections be considered a *prismoid*, then a more precise formula — the prismoidal formula — may be used. It is generally considered that, all things being equal, use of this formula gives the most accurate estimate of volume.

A prismoid is a solid made up of two end faces which must be parallel plane figures not necessarily of the same shape; the faces between them, i.e. sides, top and bottom, must be formed by straight continuous lines running from one end face to the other.

The volume of a prismoid is given by

$$V = \frac{D}{6} \, (A_1 + 4M + A_2) \tag{9.16}$$

where A_1 and A_2 are the areas of the two end faces distances D apart, M is the area of the section mid-way between.

The theory assumes that in calculating M, each linear dimension is the average of the corresponding dimensions of the two end areas. (It is no use taking the area M as the mean of A_1 and A_2, for we should only get

$$V = \frac{D}{6}\left(A_1 + 4\,\frac{(A_1 + A_2)}{2} + A_2\right)$$

$$= D \cdot \frac{A_1 + A_2}{2}$$

i.e. the 'end-areas' formula.)

As an example of the application of the prismoidal formula, we may derive the formula for the volume of a pyramid. Referring to Fig. 9.15,

$$\text{Base area} = A = x^2$$

$$\text{Mid-area} = M = \frac{x^2}{4} = \frac{A}{4}$$

$$\text{Top area } A_2 = 0$$

Therefore

$$V = \frac{D}{6}\left(A + 4 \cdot \frac{A}{4} + 0\right)$$

$$= \frac{D \cdot A}{3}$$

Similarly, for a wedge as in Fig. 9.16,

$$V = \frac{D}{6}\left(ab + \frac{4ab}{2} + 0\right)$$

$$= \frac{Dab}{2}$$

The volume of a triangular prism is also given by the product of the area of the normal section (in this case $Db/2$) and one-third of the sum of the three parallel edges. Thus in Fig. 9.16,

$$V = \frac{Db}{2} \cdot \frac{1}{3}\,(a + a + a) = \frac{Dab}{2}$$

Fig. 9.16

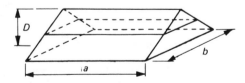

The proof of the prismoidal formula will be found in any standard textbook on solid geometry.

There is a number of alternative ways in which the prismoidal formula may be used, and some of these are given below.

(*a*) Treat each cross-section as the end area of a prismoid of length D, and estimate the dimensions of the mid-areas at the $D/2$ points as the mean of the two corresponding dimensions in the end areas. This is difficult when the sections are irregular.

(*b*) Where estimation of the mid-area is difficult, arrange for extra sections to be levelled at the mid-area positions as required. This means, of course, a large increase in the amount of field work.

(*c*) Treat alternate sections as end areas, i.e. the length of the prismoid is $2D$. Unless the ground profile is regular both transversely and longitudinally, however, it is likely that errors will be introduced in assuming that a volume of earth is in fact prismoidal over such a length. Where this method is used, by taking A_1 and A_3, A_3 and A_5, A_5 and A_7, etc., as end areas of successive prismoids, we have

$$V_1 = \frac{2D}{6} \cdot (A_1 + 4A_2 + A_3)$$

$$V_2 = \frac{2D}{6} \cdot (A_3 + 4A_4 + A_5), \text{ etc.}$$

Therefore
$$V = \frac{D}{3}(A_1 + 4A_2 + 2A_3 + 4A_4 + \ldots$$

$$+ 2A_{n-2} + 4A_{n-1} + A_n) \tag{9.17}$$

where n is an odd number.

This is Simpson's Rule for volumes. It can be applied with greater accuracy with method (*b*) in which mid-sections are levelled, but note that D in equation (9.17) will then be halved.

(*d*) Calculate the volumes between successive cross-sections by the method of end areas, and apply corrections to these volumes known as *prismoidal corrections*. Such corrections can be derived for regular sections only, e.g. consider sections level across.

If D = the spacing of cross-sections, A_1 and A_2 = the two end areas, M = the mid-area, h_1 and h_2 = the difference in level between ground level and formation at A_1 and A_2 respectively, and b = formation width, then from equation (9.1),

$$A_1 = h_1(b + mh_1)$$
$$A_2 = h_2(b + mh_2)$$

Hence
$$V_{\text{end areas}} = \frac{D}{2}(bh_1 + mh_1^2 + bh_2 + mh_2^2)$$

Assuming $h_m = \dfrac{h_1 + h_2}{2}$ (difference in level between ground level and formation at the mid-section)

$$M = \left(\frac{h_1 + h_2}{2}\right)\left\{b + m\left(\frac{h_1 + h_2}{2}\right)\right\}$$

$$= \frac{bh_1}{2} + \frac{bh_2}{2} + \frac{mh_1^2}{4} + \frac{mh_2^2}{4} + \frac{mh_1h_2}{2}$$

Hence

$$V_{\text{prismoid}} = \frac{D}{6} \left\{ bh_1 + mh_1^2 + 4 \left(\frac{bh_1}{2} + \frac{bh_2}{2} + \frac{mh_1^2}{4} \right. \right.$$
$$\left. \left. + \frac{mh_2^2}{4} + \frac{mh_1 h_2}{2} \right) + bh_2 + mh_2^2 \right\}$$

$$= \frac{D}{6} \left\{ 3bh_1 + 2mh_1^2 + 3bh_2 + 2mh_2^2 + 2mh_1 h^2 \right\}$$

$$= \frac{D}{2} \left\{ bh_1 + \tfrac{2}{3}mh_1^2 + bh_2 + \tfrac{2}{3}mh_2^2 + \frac{2mh_1 h_2}{3} \right\}$$

Hence

$$V_{EA} - V_P = \frac{D}{2} \left\{ \frac{mh_1^2}{3} - \frac{2mh_1 h_2}{3} + \frac{mh_2^2}{3} \right\}$$

$$= \frac{D}{6} \cdot m(h_1 - h_2)^2$$

$$= \textit{prismoidal correction} \text{ for a level section} \quad (9.18)$$

Since the term $(h_1 - h_2)^2$ must always be positive, the correction must be *deducted* from the volume as calculated by the end-areas formula. The correction is simple to apply since it requires no additional information to that which is already required for the end-areas calculations.

Prismoidal corrections can be similarly derived for other types of section starting from the general expression:

$$V_{EA} - V_P = \text{prismoidal correction (PC)}$$

$$= \frac{D}{2} (A_1 - A_2) - \frac{D}{6} (A_1 + 4M + A_2)$$

$$= \frac{D}{3} (A_1 - 2M + A_2) \quad (9.19)$$

For instance in the case of the section with a cross fall (Fig. 9.12) it has been proved that

$$A = \frac{1}{2m} \left\{ \left(\frac{b}{2} + mh \right)(w_1 + w_2) - \frac{b^2}{2} \right\}$$

$$= \frac{1}{2m} \left(\frac{b}{2} + mh \right)\left(\frac{b}{2} + mh \right)\left(\frac{k}{k+m} + \frac{k}{k-m} \right) - \frac{b^2}{4m}$$

If $h = h_1$, h_2 and $(h_1 + h_2)/2$ at the two end sections and mid-section respectively

$$PC = \frac{D}{3} (A_1 - 2A_m + A_2)$$

$$= \frac{D}{6m} \left(\frac{k}{k+m} + \frac{k}{k-m} \right) \left\{ \left(\frac{b}{2} + mh_1 \right)^2 + \left(\frac{b}{2} + mh_2 \right)^2 \right.$$

$$\left. -2 \left(\frac{b}{2} + m \left(\frac{h_1+h_2}{2} \right) \right)^2 \right\}$$

$$= \frac{Dk}{6m} \left(\frac{k-m+k+m}{k^2-m^2} \right) \left(\frac{m^2h_1^2}{2} - m^2h_1h_2 + \frac{m^2h_2^2}{2} \right)$$

$$= \frac{D}{6} \frac{k^2}{(k^2-m^2)} \cdot m(h_1-h_2)^2 \qquad (9.20)$$

As with the PC for a level section, this expression will always be positive, since $k > m$.

As an exercise, the student is advised to check that prismoidal corrections for sloping sections which are part in cut, part in fill, are:

$$Fill \quad PC = \frac{D}{12(k-m)} k^2 (h_1-h_2)^2 \qquad (9.21)$$

$$Cut \quad PC = \frac{D}{12(k-n)} k^2 (h_1-h_2)^2 \qquad (9.22)$$

Example 9.9 Using the data of Example 9.8 solved by the end-areas method, compute the volume by the prismoidal formula.

(a) Taking $\qquad D = 40$ metres

$$V = \frac{40}{6} (72.00 + 4 \times 103.36 + 139.84)$$

$$= \textbf{4168.5 m}^3$$

(b) Taking $D = 20$ metres, and applying the prismoidal correction to the 'end-areas' volumes,

$$V_1 = \frac{20}{2} (72.00 + 103.36)$$

$$= 1753.6 \text{ m}^3$$

$$PC = \frac{Dm}{6} (h_1-h_2)^2$$

$$= \frac{20 \times 1}{6} (6.00 - 7.60)^2$$

$$= 8.53 \text{ m}^3$$

$$V_2 = \frac{20}{2} (103.36 + 139.84)$$

$$= 2432 \text{ m}^3$$

$$PC = \frac{20}{6} (7.60 - 9.20)^2$$

$$= 8.53 \text{ m}^3$$

$$\text{Total volume} = \textbf{4168.5 m}^3$$

The answers are the same because in this example the dimensions of the middle section are the exact mean of the two outer ones. In practice, the second method is preferable, since it assumes prismoids of shorter length. A similar example, in which the volume contained between three cross-sections is required, will now be worked; in this case, however, the centre-section dimensions are not the mean of the outer ones.

Example 9.10 A cutting is to be made in ground which has a transverse slope of 1 in 5. The width of formation is 8.00 metres and the side slopes are 1 vertical to 2 horizontal. If the depths at the centre lines of three sections 20 metres apart are 2.50, 3.10 and 4.30 metres respectively, determine the volume of earth involved in this length of the cutting.

From equation (9.4),

$$A = \frac{1}{2m} \left\{ \left(\frac{b}{2} + mh \right) (w_1 + w_2) - \frac{b^2}{2} \right\}$$

From equations (9.2) and (9.3),

$$w_1 = \left(\frac{b}{2} + mh \right) \left(\frac{k}{k-m} \right)$$

$$w_2 = \left(\frac{b}{2} + mh \right) \left(\frac{k}{k+m} \right)$$

Hence $\dfrac{k}{k-m} = \dfrac{5}{3}$, $\dfrac{k}{k+m} = \dfrac{5}{7}$, since $m = 2$ and $k = 5$

Tabulating,

Section	h	mh	b/2+mh	w_1	w_2	w_1+w_2	$A(m^2)$
1	2.50	5.00	9.00	$9.00 \times \frac{5}{3}$ = 15.00	$9.00 \times \frac{5}{7}$ = 6.43	21.43	40.24
2	3.10	6.20	10.20	$10.20 \times \frac{5}{3}$ = 17.00	$10.20 \times \frac{5}{7}$ = 7.29	24.29	53.94
3	4.30	8.60	12.60	$12.60 \times \frac{5}{3}$ = 21.00	$12.60 \times \frac{5}{7}$ = 9.00	30.00	86.50

(*a*) Treating the whole as one prismoid,

$$V = \frac{40}{6} (40.24 + 4 \times 53.94 + 86.50)$$

$$= \textbf{2283.3 m}^3$$

(b) Using end-areas formula with prismoidal correction,

$$V_{EA} = \frac{20}{2}(40.24 + 2 \times 53.94 + 86.50)$$

$$= 2346.2 \text{ m}^3$$

From equation (9.20),

$$PC = \frac{D}{6}\frac{k^2}{k^2-m^2} \cdot m(h_1-h_2)^2$$

Therefore

$$\text{Total PC} = \frac{20}{6} \cdot \frac{25}{21} \cdot 2[(2.50-3.10)^2+(3.10-4.30)^2]$$

$$= 14.3 \text{ m}^3$$

Therefore $V_P = 2346.2 - 14.3$
$$= \mathbf{2331.9 \text{ m}^3}$$

Example 9.11 A road has a formation breadth of 9.00 m, and side slopes of 1 in 1 cut, and 1 in 3 in fill. The original ground had a cross fall of 1 in 5. If the depth of excavation at the centre lines of two sections 20 metres apart are 0.40 and 0.60 metres respectively, find the volumes of cut and fill over this length.

From equations (9.9) and (9.10),

$$\text{Area of fill} = \frac{1}{2} \cdot \frac{(b/2-kh)^2}{k-m}$$

$$\text{Area of cut} = \frac{1}{2} \cdot \frac{(b/2+kh)^2}{k-n}$$

Section 1:

$$\text{Area of fill} = \frac{1}{2}\frac{(4.50-5 \times 0.40)^2}{(5-3)} = 1.56 \text{ m}^2$$

$$\text{Area of cut} = \frac{1}{2}\frac{(4.50+5 \times 0.40)^2}{(5-1)} = 5.28 \text{ m}^2$$

Section 2:

$$\text{Area of fill} = \frac{1}{2}\frac{(4.50-5 \times 0.60)^2}{(5-3)} = 0.56 \text{ m}^2$$

$$\text{Area of cut} = \frac{1}{2}\frac{(4.50+5 \times 0.60)^2}{(5-1)} = 7.03 \text{ m}^2$$

Fill:

$$V_{EA} = \frac{20}{2}(1.56+0.56)$$

$$= 21.2 \text{ m}^3$$

From equation (9.21),

$$PC = \frac{D}{12(k-m)} \cdot k^2(h_1 - h_2)^2$$

$$= \frac{20}{12 \times 2} \times 25 \times 0.2^2$$

$$= 0.8 \text{ m}^3$$

Therefore $V_P = \textbf{20.4 m}^3$

Cut:

$$V_{EA} = \frac{20}{2} \cdot (5.28 + 7.03)$$

$$= 123.1 \text{ m}^3$$

From equation (9.22),

$$PC = \frac{D}{12(k-n)} \cdot k^2(h_1 - h_2)^2$$

$$= \frac{20}{12 \times 4} \times 25 \times 0.2^2$$

$$= 0.4 \text{ m}^3$$

Therefore $V_P = \textbf{122.7 m}^3$

i.e. there is an excess of 102.3 m³ of cut over fill.

Example 9.12 Obtain an expression for the total volume of a symmetrical embankment of length l with side slopes 1 in s horizontal and formation breadth b if the height varies uniformly from zero at $l = 0$ to l/n at a distance l.

A new access road to an old quarry necessitates an incline crossing an existing vertical quarry face, 9 metres high, at right angles, the original ground being level above and below the face. A length of 120 metres available on the low side is to accommodate the embankment forming the lower portion of the incline, and it is desired to use spoil from cutting the higher ground to form this embankment in such a way that cut and fill balance. The cut has vertical sides and the material from each cubic metre filled 1·1 m³ of the bank which has side slopes of 1 in 3. The formation breadth is 6 metres. Calculate the gradient required and the length of cutting.

(London)

Fig. 9.17

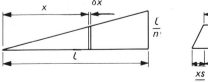
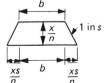

Referring to Fig. 9.17 volume of section of length δx

$$= \frac{1}{2}\frac{x}{n}\left\{b+\left(b+\frac{2xs}{n}\right)\right\}\delta x$$

$$= \frac{x}{n}\left(b+\frac{xs}{n}\right)\delta x$$

Volume of embankment

$$= \int_0^l \frac{x}{n}\left(b+\frac{xs}{n}\right)dx$$

$$= \left[\frac{bx^2}{2n}+\frac{sx^3}{3n^2}\right]_0^l$$

$$= \frac{bl^2}{2n}+\frac{sl^3}{3n^2}$$

Or, using the prismoidal formula

$$A_1 = 0$$

$$A_2 = \frac{l}{n}\cdot\frac{\left(b+b+\dfrac{2ls}{n}\right)}{2} = \frac{l}{n}\left(b+\frac{ls}{n}\right)$$

$$M = \frac{l}{2n}\cdot\frac{\left(b+b+\dfrac{ls}{n}\right)}{2} = \frac{l}{4n}\left(2b+\frac{ls}{n}\right)$$

Therefore

$$V = \frac{l}{6}\left\{0+4\cdot\frac{l}{4n}\left(2b+\frac{ls}{n}\right)+\frac{l}{n}\left(b+\frac{ls}{n}\right)\right\}$$

$$= \frac{bl^2}{2n}+\frac{sl^3}{3n^2}$$

Or, as a triangular prism

$$\text{Area of normal section} = \frac{l^2}{2n}$$

Therefore

$$V = \frac{l^2}{2n}\cdot\frac{1}{3}\left(b+b+b+\frac{2ls}{n}\right)$$

$$= \frac{bl^2}{2n}+\frac{sl^3}{3n^2}$$

In the second part, if l_1 be the length of cutting, and the gradient required be 1 in n, then the volume of cut can be deduced from the above formulae as

$$V_{\text{cut}} = \frac{bl_1^2}{2n}$$

(*Note*: The term $sl^2/3n^2$ disappears here because $s = 0$ for vertical sides.)

The embankment is to be 120 metres long, so that the height of embankment at the quarry face will be $120/n$ metres. Thus the depth of cut at the quarry face will be $(9-(120/n))$ metres.

Therefore
$$l_1 = n\left(9 - \frac{120}{n}\right)$$

$$= 9n - 120$$

Volume of material in embankment

$$= \frac{bl^2}{2n} + \frac{sl^3}{3n^2}$$

$$= \frac{6 \times 120^2}{2n} + \frac{3 \times 120^3}{3n^2}$$

Since cut and fill are to balance,

$$1.1 \times \frac{6l_1^2}{2n} = \frac{6 \times 120^2}{2n} + \frac{3 \times 120^3}{3n^2}$$

Hence
$$3.3(9n - 120)^2 = 3 \times 120^2 + \frac{120^2}{n}$$

whence
$$n = 32.8, \text{ say}$$

i.e.
$$\text{gradient} = \textbf{1 in 32.8}$$

$$\text{Length of cut} = 9 \times 32.8 - 120$$

$$= \textbf{175 m}, \text{ say}$$

Effect of curvature

In the previous calculation of volume by cross-sections, it has been assumed that the sections are parallel to each other and normal to a straight centre line. When the centre line is curved, however, the sections will no longer be parallel to each other, since they will be set out radial to the curve; a correction for curvature is then necessary and to derive this, use is made of Pappus's Theorem.

Pappus's Theorem states that the volume swept out by an area revolving about an axis is given by the product of the area and the length of the path traced by the centroid of the area. The area is to be completely to one side of the axis and in the same plane.

Thus a sphere is swept out by a semi-circle rotating about a diameter, and its volume is given by the area $(\frac{1}{2}\pi r^2)$ times the length of the path of the centroid in tracing one revolution. Since the centroid of a semi-circle is

$$\frac{4}{3} \cdot \frac{r}{\pi} \text{ from the axis,}$$

$$\text{Volume} = \frac{\pi r^2}{2} \times \left(2\pi \times \frac{4}{3}\frac{r}{\pi}\right)$$

$$= \tfrac{4}{3}\pi r^3$$

The volume of earth in cuttings and embankments which follow circular curves may therefore be determined by considering the solid as being formed by an area revolving about an axis at the centre of the curve. If the cross-section is constant round the curve, then the volume will equal the product of that area and the distance traced out by its centre of gravity.

When the sections are not uniform, an approximate volume may be obtained by the method illustrated in Fig. 9.18.

Fig. 9.18

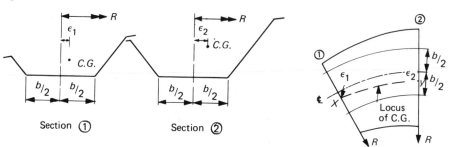

Section ①   Section ②

If the mean distance of centroids from the centre line is written as $\epsilon = (\epsilon_1 + \epsilon_2)/2$, then the mean radius of the path of the centroid will be $(R \pm \epsilon)$, the negative sign being adopted if the mean centroid lies on the same side of the centre line as the centre of the curve, and positive if on the other side.

If the distance between sections is D, then the angle θ subtended at the centre is equal to D/R.

Hence   Length of path of centroid $= XY = \dfrac{D}{R} \cdot (R \pm \epsilon)$

Then the volume is given approximately by

$$V = \tfrac{1}{2}(A_1 + A_2)D \left(1 \pm \frac{\epsilon}{R} \right) \tag{9.23}$$

The position of the centroid may change from one side of the centre line to the other if the transverse slope of the ground changes. If regard is paid to sign, however, and ϵ_1 and ϵ_2 summed algebraically, the correct sign for ϵ will be obtained.

As an alternative, each cross-sectional area may be corrected for the eccentricity of its centroid, so that instead of using A_1, an 'equivalent area' of $A_1(1 \pm \epsilon_1/R)$ is used. Each area is thus corrected by $A_1\epsilon_1/R$, $A_2\epsilon_2/R$, ... etc., and these corrected values are then used in the prismoidal formula adopted.

Example 9.13 The centre line of a cutting is on a curve of 120 metres radius, the original surface of the ground being approximately level. The cutting is to be widened by increasing the formation width from 6 to 9 metres, the excavation to be entirely on the inside of the curve and to retain the existing side

Fig. 9.19

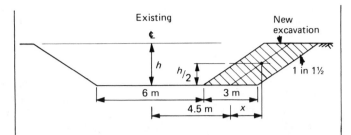

slope of $1\frac{1}{2}$ horizontal to 1 vertical. If the depth of formation increases uniformly from 2.40 metres to 5.10 metres over a length of 90 metres, calculate the volume of earthwork to be removed in this length.　　　　　　　　　　　(*London*)

Referring to Fig. 9.19, area of new excavation $A = 3h$.
For any given depth at the centre line of h,

$$x = \frac{3}{2} \times \frac{h}{2} = \frac{3}{4} h \text{ metres}$$

Therefore eccentricity of centroid of excavation will then be $4.5 + 3h/4$ from the former centre line.

Distance (m)	h(m)	A(m^2)	ϵ(m)	Mean distance of centroid from $\mathcal{L}$
0	2.40	7.20	$4.50 + 1.80 = 6.30$	6.64
30	3.30	9.90	$4.50 + 2.48 = 6.98$	7.32
60	4.20	12.60	$4.50 + 3.15 = 7.65$	7.98
90	5.10	15.30	$4.50 + 3.82 = 8.32$	

Volume of excavation between 0 and 30 metres

$$= \left(\frac{7.2 + 9.9}{2}\right) \frac{30}{120} (120 - 6.64)$$

$$= 242.31 \text{ m}^2$$

Volume of excavation between 30 and 60 metres

$$= \left(\frac{9.9 + 12.6}{2}\right) \frac{30}{120} (120 - 7.32)$$

$$= 316.91 \text{ m}^3$$

Volume of excavation between 60 and 90 metres

$$= \left(\frac{12.6 + 15.3}{2}\right) \frac{30}{120} (120 - 7.98)$$

$$= 390.67 \text{ m}^3$$

Total volume = **949.89 m^3** (950 m^3, say)

Alternatively, using equivalent areas and the prismoidal formula,

Distance (m)	h(m)	A(m^2)	ϵ(m)	$A\dfrac{\epsilon}{R}$	$A\left(1-\dfrac{\epsilon}{R}\right)$(m^2)
0	2.40	7.20	6.30	0.38	6.82
15	2.85	8.55	6.64	0.47	8.08
30	3.30	9.90	6.98	0.57	9.33
45	3.75	11.25	7.31	0.68	10.57
60	4.20	12.60	7.65	0.80	11.80
75	4.65	13.95	7.99	0.93	13.02
90	5.10	15.30	8.32	1.06	14.24

$$\text{Volume} = \frac{30}{6}\,[6.82+4\times 8.08+2\times 9.33$$
$$+4\times 10.57+2\times 11.80+4\times 13.02+14.24]$$
$$= \mathbf{950\ m^3}$$

The correction for curvature is most important when the radius of the curve is small compared to the width of the cross-section. In the above example, the uncorrected volume is

$$V = \frac{30}{6}\,[7.20+4\times 8.55+2\times 9.90$$
$$+4\times 11.25+2\times 12.60+4\times 13.95+15.30]$$
$$= 1012.5\ m^3$$

Therefore the error caused by neglecting effect of curvature is about $6\frac{1}{2}$ per cent.

---

(2) Volumes from contour lines

As an alternative to the determination of volumes by means of vertical cross-sections, it is possible to calculate volumes using the horizontal areas contained by contour lines. Owing to the relatively high cost of accurately contouring large areas, the method is of limited use, but where accurate contours are available as, for instance, in reservoir sites, they may be conveniently used.

The contour interval will determine the distance D in the 'end-area' or prismoidal formula, and for accuracy this should be as small as possible, preferably 1 or 2 metres. The areas enclosed by individual contour lines are best taken off the plan by means of planimeter. In computing the volumes, the areas enclosed by two successive contour lines are used in the 'end-areas' formula, whence

$$\begin{aligned} V =\ & \text{volume of water, earth, tipped material etc.,} \\ & \text{between contour lines } x \text{ and } y \end{aligned}$$

$$= D\,\frac{A_x+A_y}{2} \tag{9.24}$$

where D = vertical interval.

Fig. 9.20

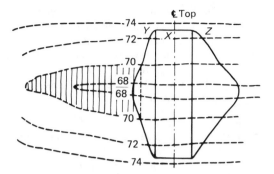

If required, the prismoidal formula can be used, either by treating alternate areas as mid-areas or by interpolating intermediate contours between those established by direct levelling.

Figure 9.20 shows, by hatching, the area enclosed by the 70 m contour line. Note that the contour line is completed across the upstream face of the dam. The plan of the dam is determined as follows. Assume the breadth at the top to be, say 8 m, and the height to be 73.0 m AOD, with side slopes 1 in 1 upstream and 1 in 2 downstream. Then at point X the 72 m contour is cut by the toe of the bank, which has thus fallen 1 m at 1 in 1, i.e. in 1 m horizontal distance from the top of the bank. Therefore XY, the distance from the centre of the dam, is $\frac{8}{2}+1 =$ 5 m. Similarly $XZ = \frac{8}{2}+2 = 6$ m, since the slope here is 1 vertical to 2 horizontal. The points of intersection of the toe with other contours are found in like manner and the outline of the dam is thus obtained.

Example 9.14 The areas within the underwater contour lines of a lake are as follows:

Contour (m AOD)	190	188	186	184	182
Area (m^2)	3150	2460	1630	840	210

Calculate the volume of water in the lake between the 182 and 190 m contours.

(*a*) By 'end areas',

$$V = \frac{2}{2} \{3150+2(2460+1630+840)+210\}$$

$$= \textbf{13 220 m}^3$$

(*b*) By prismoidal formula, using alternate sections as mid-areas,

$$V = \frac{4}{6} (3150+4(2460+840)+2 \times 1630+210\}$$

$$= \textbf{13 213 m}^3$$

With reservoirs it is often required to know the volume of water contained, corresponding to a given height. This is done by calculating the total volume contained below successive contours and then plotting volume against height to give a curve from which the volume at intermediate levels may be read. (It will be appreciated that a still-water surface is a level surface, so that the water's edge is in fact a contour line.)

(3) Volumes from spot levels

This method of volume determination is especially useful in the determination of volumes of large open excavations for tanks, basements, borrowpits, etc. and for ground levelling operations such as playing fields and building sites. It can be applied also to the determination of volumes of spoil heaps.

Having located the outline of the structure on the ground, the engineer divides up the area into squares (or rectangles), marking the corner points as described in the method of contouring by spot-levels, and then taking levels at each of these corner points.

The size of the squares will depend on the nature of the ground and the corners should be sufficiently close for the ground surfaces between lines to be considered as planes. By subtracting from the observed levels the corresponding formation levels, a series of heights can be found from which the volume within each square can be taken as the plan area multiplied by the average of the depth of excavation (or fill) at the four corners.

Fig. 9.21

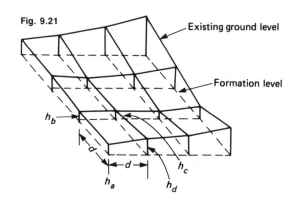

Existing ground level

Formation level

Fig. 9.22

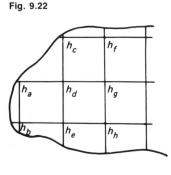

Referring to Fig. 9.21 the volume of cut under the first square

$$= \frac{d^2}{4} (h_a + h_b + h_c + h_d)$$

Since over the complete grid the heights at the internal intersection are used in the calculation of more than one square, the following type of formula can be developed (*see* Fig. 9.22)

$$\text{Vol} = \frac{d^2}{4} (\Sigma h_1 + 2\Sigma h_2 + 3\Sigma h_3 + 4\Sigma h_4) \qquad (9.25)$$

where d = length of side of square

h_1 = heights used only once (h_a, h_b)
h_2 = heights used twice (h_e, h_f)
h_3 = heights used three times (h_d)
h_4 = heights used four times (h_g)

If the sides of the excavation or fill are not vertical as in Fig. 9.19 such that the area cannot be divided exactly into squares (or rectangles), the volumes outside the grid must be estimated individually and added. As in Fig. 9.22 this will normally involve the calculation of volume of many irregular trapezia and triangles.

Difficulties may be experienced if there is a change over from cut to fill, or vice versa, within a grid square. Blind application of the standard formula can well result in large errors; so it is best to treat these grid squares as special cases and calculate their volumes individually.

If the area within a grid square cannot be considered plane, whilst reducing the grid size would result in too much work, then the squares can be split into two triangles. This generally results in better surface fits but equation (9.25) must be amended.

Fig. 9.23

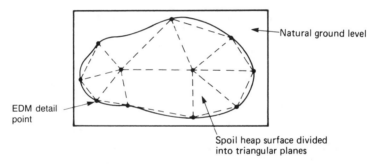

Natural ground level

EDM detail point

Spoil heap surface divided into triangular planes

It is possible to use the calculation of volumes within vertical triangular sections to estimate the volume of (1) irregular cuttings such as quarries or (2) irregular spoil or coal heaps etc. The level and position of the corners of triangles covering the area can be located by EDM (Fig. 9.23), the relevant detail pole points being selected at changes of slope, etc. such that the triangle planes closely match the surface. The volume of the cutting or fill can then be found as the sum of the product of the areas of the individual triangles times the mean difference in level of their corners from the natural ground level.

Fig. 9.24

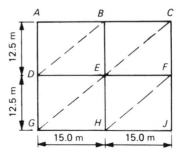

Example 9.15 Figure 9.24 shows a rectangular plot which is to be excavated to the given depths. Assuming the sides to be vertical, calculate the volume of earth to be excavated.

(*a*) Assume area is subdivided into four rectangles. Then, rather than evaluate each prism separately, since all the plan areas are the same, it is sufficient to sum the depths at the corners of all the rectangles and divide by four. Since some of the corners occur more than once, we may tabulate thus:

Station	Depth of exc. (h_n) (m)	Number of rectangles in which it occurs (n)	Product $h_n \times n$
A	3.15	1	3.15
B	3.70	2	7.40
C	4.33	1	4.33
D	3.94	2	7.88
E	4.80	4	19.20
F	4.97	2	9.94
G	5.17	1	5.17
H	6.10	2	12.20
J	4.67	1	4.67

$$\Sigma h_n \times n = 73.94$$

Hence

$$\text{Volume} = 15.0 \times 12.5 \times \frac{73.94}{4}$$

$$= \textbf{3466 m}^3$$

using equation (9.25) as a basis.

(*b*) Assume the area is divided into triangles as shown by the dotted lines in Fig. 9.24.

The tabulation is as before, but the number of times each depth of excavation occurs is different, while to obtain the mean depth we now divide by three.

Station	Depth of exc. (h_n)	Number of triangles which it occurs (n)	$h_n \times n$
A	3.15	1	3.15
B	3.70	3	11.10
C	4.33	2	8.66
D	3.94	3	11.82
E	4.80	6	28.80
F	4.97	3	14.91
G	5.17	2	10.34
H	6.10	3	18.30
J	4.67	1	4.67

$$\Sigma h_n \times n = 111.75$$

$$\text{Volume} = \frac{15.0 \times 12.5}{2} \times \frac{111.75}{3}$$

$$= \textbf{3492 m}^3$$

again using equation (9.25) as a basis.

It often happens that the topsoil and vegetable matter is removed from the surface before work begins, such soil stripping normally being paid for per square metre. If the levelling was done before the stripping and it is required to deduct this soil from the main excavation then the volume to deduct is given by the total area times the average depth of stripping.

Mass-haul diagrams

In works where large volumes of earthwork have to be handled — more especially long works such as railways and arterial roads — a mass-haul diagram is of great value both in planning and construction. The diagram is plotted after the earthwork quantities have been computed, the ordinates showing aggregate volumes in cubic metres while the horizontal base line, plotted to the same scale as the profile, gives the points at which these volumes obtain. Cuttings are taken as positive and fills as negative when evaluating the aggregate volumes, and in plotting the mass-haul curve total, positive volumes are plotted above the base line and total negatives below it.

Most materials are found to increase in volume after excavation ('bulking'), but after being re-compacted by roller or other means, soils in particular might be found to occupy less volume than originally, i.e. a 'shrinkage' has taken place when compacted in the *in situ* volume.

If the shrinkage factor of such soils is known it may be used in the mass diagram to amend the volumes required for filling. For example, if a certain material has undergone a volume shinkage of 15 per cent on final consolidation compared with its pre-excavation volume, then 100 m^3 of excavation produces 85 m^3 of fill (or 118 m^3 of excavation give 100 m^3 of bank). Volumes of cut and fill may thus be related by multiplying the excavated volumes by 0.85 to give the equivalent volumes of fill. Table 9.1 gives typical swell or shrinkage factors for

Table 9.1
Volume before Excavation = 1 m^3

Material	Volume immediately after excavation	Volume after compaction
Rock (large pieces)	1.5	1.4
Rock (small pieces)	1.7	1.35
Chalk	1.8	1.4
Clay	1.2	0.90
Light sandy soil	0.95	0.89
Gravel	1.0	0.92

certain materials. It will be seen that rock is the only material which has a larger volume after consolidation in embankment than it had before excavation.

Haul

The cost of hauling excavated material will obviously depend to some extent on the distance it must be carried. In the Bill of Quantities for a job, the unit price of excavation will include for transporting the material a certain specified distance (say 0.5 km); this distance is known as the *free haul*. If the material has to be moved a greater distance than the free haul, the extra distance is known as *overhaul*. *Haul* is the total of the product of the separate volumes of cut and the distance they are transported to the embankment. This must equal the total volume of cut multiplied by the distance between the centroids of cutting and the embankment it forms.

Consider the mass-haul diagram given in Fig. 9.25, which has been plotted from the detail contained in Table 9.2 of quantity distribution.

Table 9.2

Chainage	Centre height (m)	Volumes (m³) Cut	Fill	Shrinkage constant	Corrected volume	Accumulated volume
1000	$F1.22$					0
1040	0		230			−230
1100	$C1.52$	480		0.90	+430	+200
1200	$C3.96$	2560		0.90	+2300	+2500
1300	$C4.12$	4560		0.90	+4100	+6600
1400	$C2.74$	3940		0.90	+3550	+10150
1500	0	950		0.90	+850	+11000
1600	$F3.05$		1350		−1350	+9650
1700	$F4.27$		4010		−4010	+5640
1780	$F4.72$		4600		−4600	+1040
1820	$F4.72$		BRIDGE			+1040
1900	$F3.51$		4130		−4130	−3090
2000	$F1.22$		2370		−2370	−5460
2035	0		60		−60	−5520
2100	$C1.98$	510		0.90	+460	−5060
2200	$C3.96$	3180		0.90	+2860	−2200
2300	$C3.66$	4055		0.90	+3650	+1450
2400	$C2.44$	3860		0.90	+3470	+4920
2500	$C0.61$	1320		0.90	+1190	+6110
2530	0	100		0.90	+90	+6200
2600	$F1.07$		350		−350	+5850
2700	$F1.52$		1230		−1230	+4620
2800	0		420		−420	+4200
2900	$C1.68$	1080		0.89	+960	+5160
3000	$C3.66$	3730		0.89	+3320	+8480

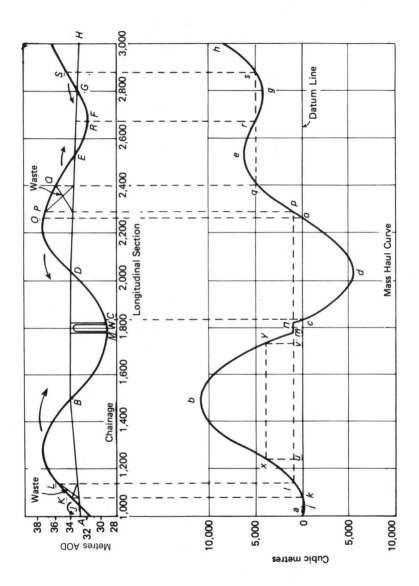

Fig. 9.25

This curve is illustrative only, and it is assumed that earthworks prior to chainage 1000 have balanced.

In plotting the curve the ordinate at j is plotted as the whole of the fill required between A and J, i.e. -230 m^3. The ordinate at chainage 1110 is the aggregate volume up to this point, i.e. $+200$ m^3, so that the mass-haul curve must cross the base line between chainage 1040 and chainage 1100. On plotting the remaining points it is seen that the curve crosses the datum line at four points, and the aggregate volume between pairs of these points is zero. The excavation and embankment between those points thus balance each other, i.e. between A and K, K and C, C and O.

The following points may also be noted in reading the mass-haul curve:

(a) The mass-haul curve rising indicates cut, since the aggregate volumes are increasing (i.e. J to B) and a maximum point occurs at the end of a cut (i.e. b, e).

(b) The mass-haul curve falling indicates fill (i.e. B to M, N to D), and a minimum point occurs at the end of a fill (i.e. d).

(c) The difference between the ordinates at two points represents the volume of cut or fill between the two points so long as there is no maximum or minimum point between the two.

(d) If any horizontal line, e.g. lm, be drawn on the curve, then the volumes of cut and fill balance because there is no difference in aggregate volume between l and m. Such a line is termed a balancing line and gives the distance of haul between the points. When the curve is above the balancing line, material must be moved to the right, i.e. LBM, and when it is below such a line the material must be moved to the left, i.e. RGS.

The base line on which the curve is drawn is a possible balancing line though not necessarily the most satisfactory one. In Fig. 9.25, using the base line as a balancing line, there is a surplus of 8480 m^3, though this may, of course, be used in the following section of the work. Any number of horizontal lines may be drawn on the curve to find balancing lines which enable the work to be done in the most economical manner, and they need not be continuous, e.g. lm and np are broken by the bridge, and np and qrs are not connected at all. When the balancing lines are not connected it means that the earthwork between those points on the profile not included by the lines will not be balanced, i.e. between K and L, P and Q, where the mass curve is rising. In these cases, material will be carried to tip as it is surplus; if the mass curve were falling it would be necessary to borrow material, since the mass curve then indicates a fill.

In considering the ways in which balancing lines may be drawn on the mass-haul curve it will be seen that to obtain the most economical scheme it is probable that the balancing lines will not be continuous. If the balancing lines were too long it would mean excessive and uneconomic haul distances. Thus a scheme involving balanced earthworks, over some lengths — having regard to the free haul — borrowing at some points and running to waste at others, is most likely to be used. It is better if material is hauled downhill, since this requires less power; thus in long hauls up steep gradients it may be worthwhile to waste material from the cutting and borrow for subsequent embankment. It is cheaper to borrow when it costs less to excavate in the cutting and run to waste, followed by excavation from a borrow pit to form the embankment, than it does to excavate in the cutting and transport this material to form the embankment.

As mentioned above, where the free haul is given, it may be plotted on the various parts of the mass-haul curve, and the extra distance for overhaul may be estimated. If the free haul distance be 500 m, then between *KBM* this would be denoted by *xy* on the mass-haul curve; elsewhere *np*, *qr* and *rs* satisfy this requirement. The area of the portions of the curve cut off by the balancing lines, i.e. *lbm*, is the haul in that section, since haul is the product of volume and distance.

One square on the diagram represents $5000 \times 200 = 10^6$ m^4. If the unit be taken as 1 m^3 hauled 100 m, then 1 square represents 10 000 units. The area giving free haul is *uxbyvu*, and the overhaul is the area *lbmn* minus area *uxbyvu*, and is given by multiplying this net area by 10 000. *xu* is the volume on which overhaul is paid.

It may be noted here that the prices in the Bill of Quantities are normally based on the unit volume of soil before excavation or the unit volume after compaction. The above analysis is mainly of use to the contractor when he is assessing plant requirements. It is also based on the assumption that all cut is directly usable as fill. In practice, the site investigation might reveal that some of the soils are unsuitable for embankment construction.

Exercises 9

1   State and prove Simpson's Rule for areas.
    Measurements made from a survey line to an irregular boundary were as follows:

Chainage (m)	0	10	20	30	40	50	60	70	80
Offset (m)		5.5	6.4	7.3	7.9	8.2	6.7	4.9	3.0	0

Calculate the area between the survey line and the boundary. *Answer*: 474 m^2.

2   Co-ordinates (E, N) of corners of a polygonal area of ground are, taken in order, as follows, in metres:

A (0, 0); B (-32, 40); C (-41, 126); D (14, 200); E (80, 144); F (108, 62); G (27, -19), returning to A:

Calculate the area enclosed, and the co-ordinates (to the nearest metre) of the far end of a straight fence from A which just cuts the area in half.

How far from A (to the nearest metre) would you position a fence running north–south so that the area to the west is 1 hectare? *Answer*: 2.033 hectares; 52, 168; 25 m east.

3   The whole circle bearings of the straight sides of a plot of land *ABCD* are:

$$AB\ 352°26' \qquad BC\ 111°04'$$
$$CD\ 195°56' \qquad DA\ 242°15'$$

The side CD is 175.82 m long and the area of the plot is 8.094 hectares. Calculate the length of side AB.   (*London*)

Answer: 418.23 m.

4   Levelling carried out at an open-cast coal site yielded the following results:

Grid co-ordinates (m)		Ground level
Easting	*Northing*	(m)
100	100	87.6
200	100	89.0
300	100	90.0
100	200	88.4
200	200	89.7
300	200	90.8
100	300	89.3
200	300	90.6
300	300	91.9

A borehole at co-ordinates (200, 200) has revealed that the top of a coal seam 1.68 m thick is located 8.4 m below ground level. The seam is known to dip towards the north at a gradient of 1 in 50.

(i) Calculate the volume of overburden contained in the gridded area.

(ii) If the ground level rises to the north at a mean gradient of 1 in 80 from 300 mN and the maximum ratio of overburden thickness to coal-seam thickness for economic working is 15 to 1, estimate the grid northing at easting 200 m to which an east–west working face may be advanced before the site becomes uneconomic.

Answer: (i) 336,000 m³; (ii) 727.7 m.

5   Tabulated below are the areas within the contour lines at the site of a reservoir, obtained by planimeter from a plan of scale 1:1000.

Contour level (m AOD)	Area (mm²)
60	85 160
58	79 355
56	70 970
54	61 835
52	56 775
50	40 290
48	16 900

If the lowest draw-off level is 48 m AOD and the maximum top-water level 60 m AOD, estimate (*a*) the full storage capacity,

(b) the top-water level for 50 per cent of full storage capacity.
Answer: Capacity 722 313 m³, say.

6 Figure 9.26 shows the straight centre line *AB* for a road of formation width 6 metres, set out on a hillside which may be considered as a plane surface of uniform slope 1 in 5.

Show that $\tan \omega = \tan \theta \sin \phi$
and $\tan \lambda = \tan \theta \cos \phi$

Hence determine the quantity of excavation obtained in a length of 100 m if side slopes of 1 vertical in $1\frac{1}{2}$ horizontal are provided and there is neither cut nor fill along *AB*, which has a uniform gradient of 1 in 10.
Answer: 105.3 m³, say.

Fig. 9.26

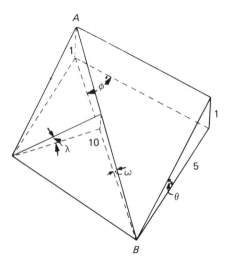

7 The centre line of a certain section of highway cutting lies on a circular curve in plan.

This cutting is to be widened by increasing the formation width of 20 m to 26 m, the excavation being on the inside of the curve and retaining the original side slopes of 2 horizontal to 1 vertical. The ground surface and the formation are each horizontal, and the depth to formation over a length of 400 m increases uniformly from 3 m to 5 m. Determine the radius of the centre line, if the volume of excavation is overestimated by 5 per cent when the influence of curvature is neglected.   (*Salford*)
Answer: 342 m.

8 The following earthwork quantities for a new road were based upon station intervals of 100 m, negative volumes denoting fill

Station	Cumulative vol. (m³)	Station	Cumulative vol. (m³)
0	0	11	14 000
1	− 3000	12	13 000
2	− 7000	13	9000
3	− 9000	14	1000
4	− 4000	15	− 2000
5	800	16	− 6200
6	5000	17	− 9000
7	10 000	18	− 10 000
8	17 000	19	− 10 000
9	19 200	20	− 8000
10	15 000	21	− 4000
		22	1000

(a) Draw a mass-haul curve and deduce the longitudinal profile assuming that the road formation is horizontal.

(b) The following excavation rates apply:

Excavation, cart and deposit, not exceeding. 3 stations
$£1.50/m^3$

Extra over for overhaul $60p/m^3$ station

Borrow and place in embankment $£2.40/m^3$

Considering the loop of the mass-haul curve contained by stations 5 and 14 approximately, estimate the difference in costs of the earthwork when (i) no borrowing is allowed and (ii) when 6500 m³ of borrow material is available. In each case, running to waste is impossible.

Answer: (i) £59 880; (ii) £56 280.

9 The following are notes from a closed traverse survey of an abandoned quarry. Calculate the area of the land bounded by the survey lines from co-ordinates.

Line	Horizontal length (m)	Bearing
AB	120	9°12′
BC	240	127°30′
CD	360	185°40′
DE	420	258°20′

Check your answer by means of an accurately scaled plan.

(Eng. Council)

Answer: 137 466 m²

10
(a) Show that the area, Δ, of a triangle, *ABC*, which lies in a horizontal plane (*E*, *N*), may be determined from the formula:

$$\Delta = \tfrac{1}{2}(N_aE_b+N_bE_c+N_cE_a) - \tfrac{1}{2}(N_aE_c+N_bE_a+N_cE_b)$$

Where E_a = Eastings of point A, N_a = Northings of point A, etc.

(b) If the vertices of a triangle, $A'B'C'$, are at elevations, h_a, h_b,

Fig. 9.27

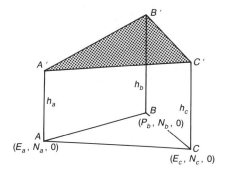

h_c, above a horizontal datum plane show that the volume of the figure $ABCA'B'C'$ (Fig. 9.27) is given by the expression:

$$V = \frac{\Delta}{3} (h_a + h_b + h_c)$$

(c) Digital Terrain Models (DTM) derived from large scale detail surveys often take the form of Triangulated Irregular Networks (TINs). Describe how such networks may be derived from detail observations indicating any additional information which may be required in order to form a sensible elevation model.

(d) Explain how a surveying company might utilize such models in the production of conventional large scale mapping products and how clients of that company might benefit from the supply of such data in a digital format.

(London)

10 Setting Out

Setting out is the process of using the surveying instruments and techniques described in preceding chapters to transfer information from a plan to the ground. As such it is the opposite of surveying and is the task most commonly performed by site engineers supervising the construction of new works.

There are three distinct elements to the task: horizontal control to ensure that the new works are in the correct place, vertical control to ensure that they are at the correct level, and vertical alignment to ensure that multi-storey or underground construction is plumb. Each stage in the process will be discussed in detail.

Accuracy

It is important that the engineer is aware of the accuracy achievable with various items of equipment at his disposal and that techniques appropriate to the job in hand are utilized. Table 10.1 shows the expected accuracy for equipment in good working condition and adjustment as recommended in BS 5606: 1990 *Code of Practice for Accuracy in Building*.

Many of the values quoted in Table 10.1 are substantiated in the relevant preceding chapters. However, some general observations can be made. In respect of linear measurements, it is worth noting that EDM is not recommended for distances less than 30 m where a steel tape used with precision is more accurate. For precise work the steel tape must be used under an applied tension and the Building Research Station have developed a constant tension handle to make this task simpler. In this device a slide, carrying a grip which can be attached to the tape, projects from a handle containing coiled spring steel bands to which the slide is connected. When the handle is pulled the slide extends from it and a pre-determined tension is applied to the tape. From the equations in Chapter 2 it will be evident that a steel tape of 3.24 mm^2 cross-sectional area, which has been standardized on the flat at 20°C under a tension of 89 N, should be used on the flat at a tension between 38.8 N and 139.2 N to maintain an accuracy of length to within 1/10 000. It will be appreciated that compensation for tape temperature still remains when we are considering overall accuracy.

In so far as angular measurement is concerned, Table 10.1 shows that procedures such as the 3:4:5 rule can be used to set out right angles by tape but that this is inferior by far to the accuracy given by a 20-second theodolite. Such instruments are commonly used on site but

Table 10.1 (*Courtesy:* BSI)

Measurement	Instrument	Deviation
Linear	30 m steel tape, general use	± 5 mm up to 5 m ±10 mm, 5 to 25 m ±15 mm, above 25 m
	30 m steel tape, precise use	± 3 mm up to 10 m ± 6 mm, 10 to 30 m
	EDM (standard infra-red type)	±10 mm, 30 to 50 m† ±10 mm + 10 ppm, above 50 m
Angular	30 m steel tape on uneven ground	± 5′ (±25 mm in 15 m)
	30 m steel tape on flat ground	± 2′ (±10 mm in 15 m)
	20″ Glass-arc theodolite	±20″ (±5 mm in 50 m)
	1″ Glass-arc theodolite	±5″ (±2 mm in 80 m)
Verticality	Spirit level	±10 mm in 3 m
	Plumb bob, freely suspended	± 5 mm in 5 m
	Plumb bob, damped in oil	± 5 mm in 10 m
	Theodolite with optical plummet and diagonal eye-piece	± 5 mm in 30 m
	Laser (visible)	± 7 mm up to 100 m
	Optical plumbing device	± 5 mm in 100 m
Levels	Spirit level	± 5 mm in 5 m distance
	Water level	± 5 mm in 15 m distance
	Laser (visible)	± 7 mm up to 100 m
	Laser (invisible)	± 5 mm up to 100 m
	Optical level (builders)	± 5 mm per sight *
	Optical level (engineers)	± 2 mm per sight * ±10 mm per km
	Optical level (precise)	± 2 mm per sight * ± 8 mm per km

† not recommended for distances less than 30 m
* up to 60 m

it is recommended that the 1-second instrument be adopted for precise work.

Basic principles and point marking

Information on the drawing concerning the proposed works must be transferred to the ground in a format that can be understood by all levels of the workforce on site. With this in mind it is normal to provide a recognized system of pegs and markers that are similar from one site to another. Effectively three orders of points can be defined, primary stations, secondary stations and detail points.

(a) Primary setting out points are stations on the control traverse or control triangulation system (see Fig. 6.18) and they can be referenced to the National Grid for orientation and co-ordination if so required by the client. The station should be permanent for the life of the works and typical construction could range from a brass plate or stud set in concrete to a road nail driven into a carriageway, it is also possible to purchase proprietary forms of plastic and metallic marking system for the purposes of monumentation. The stations should be clearly marked and protected so that they are not disturbed by construction traffic and should have sufficient space for the instrument to be set up and freely accessed. A typical construction is shown in Fig. 10.1.

Fig. 10.1 Setting out station

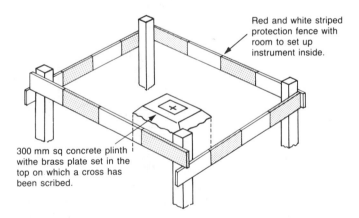

Red and white striped protection fence with room to set up instrument inside.

300 mm sq concrete plinth withe brass plate set in the top on which a cross has been scribed.

(b) Secondary setting out points are established closer to points of detail on the proposed works and are referenced by measurements from the primary points. They could, for example, be points on a grid surrounding a building as discussed below. They must be robust and rigid, designed to survive the construction period of the element of the works that they provide control to. They are thus constructed in a similar manner to primary control points.

(c) Detail points mark the location of features on the works such as the centre of a pile, corner of a building, etc. Any marker will be lost in the excavation as soon as construction starts and a temporary format such as a 50 mm square timber peg with a nail in the top, or a road pin, will be quite adequate.

(d) Temporary bench marks (TBM) are used for vertical control as mentioned in Chapter 3, their purpose being to provide a bench mark adjacent to the works and thus avoid substantial runs of flying levels to an OBM each day. It will be realized that they are analogous, in heighting terms, to the secondary control points defined above. In an ideal situation daily levelling should simply involve a backsight onto a TBM and foresight to the works.

The bench mark should be a rigid 'permanent' construction like the primary and secondary points for horizontal control. Typical construction ranges from a 600 mm long steel pin driven into the ground and set in a 300 mm cube of concrete, Fig. 10.2, to a road nail or mark on a completed part of the works. In all cases the TBM must be accurately levelled to an Ordnance Bench Mark, or Primary Site Bench Mark if a site datum is being used, to a closing error not exceeding 5 mm. Like the primary and secondary points for horizontal control their integrity needs to be checked frequently and they need to be protected from site traffic, etc.

Fig. 10.2 Typical TBM construction

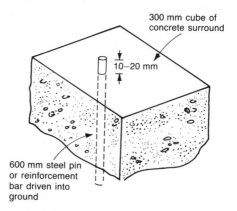

300 mm cube of concrete surround

10–20 mm

600 mm steel pin or reinforcement bar driven into ground

The reader will find further detail on the location and construction of markers in the Construction Industry Research and Information Association guide *Setting out Procedures*, B.M. Sadgrove (Butterworth) and in BS 5964 (1989) *Measurement Methods for Building — Setting out and Measurement*.

Example 10.1 Two primary control points, *A* and *B*, have co-ordinates 1000.00 mE, 2000.00 mN and 986.72 mE, 1897.46 mN respectively. Calculate data to set out point *S*, 1025.00 mE, 1950.00 mN by two measurements.

$$\text{Angle } BAS = \tan^{-1} \frac{1025.00 - 1000.00}{2000.00 - 1950.00}$$

$$+ \tan^{-1} \frac{1000.00 - 986.72}{2000.00 - 1897.46}$$

$$= 33°56'40''$$

i.e. $B\hat{A}S$ is $326°03'20''$ clockwise from direction AB.

$$\text{Angle } ABS = \tan^{-1} \frac{1025.00 - 986.72}{1950.00 - 1897.46}$$

$$- \tan^{-1} \frac{1000.00 - 986.72}{2000.00 - 1897.46}$$

$$= 28°41'50''$$

i.e. $A\hat{B}S$ is $28°41'50''$ clockwise from direction BA.

$$AS = \sqrt{25.00^2 + 50.00^2} = \textbf{55.90 m}$$
$$SB = \sqrt{38.28^2 + 52.54^2} = \textbf{65.01 m}$$

Note also that the standard expressions for intersection, Chapter 6, can be used to solve for $B\hat{A}S$ and $A\hat{B}S$, the co-ordinates of the three points being known.

Horizontal control

Having established the primary and secondary points by traverse or triangulation, as described in Chapter 6, the detail has to be set out from them. Various approaches can be adopted and in Chapter 1 (Fig. 1.1) four methods were described for the location of a point; these are equally applicable in the setting out of a point from control stations and each will be illustrated in this section.

Site grid

One of the most common ways of controlling works on a compact site is to establish a grid of secondary points in the form of a rectangle around the perimeter as shown on Fig. 10.3. The spacing of the grid

Fig. 10.3 Site grid

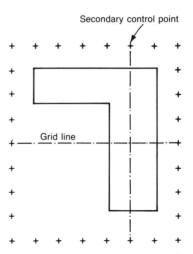

Secondary control point

Grid line

can be varied to suit the site and typically it will be arranged to coincide with the spacing of piles or columns in the permanent construction. However, the grid will normally be offset from the centre line of such columns to prevent the construction work from obscuring sight lines. Detail setting out can be carried out by offsets perpendicular to the grid lines.

On very large industrial sites where structures, roads, sewers, etc. have to be located the survey data can be used to establish two conveniently situated lines, mutually at right angles, and to fix thereon secondary points at suitable intervals. In effect the basis of a site grid has now been established and by setting out perpendiculars by theodolite at relevant secondary points a 'local' grid can be constituted as required for individual buildings.

Co-ordinates from control points

On linear sites, such as motorways, it is normal to have a system of control points remote from the line of the works, probably on a traverse that was established at the survey stage. The works will have been designed by computer and hence co-ordinates are available for all detail points. These can thus be set out by bearing and distance methods using EDM equipment from the control point. Examples of the calculations needed for this method of setting out are given in Chapter 11 in respect to curves.

Fig. 10.4

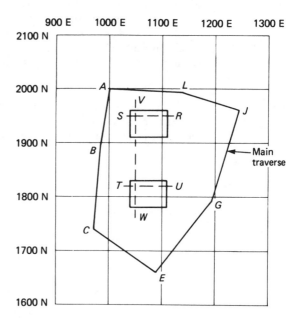

In the context of industrial sites or shopping centres, etc., secondary setting out stations i.e. S and R, V and W (Fig. 10.4) can be located from the primary control traverse by intersection methods, each pair being related to one or more structures. The secondary stations can of course be placed in a more random or scattered interrelationship to cater for a number of construction features. Swedish Standard 602.5 suggests that check lengths, such as *SR*, should not differ from the theoretical length by more than $2\sqrt{L}$ mm, where L is the length in metres. This value also applies to the relationship between the secondary setting out station and the main traverse.

In Chapter 6 we discussed resection and free station methods of fixing the arbitrary location of an instrument from primary or secondary setting out points. This allows the instrument to be set up at a convenient point close to the construction work and by readings onto two, or three, stations the co-ordinates of the instrument can be obtained. Setting out data can then be produced for the detail points and since the distances are reduced the accuracy is enhanced.

Local control

The most common local control activity is to ensure that the corners of a building are square, this involves setting out 90° angles and a variety of methods are discussed in Chapter 2. The importance of checking setting-out detail cannot be over-emphasized and all obvious check measurements should be carried out by the engineer as a matter of course. For example, for a rectangular structure the opposite sides should be of equal length, but this does not necessarily mean that it is correct. In addition the diagonals should be checked to ensure that they are equal to the theoretical value given by a simple Pythagoras calculation; if this is not true then the shape has been set out as a trapezium or parallelogram and should be corrected.

In many cases it will be necessary to locate detail points more than once, for example if the detail point is the corner of a brick building it will need to be set out for the centre line of excavation, foundation and brickwork. At each stage the marker will be destroyed by the construction process. Offset pegs can be located close to the works but outside the immediate working area such that the detail peg can be replaced by the minimum of effort. Figure 10.5 shows typical offset pegs for setting out a pile cap. Offset markers may be constructed from 50 mm square timber pegs with a nail in the top, or road nails driven into a carriageway. They should be brightly painted both to facilitate their relocation and to limit the risk of accidental damage by site traffic.

Fig. 10.5 Offset pegs for a pile cap

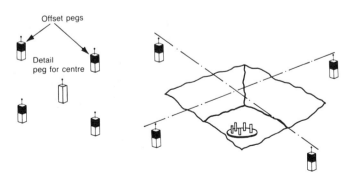

Ideally, safe locations against boundary fences, etc. should be chosen but where this is not possible the peg should be adequately protected.

Since it is important that the overall structure is correct it is good practice to establish setting out markers that are useful to tradesmen. This can be achieved by using profile boards instead of offset pegs, Fig. 10.6(*a*) and (*b*). These are horizontal timber rails, long enough to cover the respective excavation width, in which saw cuts or nails can be placed to represent lines on the construction. In Fig. 10.4, let us say that ten columns are to be placed in a building near *SR* and that for the sake of simplicity one of the lines is collinear with *SR*, Fig. 10.6(*c*). A theodolite established at *S* and sighting along *SR* allows the location of A_1, B_1, etc. by direct measurement with checks from *R*. A_2, B_2, etc. can then be established by setting off perpendiculars to *SR* by theodolite, at the same time profile boards can be established on the lines, such as A_1A_3, extended clear of the excavation. A study

Fig. 10.6 Profile boards

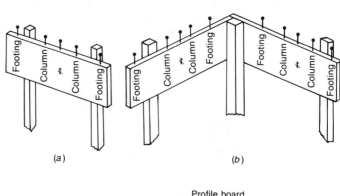

(*a*)　　　　　　　　　(*b*)

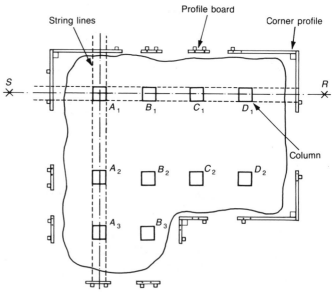

(*c*)

of the diagram will show that a number of check linear measurements can also be made.

The markers for A_1, B_1, etc. will be disturbed as construction proceeds but the positioning of footings and reinforcement cages can be facilitated by suspending string lines from appropriate pairs of nails on the top of the profile board and then plumbing down from the string. Since string lines do not normally exceed 30 m in length, intermediate profile boards are required if the building dimension exceeds this. In some cases it is also possible to set the top of the profile board at a known level so that the tradesman can transfer levels onto the proceeding construction using a long spirit level. At corners such as A_1, D_1, etc., profile boards can be brought together as in Fig. 10.6(b).

Setting out a rectangular structure

When setting out a small building by theodolite and tape, such as a rectangular structure, the following stages would occur (Fig. 10.7):

(a) Establish the line of the building from control information;
(b) Set up the theodolite on this line and locate two corners of the building and two offset pegs or profile boards (Fig. 10.7(a));

Fig. 10.7

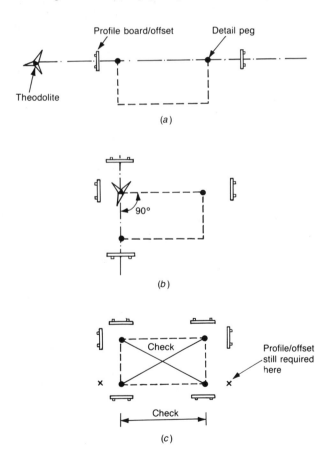

(a)

(b)

(c)

(c) Set up the theodolite at one corner then sight onto the other corner, or the original instrument location, whichever is furthest away. Turn the theodolite through 90° and establish a third corner of the building the correct distance away along the line of sight, plus further offset pegs or profile boards (Fig. 10.7(b));

(d) Repeat (c) at the second corner;

(e) Check the length of the furthest side and the diagonals (Fig. 10.7(c));

(f) Set up the theodolite on the third corner, sight the fourth corner and establish the remaining offset pegs or profile boards.

Setting out bridges

The setting out and construction of bridges on major road schemes often takes place in advance of the linear construction of the road and it is important that secondary control stations are placed so that they will not be disturbed by the latter operation. Figure 10.8(a) shows

Fig. 10.8

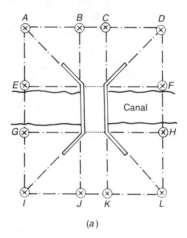

(a)

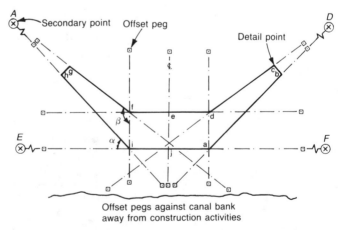

Offset pegs against canal bank
away from construction activities

(b)

one possible method of establishing a grid of secondary setting out points around a road bridge crossing a canal, only the four stations *B, C, J* and *K*, defining the parapet, are in the way of other construction operations and these could be replaced if needed for later phases of the construction.

Figure 10.8(*b*) illustrates the use of the secondary control points in establishing the footings for an abutment. Pegs have been placed to define the road centreline and eight detail pegs have been located at the corners of the foundation. In addition, offset pegs have been erected back from the working space to allow these corners to be quickly relocated once excavation has been completed. The task could be carried out by theodolite and tape or with a total station instrument.

A typical setting out sequence would be:

(*a*) Set up the instrument at secondary control point *E* and sight onto *F*. Hence establish detail points *a* and *i* and centre line peg *j* at a measured distance from *E*; check the measurements from *F*.

(*b*) Move the instrument to *i* and set the horizontal circle to zero sighting onto *F*. Turn the telescope through 90° and set out detail point *f* at a measured distance from *i*, plus two offset pegs.

(*c*) Set the horizontal circle to angle α and check that the telescope is pointing at secondary control point *A*. Set out detail point *h* at a measured distance from *i*, plus two offset pegs.

(*d*) Repeat at detail point *a*, with an initial sighting onto *E*. Check pointing at secondary control point *D* and that dimension *fd* is equal to *ai*.

(*e*) Set up the instrument at centre line point *j* and set the horizontal circle to zero pointing onto *F*, turn the telescope through 90° and establish point *e* at a measured distance from *j*, plus other points on the centre line as required. Check that *ed* is equal to *ef*.

(*f*) Move the instrument to point *f* and set the horizontal circle to zero pointing on *i*. Turn the telescope through 90° and check that it is pointing at *d*. Establish two offset pegs.

(*g*) Set the instrument to angle β and establish detail point *g* and two further offset pegs. Check dimension *gh*.

(*h*) Repeat at detail point *d*.

(*i*) If possible check dimensions *gc* and *bh*.

(*j*) Offset pegs may be established for lines *cb* and *gh*, if desired, by extending the lines either by tape or with the instrument.

Vertical control   The first stage in vertical control is to establish a series of TBM as discussed above, the next stage being to transfer the information to the workforce. The traditional method of doing this was to use sight rails (often called 'profiles' but different to profile boards) and travellers. The sight rails are horizontal timbers established by the engineer so that they define the plane of the finished works at a level above the ground. The traveller is a portable sight rail with a vertical support whose length defines the level of the finished works below the sight rail plane.

As an example, consider a length of sewer being laid from manhole
A, with an invert level of 30.02 m, to manhole B, 60 m away, the
gradient from A to B being 1 in 100 and falling from A to B. Figure
10.9 shows the proposed sewer, and it will be seen that the general
depth of the sewer is below 3 m. Thus if two rails are fixed over stations
A and B about 1 m above ground level, and each a fixed height above
invert level, then an eye sighting from rail A to rail B will be sighting
down a gradient equal to that of the proposed sewer. In the example,
a convenient height above invert would be 3.75 m, so that a traveller
(a timber with a sight bar across one end, and looking like a T-square)
of this length, held vertically so that its sight bar just touched the line
of sight between sight rails A and B, would give at its lower end a
point on the sewer invert line. (*Note*: The invert of a sewer is the lowest
point on the inside surface.) Staff readings might well be required to
0.001 m when establishing sight rails even though the nearby bench-
mark heights are known only to 0.01 m.

Fig. 10.9

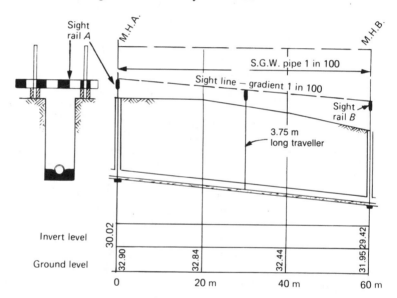

To fix these sight rails for use with a 3.75 m-long traveller, therefore,
we drive two posts on either side of the manholes and nail the rails
between these at the following levels:

$$\text{Sight rail } A, \text{ RL } = 30.02 + 3.75$$
$$= 33.77 \text{ m AOD.}$$
$$\text{Distance } AB = 60 \text{ m}$$

Therefore
$$\text{Fall} = \frac{60}{100} = 0.60 \text{ m}$$

Hence
$$\text{Invert level } B = 30.02 - 0.60$$
$$= 29.42 \text{ m AOD}$$

and
$$\text{Sight rail } B, \text{ RL } = 29.42 + 3.75$$
$$= 33.17 \text{ m AOD};$$

If a level set up nearby has a height of collimation of, say, 34.845 m AOD, then the staff is moved up and down the posts at MHA until a reading of $34.845 - 33.770 = 1.075$ m is obtained. Pencil marks are made on each post and the black and white sight rail is nailed in position as shown in Fig. 10.9. For rail B, the staff reading would, of course, be

$$34.845 - 33.170 = 1.675 \text{ m}$$

Frequent checking of sight rails is required, as they are liable to be disturbed by excavators, dumpers, lorries, etc.

In the above example the sight rail was attached to two posts straddling the trench. For deep sewers constructed by mechanical excavator this may be a hindrance to construction, in which case the sight rails may be placed upon one side of the trench, or duplicated on either side of it. Whilst it is common practice to construct sight rails in a cross shape with only one upright this is not good practice since the method relies upon the cross rail being horizontal and a single post can not guarantee this.

The task of viewing across the sight rails to estimate the degree of additional cut or fill needed at the traveller is called 'boning' and experienced operators claim to be able to carry out the process to within a few millimetres, although a tolerance of ± 15 mm is all that could reasonably be expected. The engineer should note that he must erect at least three sight rails to define any particular plane, since it is possible to bone any two sight rails in with the traveller. A mistake in the level of a sight rail, either due to a setting out error or because it has been disturbed, would not be detected without a third and subsequent sight rails.

For road works, sight rails are normally provided at round chainage intervals in pairs on either side of the working area. On even gradients the pair of sight rails may be at the same level but often they will reflect a degree of superelevation. Levels on the drawings will be given at the kerb line and if the sight rails have been set up some distance behind the kerb, clear of construction traffic, the level of the horizontal rail should represent the correct degree of cross fall across the road width, i.e. the low rail will be lower than the kerb level by y/x where y is the dimension back from the kerb and 1 in x the crossfall. The high rail will be raised by a similar amount.

Batter rails

The slope of cuttings and embankments could be defined by sight rails up the slope, but it is normal practice to erect batter rails (Fig. 10.10) which define the excavation surface with a sloping rail. These are set up as follows:

(a) Establish the plan location of the top of the cutting or bottom of the embankment.
(b) Measure 1 m away from the excavation and drive in a stake.
(c) Measure a further 1 m and put in a second stake.
(d) Decide upon the traveller length such that the top edge of the batter

rail will not be too high for cuttings (max 1.5 m) or too low for embankments (min 0.3 m) to be sighted along.

(e) If the excavation slopes at 1 in x,

for a cutting:

On the first stake mark the level corresponding to RL of cutting top + traveller height + $1/x$. On the second stake mark the level corresponding to RL of cutting top + traveller height + $2/x$.

for an embankment:

On the first stake mark the level corresponding to RL of base + traveller height $-1/x$. On the second stake mark the level corresponding to RL of base + traveller height $-2/x$.

(f) Nail on the sloping rail so that its top edge is in contact with the two marks and this sloping timber will define a plane at the traveller height above the finished formation.

Fig. 10.10

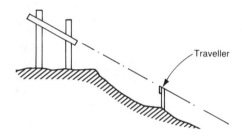

Traveller

Example 10.2 Calculate data to establish batter rails that define a cutting that is to slope downwards at 1 in 2 from a point of reduced level 50.457 m.

If we select an 0.4 m long traveller, and drive in two stakes at 1 m and 2 m from the cutting top then the level mark on:

Stake 1 = 50.457 + 0.4 + 1/2 = **51.357 m**
Stake 2 = 50.457 + 0.4 + 2/2 = **51.857 m**

Now check the height of the upper edge of the batter rail, assuming that the ground is level at 50.457 m. Height of upper edge = 51.857 − 50.457 = 1.4 m, which is satisfactory.

Construction lasers

Whilst sight rails are still used for road works their use has diminished for drainage work and work on compact, level surfaces like car parks and floor slab construction, due to the advent of construction lasers. There are two types in common usage, the rotating laser where the laser beam spins out a horizontal, or on some units a sloping, plane and the pipe laser where a beam of laser light defines the line and gradient of the pipeline or tunnel. Nowadays some laser units can be used for either function.

Some devices, such as the AMA SL-80 COMBI, use a 1.4 mW Helium Neon laser tube to produce a visible beam of red laser light with a range of 300 m, the laser having a Class 3A safety rating as discussed below. The laser tube is mounted vertically in the unit and shines at a prism head that can be rotated at controlled speed from 0 to 600 rpm. This produces a plane of laser light in a 300 m diameter circle. The unit is self levelling provided that the casing is standing within 15° of horizontal: the rotating head stops and the beam flashes if this range is exceeded. The instrument is operated from a 12 volt rechargeable battery.

More recent, lower powered 6 volt models, such as the Sokkisha LP3A (Fig. 10.11), use an infra-red laser diode, which produces an invisible beam of laser light with a Class 1 safety rating as discussed below. The prism head can rotate at up to 300 rpm and the beam has a range of 100 m with a quoted level accuracy of 10″, i.e. 2.4 mm at 50 m. The instrument is also self-levelling and flashes if the instrument is knocked. Since the beam is invisible the instrument has to be used with a photoelectric sensor which is also shown in Fig. 10.11. This particular type is attached to a graduated staff and manually slid up and down until the sensor falls within the path of the beam, at which point visual liquid crystal display arrows and, if selected, audio indicators are activated that facilitate accurate alignment upon the centre of the beam. The staff is then read against a pointer.

Photoelectric sensors may also be used with visible-beam lasers and some manufacturers produce 'laser wands' that are special staffs with driven sensors that automatically seek and find the level of the laser plane. It is further possible to attach sensors to construction plant, such as the blade of a motor grader, to allow the driver to read the sensor

Fig. 10.11 Sokkisha LP3A infra-red rotating laser (*Courtesy*: Sokkisha (UK) Ltd)

from his cab and accurately control the finished formation without the need for a banksman boning across sight rails and giving verbal instructions.

As well as being a quick one-man method of obtaining or setting out levels across open sites the rotating beam can be of benefit to tradesmen, particularly those responsible for positioning suspended floors or ceilings which have to be horizontal. In addition some rotating lasers can be used vertically so that when correctly level they define a vertical plane. This can be used, for example, to set out partition walls.

Pipe lasers

The AMA SL-80 COMBI was quoted above as an example of a visible light rotating laser, but by removing the rotating prism it can be used as a pipe laser as well. This type of laser is always visible light and the unit is set up in a manhole or the end of the pipe so that the beam is horizontal. The above unit is self levelling within a range of 22° from the horizontal, although some older devices have to be manually levelled with a traditional three-screw tribrach. The required gradient for the pipeline, in the range +20 per cent to −10 per cent, is then entered digitally into the instrument and the beam automatically corrects to this gradient. The unit can be aligned for direction using push button controls within a range of 26 m/100 m (i.e. 15°) and the latest instruments such as the Spectra-Physics Dialgrade 1165 (Fig. 10.12) can be aligned by remote infra-red hand controllers at distances up to 150 m from the front and 30 m from the back of the laser unit. The 12 mm diameter visible beam of light shines up the pipeline with a quoted accuracy of gradient of ±3 mm per 30 m, although users should be aware that the beam can be refracted by non-uniform temperature gradients or if it passes too close to a solid object, for example a pipe or tunnel wall. Temperature gradients can produce serious errors if the laser is used inside a run of pipes that are exposed to the sun on

Fig. 10.12 Spectra-Physics Dialgrade 1165 pipe laser, target and remote control unit (*Courtesy*: Spectra-Physics)

a hot day. The red beam is displayed on the end wall of the trench for initial excavation and purpose designed targets are used for the levelling of bedding and the invert of the pipe. Similar devices are used for alignment of tunnels where the laser is normally fixed to a bracket on the tunnel wall; this is discussed in more detail in the topic of underground control.

Laser safety

Laser light can be either visible or invisible, and can enter the eye in very narrow and virtually parallel beams. Hence even laser light of low power can be focused by the eye to cause burning of the retina.

The eye compensates to some degree against the visible radiation of the sun by the reflex action of blinking which causes the iris diameter to contract and thus reduce the power entering the eye. Invisible laser light does not induce any such compensating reflexes and accordingly its intensity must be kept low if damage is to be avoided.

Laser safety is covered by BS 4803: 1983, *Radiation Safety of Laser Products and Equipment* in which five categories are specified, Class 1, 2, 3A, 3B and 4. The higher the classification the more powerful the unit and hence the greater degree of safety required. The classification is based upon the unit in its most dangerous operating condition, i.e. for a rotating laser when it is stationary. For Class 1 and 2 lasers, which are typically the invisible infra-red devices, the safety requirements are not onerous, mainly common sense things like not deliberately staring into the beam, provide safety warning signs on the unit and not terminating the beam at a reflective surface.

For Class 3A lasers trained operators are required and the instrument must have a key control to prevent unauthorized use. There must be an emission indicator when the beam is energized and this is normally achieved by making the units visible light. In addition, sensible safety precautions must be observed in operation like not setting the beam up at eye level or allowing it to unintentionally reflect off surfaces. Care also has to be taken when working with surveying instruments in the vicinity of the beam since the magnification of the telescope increases the safety risk. Class 3B and 4 lasers are not normally found on construction sites because the safety requirements become onerous including the use of eye protection and protective clothing.

Vertical alignment

In addition to constructing works to the correct line and level it is necessary to ensure that there is accurate control of vertical alignment during the construction of multi-storey structures. For buildings this applies not only to the framework but also to the transfer to higher floors of the horizontal control established at ground level. A number of approaches are listed in Table 10.1, namely:

(*a*) plumb bob or spirit level
(*b*) theodolite
(*c*) optical plumbing device
(*d*) laser plumbing device

Vertical alignment by spirit level and plumb bob

In Table 10.1 it is suggested that a long builder's spirit level could be used to achieve vertical alignment of ± 10 mm in 3 m. This degree of accuracy is adequate for domestic house building and other low rise dwellings. An improved accuracy can be achieved with a plumb bob, but as the length of the suspension wire increases so the bob tends to swing as a result of the Earth's rotation. Some of this effect can be reduced by immersing the bob in a bucket of oil or other viscous fluid and then an accuracy of ± 5 mm in 10 m is attainable. This technique could be used to align concrete formwork for retaining walls and it has uses in tunnelling as mentioned later.

Vertical alignment by theodolite

For more accurate vertical alignment a theodolite should be used. For one or two floors the theodolite could be set up over a hole in the floor and its optical plummet used to set up over a station on the ground floor. However, the quality of the telescope on the optical plummet is inadequate to perform this operation accurately from any great height.

A typical working method is to set up the theodolite on the ground floor over the control station and then sight the telescope vertically upwards. A diagonal eye-piece must be fitted to the telescope because the normal eyepiece is obscured by the body of the horizontal circle and such an attachment is available for most makes of theodolite. For accuracy the instrument must be used in both the face left and face right positions. The vertical sighting should be through a space in the building, for example a lift or service shaft, sighting onto a temporary target at the appropriate level (Fig. 10.16). Alternatively the sight can be up the outside of the building onto a target suspended over the side at the appropriate floor.

An alternative approach is to use a single special prism, a pentagonal prism (Fig. 10.13) attached at the objective end of the telescope of a theodolite. These are produced by most manufacturers and facilitate the transfer of a given bearing to different levels as in the connection of surface and underground lines as well as upward and downward plumbing. Providing the telescope's line of sight is truly horizontal,

Fig. 10.13

The pentaprism housing can be rotated after its circular flange has been clamped to the telescope, the line of sight thereby generating a plane perpendicular to the telescope axis

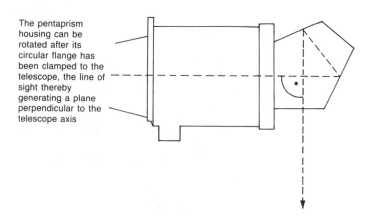

then a vertical sighting line is obtained so that a point can be suitably located at a required level on that line. Turning the instrument through 180° in azimuth gives a second position and the mean of the two positions gives the plumb position with respect to the theodolite axis. This set of observations can be repeated for rotations of 90° and 270° in azimuth to the original bearing, and also after changing face.

The vertical circle on a theodolite can be used to plumb columns as illustrated in Fig. 10.14. Two theodolite stations are established at 90° to each other, often utilizing points on a control grid. The theodolite is sighted along the grid line and the offset to the base of the column measured. If the column is vertical in the plane at 90° to the theodolite line of sight, then the same offset will be observed in the theodolite telescope at the top of the column. The instrument is then moved to, or a second instrument set up on, the grid line at 90° to the previous one and the procedure repeated. If an adjustment needs to be made to the column position then it may be necessary to repeat the process until both directions are correct.

Fig. 10.14

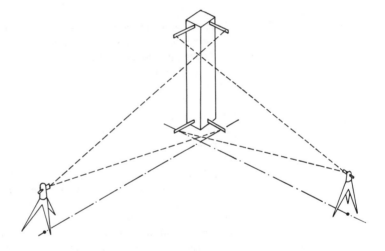

Vertical alignment by optical plumbing

Various types of plummet are available for upwards and downwards sighting to allow the establishment of a vertical line, and these are normally manufactured so as to be interchangeable with theodolites on their tripods. Figure 10.15 shows one such instrument, the Watts Autoplumb, which has two telescopes, one low power for sighting downwards and locating the instrument over the ground mark, and the other of high power for sighting upwards onto a target. In this instrument the upwards-sighting telescope is fitted with the same type of compensator which was formerly used in the Watts Autoset levels and this automatically ensures a vertical line of sight, even if the instrument is tilted by several minutes of arc.

The instrument is attached to a three-screw levelling base which can be mounted on a tripod, and a centring motion is available for

Fig. 10.15 Optical system of the Watts Autoplumb

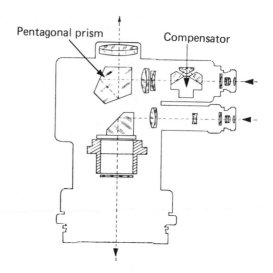

positioning over ground points. Since the downward-sighting telescope is not compensated, it is essential that the instrument be levelled using a plate bubble, positioned parallel to the telescope axes, and a circular bubble, found at the top of the case.

A pawl engaging with vee-slot locators, positioned at 90° to each other on the bevelled fitting just above the instrument base, allows the instrument to be readily placed in two positions mutually at right angles.

The upward-sighting telescope is, as mentioned earlier, provided with a compensator and the sighting line is deflected upwards by means of a pentagonal prism which can be tilted by a micrometer movement. In the field of view, a horizontal line defining a vertical plane, together with three concentric circles, will be observed, and the line of sight is vertical when the micrometer reading is 10.00. Thus if the Autoplumb be now rotated through 90° in azimuth the intersection of the second vertical plane with the first plane gives a vertical line.

Figure 10.16 shows the plan of four ground stations in a building with holes left for the vertical sighting line in the falsework etc. of the floors, together with a target, 250 mm side say, on which line AB has been scribed. An assistant arranges that line AB is brought into coincidence with the horizontal line in the field of view on instructions from the surveyor at the instrument. Extension lines AA_1 and BB_1 are then scribed on to the falsework or on to a suitable plate. The Autoplumb is swung through 90° and line AB is now positioned at $A'B'$ to lie on the horizontal line in the field of view. Extension lines $A'A_1'$ and $B'B_1'$ are again scribed and the target removed.

Lines can now be set on the pairs of scribed lines to give the intersection or, alternatively, if lines PQ and $P'Q'$ mutually at right angles have been previously scribed on a perspex or metal plate larger in dimension than the hole in the falsework, then these lines can be brought into coincidence with lines A_1ABB_1 and $A_1'A'B'B_1'$, and point O is now located to correspond with its respective ground station.

Fig. 10.16

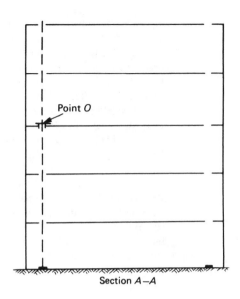

Section A—A

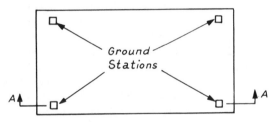

Plan at lowest floor level

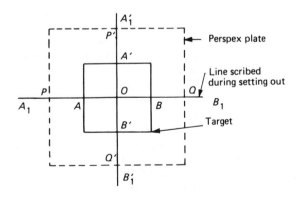

Measurements can be related to point O for setting-out purposes. Forms of target to suit the problem in hand can be designed by the surveyor, e.g. the target could be mounted on an adjustable offset arm for external use.

It has, of course, been assumed above that the instrument is in adjustment and that the line of sight is vertical with the micrometer set at 10.00. To check the adjustment of the instrument, sightings can be made near to a target line with the instrument set at 10.00. The target line is moved until it is close to the horizontal sighting line in the field of view, and then, using the micrometer drum, correct lining-in can be effected. Reading M_1 is now noted. The instrument is rotated through 180° in azimuth and a further reading M_2 of the micrometer taken for coincidence. If the instrument is in adjustment, $\frac{1}{2}(M_1 + M_2)$ = 10.00: if not, then $\frac{1}{2}(M_1 + M_2)$ is set as the micrometer reading and the target moved for coincidence with the line of sight which is now vertical.

The micrometer can now be set at 10.00 and the line of sight moved back into position by adjustment of the reticule. The adjustment can be rechecked, but the manufacturer states that the error should not exceed 0.50 revolution of the drum for this correction to be carried out. The accuracy of a single sighting is of the order of 1 mm in 40 m and this can be improved to about 0.3 mm in 40 m for two opposite sightings with the instrument turned 180° between each.

One revolution of the micrometer drum displaces the line of sight by an angle of $\tan^{-1} 0.001$; one division of the drum scale represents one hundredth of this movement, and a movement of $\tan^{-1} 0.002$ is possible in either direction from the vertical. It will be appreciated therefore that if the zeroing of the micrometer drum is correct, and the micrometer drum reads M_1 on lining on to the target, then the relevant displacement $= [(10.00 - M_1)/1000]H$ where H is the height of the target above the instrument. Errors in verticality can be established by such measurements and, similarly, displacements imparted by the micrometer can be measured on a horizontally held levelling staff so that H can be deduced.

Vertical alignment by laser

Any rotating laser unit in which the rotating prism is removable can be used for vertical plumbing, the beam of laser light being projected upwards with a typical accuracy of ±3 mm in 30 m.

In addition to the commercial laser devices described earlier, attachments can be obtained to project a beam of laser light through the optics of a theodolite. This allows alignment at angles other than vertical or near horizontal and can also be useful for fixing points in confined space or for surveying at night.

The Kern LO Laser attachment is one such instrument using this principle (Fig. 10.17). A laser is mounted on the tripod of a Kern theodolite and its light is passed through a condenser lens whereby its diameter is reduced sufficiently for it to pass along a light-carrying fibre of diameter 0.08 mm. This fibre passes the light to a reticule in the telescope of the theodolite and the beam is placed on the telescope axis by means of a light-dividing cube, whilst a filter absorbs any laser light which might be reflected towards the eyepiece. The line of sight is defined by a cross produced by the illuminated reticule and accordingly can be directed to a target since the red laser light, of wavelength 632.8 nm, is visible. Accessories such as diagonal eyepieces

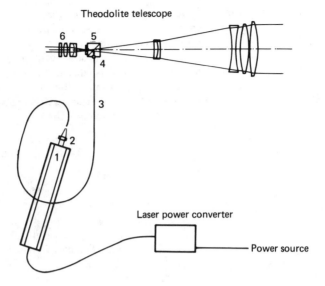

Fig. 10.17 Principle of Kern Laser attachment

Theodolite telescope

6 5 4 3 2 1

Laser power converter

Power source

and pentaprisms can be used to establish lines either directed towards zenith or for plumbing and most commercial lasers can be used as the light source providing that the necessary connectors are available. Under favourable atmospheric conditions, targets can be aligned fairly precisely to within a few millimetres within a range of 400 m; the theodolite can be used independently of the laser beam if required.

Sokkisha's SLT20 Laser Theodolite has the laser unit positioned around the theodolite eyepiece to give a compact single instrument. The theodolite has conventional glass-arc circles which have graduations and digital reading every 20″, with dots every 10″, whilst the laser is a 2 mW helium-neon type, with Class 2 safety rating. The laser beam is projected along the theodolite axis to a quoted accuracy of ±5 mm, at 100 m the beam itself has negligible divergence, being 3 mm diameter at 20 m and 10 mm diameter at 100 m.

In addition to their use for plumbing devices, laser theodolites may be used in inclined alignment to monitor movements of surfaces over a long period of time, by arranging pointings at fixed bearings from a base line or base lines and at known angles of elevation or depression. The lines of sight of a second theodolite are then directed in turn on to the spot positions indicated by the laser beams on the surface in question and after measurement of the relevant horizontal and vertical angles at the station(s) at the other end(s) of the line(s), movement of the surface can be deduced.

It can be mentioned here that Kern and the RCA Corporation have developed a system which allows bearings or directions to be measured between points which are not mutually intervisible. A laser beam is transmitted vertically upwards at one station and 'sighted' by an adapted one-second theodolite at the other station.

As well as transferring the horizontal control to other levels in the construction it is necessary to transfer a TBM. For multi-storey buildings this may simply involve conventional levelling up a staircase. However, for shafts on tunnels and other restricted spaces it is common practice to measure depth directly by steel tape. Ideally the tape should be longer than the depth to be measured and in this case the tape is suspended at the surface and a weight applied. Elongation of the tape is given by the equation

$$s = \frac{gl}{AE} \left[M + 0.5M\ (2L-l) - \frac{T_s}{g} \right]$$

where:
- s = elongation
- g = gravitational acceleration
- l = length of suspended tape
- L = total length of tape
- A = cross-sectional area of tape
- E = Young's Modulus for tape
- M = attached mass
- T_s = standard tension

If the depth is deeper than the length of a tape then the surveyor could work down the shaft in tape lengths using the hoisting device, or an EDM could be set up on the surface and the carrier wave directed down the shaft by means of a mirror mounted at the top.

Examples of the calculations for both tape and EDM measurement of shafts are given in *Solving Problems in Surveying*, Bannister and Baker (Longman).

Connection of surface and underground lines

Where a sewer or pipeline has to be laid in a tunnel, its alignment must be established accurately. Plumb lines are used to transfer lines below ground and two possible cases may be met when setting out, namely the case in which the underground line lies directly below the surface traverse line and, secondly, the case in which it does not. This second case means that the proposed underground line must be initially established from the overground control and then the plumb lines put down. Typical calculations are as follows:

Example 10.3 A length of sewer RQS is to be constructed in heading, the straights QR and QS having whole circle bearings of $202°46'$ and $20°14'$ respectively, whilst manhole Q has co-ordinates of 127.05 mE, 448.62 mN. If the co-ordinates of a nearby station A on a street traverse are 60.00 mE, 300.00 mN and the bearing of a traverse line AB is $21°33'$ obtain data for setting out the two straight lengths of sewer.

Assuming ground conditions allow a line PQ to be set out, P can be located along AB, so that PQ may be set off at right angles to AB and Q then located by direct measurement (Fig. 10.18(a)).

Fig. 10.18

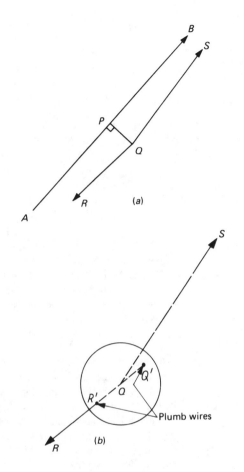

(a)

(b)

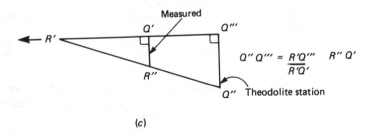

Measured

$$Q'' Q''' = \frac{R'Q'''}{\overline{R'Q'}} \quad R'' Q'$$

Q'' Theodolite station

(c)

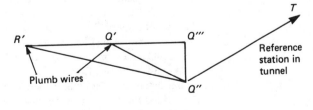

T

Reference station in tunnel

Plumb wires

(d)

Bearing $AQ = \tan^{-1} (127.05 - 60.00)/(448.62 - 300.00)$
$$= 67.05/148.62$$
$$= 24°17'$$
$AQ = \sqrt{(67.05^2 + 148.62^2)} = \mathbf{163.08 \ m}$
$AP = 163.08 \cos (24°17' - 21°33') = \mathbf{162. \ 90 \ m}$
$PQ = 163.08 \tan (24°17' - 21°33') = \mathbf{7.79 \ m}$

Note that writing the co-ordinates of A, B and Q as E_A, N_A etc., then AP and PQ could be determined directly from the two equations:

$$\frac{E_B - E_A}{AB} \cdot AP + \frac{N_B - N_A}{AB} \cdot PQ = E_Q - E_A$$

$$\frac{N_B - N_A}{AB} \cdot AP + \frac{E_B - E_A}{AB} \cdot PQ = N_Q - N_A$$

Either approach allows Q to be positioned on the surface.

The bearing of PQ is $111°33'$ and thus angles $P\hat{Q}R$ and $P\hat{Q}S$ equal to $88°47'$ and $88°41'$ respectively, can be set out on the ground and offset pegs or reference points can be established clear of the shaft excavation.

Coplaning

If the sewer is not to be laid at great depth plumb lines might well be established directly on QR or QS after sinking a shaft at Q (Fig. 10.18(b)). In this case a theodolite would be set up near Q such that its line of sight would lie on line RQ say, using the centring device if fitted. Fine marks could be scribed on timbers, or metal plates thereon, across the top of the shaft so that thin plumb lines or wires, R' and Q', could be suspended so that they coincide with those marks and therefore lie on line RQ. For relatively shallow depths, such plumb line positions might be defined by marks scribed on dogs driven into the shaft lining and the wires could be suspended directly from these points.

A plumb plane $R'Q'$ can thus be established compatible with line RQ, the co-ordinates of R' and Q' being readily determined from the surface survey. One must ensure that the lines are vertical so that the plumb plane $R'Q'$ is vertical. Overall, this is not an easy task to carry out since, for instance, the displacement of R' by the amount $R'Q'/10\ 000$ perpendicular to $R'Q'$ causes an error of about 20 seconds in the bearing of plane $R'Q'$. Repeated observations are needed to fix R' and Q', and it would be advisable to carry out a further set from a theodolite set up on the opposite side of Q.

Visual alignment will allow the tunnel to be commenced along $R'Q'$ at the shaft bottom, but at the earliest opportunity the theodolite is taken below and lined in on the two wires. This alignment procedure, or technique, is known as 'coplaning'. Should the wires oscillate, scales can be placed behind them and the mean positions of swing deduced from the scale readings given when the theodolite's line of sight has followed the oscillations. Alternatively these oscillations could be damped by immersing the plumb bobs in a liquid of suitable viscosity: their weights should be such as to stress the wires to say half ultimate stress. The reader's attention is drawn again to the importance of ensuring verticality of the plumb-plane, and in addition careful setting up of the theodolite is required during the coplaning procedure. It is unlikely that the first estimate of alignment will be correct and so $R'Q'$ will subtend a small angle at the theodolite, i.e. such as $R'Q''Q'$ in Fig. 10.18(c). The amount of shift $Q''Q'''$ to place the vertical axis of the theodolite on $Q'R'$ can be calculated after measuring on a scale the distance from sight line $Q''R'$ to Q', the scale being placed perpendicular to $Q'R'$. Repetitions, with changes of face, are required to satisfy the engineer that he has an underground sighting line related to the surface line. After shifting the vertical axis with the aid of the centring device the theodolite station can be established by using the optical plummet. Tunnel centre line points can be fixed by means of the sighting line: these can be recovered as required for prolonging the centre line and for fixing new instrument stations as applicable. Throughout, the theodolite needs to be well adjusted and its setting up again meticulously carried out; it is advisable to 'coplane' with the theodolite set up on the opposite side of the plumb wires near R' both to check the previous work and to fix further reference points within the tunnel.

Weisbach Triangle   Alternatively, the Weisbach Triangle (Fig. 10.18(d)) can be adopted since coplaning is not an easy procedure for the observer, particularly when the tunnel is at appreciable depth. The theodolite, reading to one second of arc, is established just off line $R'Q'$ given by the plumb wires, length $Q'Q''$ being preferably less than $R'Q'$. Both these lengths are measured along with angle $R'Q''Q'$, the instrument being at Q''. This allows angle $Q'R'Q''$ to be estimated, since by the sine rule

$$\frac{R'Q'}{\sin R'Q''Q'} = \frac{Q'Q''}{\sin Q'R'Q''}$$

For small angles we can put $\sin R'Q''Q' = \sin 1'' \times (R'Q''Q')$
and $\sin Q'R'Q'' = \sin 1'' \times (Q'R'Q'')$

where $(R'Q''Q')$ and $(Q'R'Q'')$ are expressed in seconds of arc. Thus

$$\frac{R'Q'}{(R'Q''Q')} = \frac{Q'Q''}{(Q'R'Q'')}$$

Although $Q''Q'''$ can be calculated now and Q''' placed on the tunnel centre line, since the bearing of $R'Q'$ is known we can also determine the bearing of $R'Q''$ and hence the bearing of a line joining Q'' and a suitably positioned reference station T within the tunnel (Fig. 10.18(d)). The co-ordinates of R' and Q' being known, all such stations as T, which give underground reference lines, can be similarly co-ordinated following suitable angular and linear measurements. It is recommended that two or three positions be adopted for Q'' on each side of the line $R'Q'$ and that in addition theodolite stations should be adopted near R'. For deep shafts a few pairs of plumb lines could be put down, not necessarily on the centre line, their bearings being established at the surface by Weisbach Triangles linked to known ground points (*see* Question 4 in Exercises 10).

Example 10.4 The centre-line of the tunnel AB shown in Fig. 10.19 is to be set out to a given bearing. A short section of the main tunnel has been constructed along the approximate line and access is gained to it by means of an adit connected to a shaft. Two wires, C and D, are plumbed down the shaft, and readings are taken on to them by a theodolite set up at station E slightly off the line CD produced. A point F is located in the tunnel, and a sighting is taken on to this from station E. Finally, a further point G is located in the tunnel and the angle $E\hat{F}G$ measured.

From the surface survey initially carried out, the co-ordinates of C and D have been calculated and found to be 375.78 mE, 1119.32 mN, for C, and 357.37 mE, 1115.70 mN for D.

Calculate the co-ordinates of stations F and G. Without making any further calculations describe how the required centre-line could then be set out. (*I.C.E.*)

In the Weisbach Triangle CDE (Fig. 10.19)

$$D\hat{C}E = \frac{4.46}{3.64} \times 38'' = 46.6''$$

or external angle $\qquad C\hat{D}E = \frac{4.46+3.64}{3.64} \times 38''$

$$= 84.6'' = 01'24.6''.$$

assuming $\qquad CD+DE = CE$ since $C\hat{D}E \simeq 180°$.

The bearing of $DC \quad = \tan^{-1}\frac{0.41}{3.62}$

$$= 6°27'42.4''$$

Hence bearing of $ED = 6°27'42.4'' - 01'24.6''$

$$= 6°26'17.8''$$

i.e. $\qquad\qquad\qquad = 6°26'18''$

The reader can now check that

(1) bearing of $EF = 173°37'16''$;
(2) bearing of $FG = 81°00'57''$;
(3) co-ordinates are

$\qquad\qquad F$ **376.33 E, 1098.23 N**
$\qquad\qquad G$ **433.12 E, 1107.22 N**

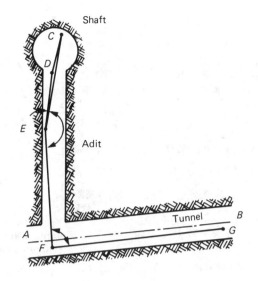

Fig. 10.19
CD = 3.64 m
DE = 4.46 m
EF = 13.12 m
FG = 57.50 m
Angle DEC = 38″
Angle CEF =
　　　167°10′22″
Angle EFG =
　　　87°23′41″
The drawing is not to
scale

Shaft

Adit

Tunnel

Underground control

Once the line has been transferred underground it is then necessary to project this line forward for the length of the tunnel. For short urban tunnels, such as those met on sewerage schemes, it will be sufficient to simply align a pipe laser with the underground control points at the shaft. Due to the small size of the tunnel and shaft the laser would normally be set up in a back extension of the tunnel line away from construction activities. If there is a change in line at the shaft this may involve the excavation of a special adit to house the laser. The line and level of the tunnel should be checked at regular intervals by theodolite and level.

On larger tunnels, which will probably include curved sections, more accurate internal control will be required. Pipe lasers will be used for local alignment and the directional control of tunnelling machines and these are typically attached to brackets mounted high on the tunnel lining out of the way of construction operations. The laser position, and the line and level given by the beam, need accurate reference to the surface control and this will probably include a traverse within the tunnel bore with position checks using a gyro-theodolite as discussed in Chapter 7.

In Chapter 6 we have already mentioned the surface control network for the Channel Tunnel and the reader is directed towards an article in *New Civil Engineer*, 29 November 1990, in which the survey and control aspects of this project are discussed. Due to difficulties of access the survey team's initial idea was to set up a traverse along one side of the tunnel only. Stations were marked by brackets at about 75 m intervals high on the tunnel walls, but this meant that theodolite sights had to be made close to the cold concrete lining of the tunnel, whilst the general body of air in the tunnel was warm from plant and machinery. Checks by gyro-theodolite showed much larger closing errors than implied by the theodolite and EDM data and it was

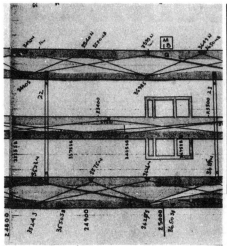

Section of the main survey.

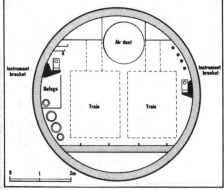

Space within the service tunnel was very restricted; surveyors had to take refuge as each works train passed.

Fig. 10.20 Internal control traverse for the Channel Tunnel (*Courtesy*: Editor, *New Civil Engineer*)

concluded that refraction due to the temperature differential was inducing errors that turned a straight tunnel length into a 3000 km radius curve. The traverse was changed to a zig-zag arrangement as shown on Fig. 10.20 and this allowed the two ends of drives from France and England to meet within 0.5 m, a satisfactory achievement when it is considered that a 1″ error corresponds to a deviation of 200 mm in horizontal alignment over the 37 km length of the tunnel.

Exercises 10

1 (i) Explain, with the aid of diagrams, how sight rails may be used to control the excavation of a trench to accommodate the footing for a new building detailing how you would ensure that your setting out was free from any blunder or mistake.

(ii) The booking form below forms part of the level sheet of a site engineer engaged in such a procedure. If the formation level for the excavation is 92.4 m and the general ground level in the vicinity of the site is given by the level of position *C*, compute a height for the sight rails which would allow a traveller of a suitable length to be used, and hence the staff intercept which should be set out from the current instrument position in order to attain this level.

(London)

Answer: 3.3 m traveller = 1.169 m sight rails
staff reading = 0.769 m

2 A sewer is to be laid at a uniform gradient of 1 in 200, between two points *X* and *Y*, 240 m apart. The reduced level of the invert at the outfall *X* is 150.82. In order to fix sight rails at *X* and *Y*,

Level Booking Form

From: TBM A   To: Observer: DPC

Point	Backsight	Intermediate sight	Foresight	Height of collimation	Reduced level	Remarks
TBM A	1.534				95.312	TBM
	1.456		1.235			
	1.279		1.472			
	1.043		1.432			
C		1.974				Ground Level

readings are taken with a level in the following order:

Reading	Staff Station
BS 0.81	TBM (near X), RL 153.81
IS 'a'	Top of sight rail at X
IS 1.07	Peg at X
FS 0.55	CP between X and Y
BS 2.15	CP between X and Y
IS 'b'	Top of sight rail at Y
FS 1.88	Peg at Y

(i) Draw up a level book and find the reduced levels of the pegs.

(ii) If a boning rod of length 3 m is to be used, find the level readings a and b.

(iii) Find the height of the sight rails above the pegs at X and Y.

(*London*)

Answer: (iii) 0.27 m, 0.68 m.

3   Calculate the data that is required to establish a batter rail, for use with a 1.3 m long traveller, that will define an embankment which is to slope upwards at 1 in 2.5 from a level of 72.509 m.

Answer: stake 1 m from toe, level = 73.409 m

stake 2 m from toe, level = 73.009 m

4   During the setting out of a tunnel, the following observations were made at a shaft, the instrument stations being established eastwards of the plumb wires and with X the nearer in each case. The plumb wires were 6.245 m apart and the whole circle bearing of PQ was 194°15′40″.

Theodolite at Surface Station

Pointing	Circle reading	Distance from theodolite (m)
Reference Station P	277°41′10″	139.65
Reference Station Q	97°42′19″	475.30
Plumb wire X	185°18′18″	8.785
Plumb wire Y	185°19′02″	

Theodolite at Underground Station

Pointing	Circle reading	Distance from theodolite (m)
Plumb wire *X*	43°42′30″	7.233
Plumb wire *Y*	43°43′02″	
Reference Station *A*	223°44′34″	27.58
Reference Station *B*	43°45′17″	24.45

Determine the bearing of *BA*.
Answer: 101°54′56″.

5   During the setting out of a tunnel the following observations were made by theodolite at a surface station *I*, near to both a survey line *AB* and a vertical shaft.

Pointing	Horizontal circle reading	Distance from theodolite (m)
Station *A*	121°57′20″	125.38
Station *B*	301°57′38″	424.59
Plumb line *X*	214°20′12″	9.932
Plumb line *Y*	214°20′36″	

The plumb lines down the shaft were 5.297 m apart. If the whole circle bearing of *XY* was determined to be 214°57′50″ what was the whole circle bearing of *AB*? It may be assumed that *X* was the nearer of the two plumb lines to *I*.

How would the line *XY* be continued in the tunnel by means of co-planing?

(*Salford*)

Answer: 302°34′03″.

6   In Weisbach Triangle *ABC*, *A* and *B* denote the plumb wires and *C* denotes the theodolite station, with *B* the nearer wire thereto. If $ACB = \theta$ and $CAB = \phi$ show that the standard error $s_{\overline{\phi}}$ (rad) in ϕ is related to the corresponding standard errors in θ, *AB* and *BC* by the expression

$$\cot^2 \phi \cdot s_{\overline{\phi}}^2 = \cot^2 \theta \cdot s_{\overline{\theta}}^2 + \frac{s_{\overline{AB}}^2}{AB^2} + \frac{s_{\overline{BC}}^2}{BC^2}$$

The following data apply to a certain Weisbach Triangle:

AB = 2.958 m	standard error 3 mm
BC = 2.105 m	standard error 2 mm
θ = 06′20″	standard deviation 10″

A reference mark *D* was established to the requirement that the standard error in bearing of *CD* would not exceed 5″, assuming that the bearing of *AB* was free of error in this instance. If angle *ACD* was measured with a standard error of 4″ how many times would angle *ACB*, whose mean value is quoted above, have been measured to meet that requirement?
Answer: 6.

7   A length of sewer *CD* is to be constructed in a tunnel. Two plumb
    wires, *X* and *Y*, are put down the shaft at *C*. The plumb wires
    were observed from control stations *A* and *B* to the south of *C*
    and the following angles were recorded:

$$B\hat{A}X = 61°31'22''$$
$$B\hat{A}Y = 60°18'15''$$
$$A\hat{B}X = 55°30'32''$$
$$A\hat{B}Y = 56°36'51''$$

The control stations have co-ordinates

	E	N
A	850.36 m	1360.21 m
B	990.37 m	1401.47 m

A theodolite was then established below ground, at *U*, close to
but to the west of line *XY* produced and 5.631 m from nearest
wire *X*. Horizontal circle readings for pointings on *X* and *Y* were
32°18′10″ and 32°12′18″ respectively. What angle should be
set on the theodolite at *U* to locate underground control station
R, co-ordinates 966.75 mE, 1267.25 mN and how far is *R* from
U?

(Salford)

Answer: 39°2′29″, 26.82 m

8   The centre line of a tunnel heading in an Easterly direction is to
    be set out between two points *A* and *B*. Both points are visible
    from a third point *C* overlooking the site of the tunnel to the South-
    West generally of the line *AB*. The horizontal angle *A*Ĉ*B* is
    34°34′55″. The mean of reciprocally observed vertical angles
    on the lines *CA* and *CB* are −02°56′05″ and −08°29′26″ and
    the slope distances *CA* and *CB* are 2443.902 m and 1923.433 m.

Compute
(*a*) the horizontal distance *AB*
(*b*) the difference in level between points *A* and *B*
(*c*) the horizontal angles required at *A* and *B* for setting out the
       centre line of the tunnel.

(Eng. Council)

Answer: 1389.436 m; 158.865 m; *A*B̂*C* = 94°25′17″,
*B*Â*C* = 50°59′48″

11 Curve ranging

Mention has already been made in Chapter 10 of the techniques of setting out for certain construction works defined by straight lines, such as buildings. However, in many types of construction, such as roads and railways, two straights will normally be connected by a curve wherever there is a change in direction. The setting out, or ranging of curves, will be dealt with at some length in this chapter, treatments of the circular curve and the transition curve in plan being given, along with the vertical curve. In this last type, the setting out operations consist of fixing level pegs, and since this has already been dealt with, only the design method will be described.

The term *chainage* is used in linear construction projects such as roads, railways and sewers to distinguish a unique point on the works. Chainage is a cumulative increase of distance measured from a starting point normally at one end of the scheme, i.e. the point of zero chainage. Thus a point with chainage 2000 m would be 2 km along the scheme from the starting point.

Circular curves

In Fig. 11.1, two straights meet at the point of intersection I, and a circular arc of radius R runs between the straights, meeting them tangentially at the tangent points T and U.

The angle of total deflection θ of the straights is also shown. This angle is sometimes referred to as the angle of deviation, or as the angle of intersection.

Then, since
$$I\hat{T}O = I\hat{U}O = 90°$$
$$T\hat{I}U + T\hat{O}U = 180°$$
but
$$T\hat{I}U + T'\hat{I}U = 180°$$
Therefore
$$T\hat{I}U = \theta = T\hat{O}U.$$

Thus the long chord TU subtends an angle equal to the deflection angle at the centre of the curve.

Fig. 11.1

Length of tangents

Triangles IOT, IOU, are congruent and so
(i) $IT = IU$, i.e. the tangent lengths are equal.

Therefore
$$ITU \text{ is isosceles}$$
but
$$I\hat{T}U + I\hat{U}T = T'\hat{I}U = \theta$$

Therefore
$$I\hat{T}U = \frac{\theta}{2}$$

(ii) Also,
$$T\hat{O}S = U\hat{O}S$$
$$= \frac{\theta}{2}$$

Therefore
$$IT = \text{tangent length}$$
$$= R \tan \frac{\theta}{2}$$

Thus, if the location of the intersection point I be known, then by measuring a distance $R \tan \theta/2$ back along the straight from I, the tangent point T can be located. Similarly, measuring the same distance along the other straight will locate the second tangent point U.

Length of curve

Circumference of a circle of radius $R = 2\pi R$. This subtends an angle of $360°$ at the centre of the circle,

hence the length L of the arc TU which subtends θ at the centre

$$= 2\pi R \frac{\theta}{360} = \frac{\pi}{180} R\theta$$

Chainage of tangents

Knowing the chainage of I,

then   chainage of T = chainage of $I - IT$,

and   chainage of U = chainage of $T + \dfrac{\pi}{180} R\theta$

(*Note*: The chainage of U is *not* the chainage of $I+IU$.)

Setting out

One approach for setting out circular curves is to locate the tangent points and intersection points that define the line of the construction and then to proceed with the setting out of the pegs along the curve by taking measurements from these positions. The tangent and intersection points can be set out from control points, established during previous surveys, provided the co-ordinates of all points are known.

For instance T (Fig. 11.2) could be positioned after setting off angle γ and then measuring l along that ray, whilst I could be located by measuring lengths m and n. The theodolite can of course be established at a totally independent station so long as at least three suitable points of known co-ordinates can be observed therefrom. The pegs that mark the centre line of the curve can be set out from the tangent points

(1) by tape and offset
(2) by angle and distance measurement
(3) by angle measurement only

as described in subsequent sections.

A second approach to the setting out of curves is to calculate the co-ordinates of the pegs along the centre line in the same co-ordinate

Fig. 11.2

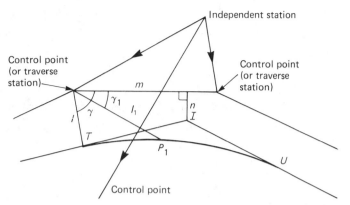

system as the local control points. The pegs can then be located directly from the existing control without setting up at the tangent or intersection points. For example, in Fig. 11.2 peg P_1 could be located by setting off angle γ_1, and length l_1.

This second approach involves more calculation than the first; but it is standard practice when computing facilities and electromagnetic distance measuring equipment are available.

(1) Methods using tapes

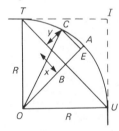

Fig. 11.3

(a) Offsets from the long chord This method is suitable for curves of small radius such as kerb lines at road intersections, and boundary walls, but it can also be used for longer curves providing AB (Fig. 11.3) is not excessively long. To set out the curve, points such as C must be located. The offsets y are calculated for corresponding distances x from origin B, which is the mid point of TU, and are set off at right angles, using any of the methods already described. TU is the long chord of length L, T and U being the tangent points. AB is termed the versed sine of the curve.

In Fig. 11.3
$$\begin{aligned} AB &= AO - OB \\ &= AO - \sqrt{OU^2 - UB^2} \\ &= R - \sqrt{R^2 - \left(\frac{L}{2}\right)^2} \end{aligned}$$

Draw CE parallel to TU

Then
$$y = EB = EO - BO$$
$$EO^2 = CO^2 - CE^2 \therefore EO = \sqrt{R^2 - x^2}$$

whence
$$y = \sqrt{R^2 - x^2} - \sqrt{R^2 - \left(\frac{L}{2}\right)^2}$$

Or, since
$$OB = R - AB = R - \text{versed sine},$$
$$y = \sqrt{R^2 - x^2} - (R - \text{versed sine})$$

Example 11.1 Derive data for setting out the kerb line shown in Fig. 11.3 if the radius be 12 m, and $T\hat{O}U = 90°$. Offsets are required at 2 m intervals.

Now $T\hat{O}U = 90°$ so $TU^2 = TO^2 + OU^2 = 12^2 + 12^2 = 288$

Therefore $TU = L = 16.97$ m

$$\sqrt{R^2 - \left(\frac{L}{2}\right)^2} = 8.49 \text{ m}$$

Note:

R − versed sine = 8.49 m, hence versed sine AB = 3.51 m

x (m)	x^2	$(R^2 - x^2)$	$\sqrt{R^2 - x^2}$	Offset (y) (m)
0	0	144	12.00	3.51
2	4	140	11.83	3.34
4	16	128	11.31	2.82
6	36	108	10.39	1.90
8	64	80	8.94	0.45

Points T and U would be located by measuring $IT (= IU)$ from the intersection point I.

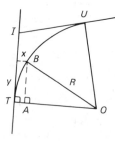

Fig. 11.4

(*b*) *Offsets from the tangent* This method is also suitable for short curves and, as in the previous method, no attempt is made to keep the chords of equal lengths. In Fig. 11.4

$$R^2 = AB^2 + AO^2 \text{ (Pythagoras' theorem),}$$
$$= y^2 + (R-x)^2$$

Therefore $x = R - \sqrt{(R^2 - y^2)}$

$$= R - R\left(1 - \frac{y^2}{R^2}\right)^{1/2}$$

Expand $\left(1 - \frac{y^2}{R^2}\right)^{1/2}$ using Binomial theorem,

then $x = R - R\left(1 - \frac{1}{2}\frac{y^2}{R^2} + \ldots\right)$

$$= \frac{y^2}{2R} \text{ approximately}$$

The curve is set out in two parts, starting from each tangent.

(*c*) *Offsets from chords produced* For longer curves of larger radius, the curve can be set out by offsets from chords using a tape. This method has the advantage that not all the land between tangents and curve need be accessible, and can be used when the accuracy attainable with the theodolite is not required, e.g. giving the line for soil stripping and excavation in road works. In order that the assumptions made in the derivations of formulae are not invalidated, namely, that the length of the arc is very nearly equal to the length of the corresponding chord, the length of chord chosen should not exceed $R/20$. It is usual to peg the curve at regular intervals with cross-sections at perhaps 100 m intervals. Thus a convenient interval for the pegs is 20 m thereby allowing a curve of radius 400 m to be so treated. If the chainage of T be, say, 22 186, then the first chord can be made 14 m long (assumed to agree with the $R/20$ concept) in order that the next peg will be placed at the even chainage of 22 220 m.

Fig. 11.5

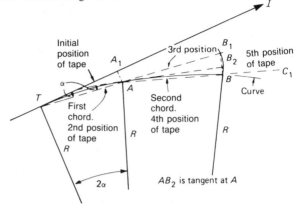

Procedure: Taking chainage of T as 22 186 (referring to Fig. 11.5),

(a) Hold 14 m mark at T with the tape arranged on the line TI. The zero end of the tape at A_1 is swung round through a calculated offset A_1A, with T as centre, thus locating peg A (chainage 22 220) on the curve. (Note, however, that chainages are often not kept continuous in this method, i.e. chainage of T is assumed to be zero, especially if cross-sections are not required.)

(b) Pull tape forward along TA produced until forward end is at B_1, and 20 m mark is at A (if a length of 10 m is required for AB, then the 10 m mark will be held at A). The end B is now swung round through a calculated offset B_1B to locate B.

Fig. 11.6

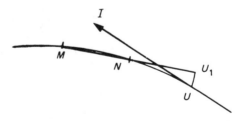

(c) Repeat for all other points and the final offset should be such that U is located on the tangent (*see* Fig. 11.6). As a check, U can be located by measuring tangent length IU from I and the final offset should then agree with this position of U. If not, then it will be necessary to check all points such as A, B, adjusting their positions until coincidence at U is obtained.

Calculations: Referring to Fig. 11.5, let

$$A_1\hat{T}A = \alpha$$

Then for small angles

$$\text{chord } AA_1 = \text{arc } AA_1$$

Hence
$$AA_1 = TA_1 \cdot \alpha = TA \cdot \alpha$$

If $A_1\hat{T}A = \alpha$, then the angle subtended at the centre of the curve will be 2α,

$$\text{Arc } TA \simeq \text{chord } TA \left(\text{so long as } TA \not> \frac{R}{20} \right)$$

$$= R \cdot 2\alpha$$

Thus
$$\alpha = \frac{TA}{2R}$$

Hence
$$AA_1 = TA \cdot \alpha = \frac{TA}{2R} \cdot TA = \frac{TA^2}{2R}$$

which is the first offset.

$$\text{Also} \quad B_1B_2 = AB_1 \cdot \alpha$$

$$= AB_1 \cdot \frac{TA}{2R} = AB \frac{TA}{2R}, \text{ since } AB_1 = AB$$

Since B_2B is the offset from the tangent AB_2, then exactly as for AA_1 which was offset from TI we may derive B_2B to be $AB^2/2R$

Therefore
$$B_1B = B_1B_2 + B_2B = AB \cdot \frac{TA}{2R} + \frac{AB^2}{2R}$$

$$= \frac{AB}{2R}(TA + AB)$$

For all subsequent offsets but the last, the chord lengths will now be equal and so

$$C_1C = \frac{BC}{2R}(AB + BC) = \frac{BC^2}{R}$$

It may be shown that the final offset U_1U will be

$$[NU(MN + NU)]/2R$$

(*see.* Fig. 11.6).

Example 11.2 Tabulate data needed to set out, using two tapes, a circular curve of radius 600 m to connect two straights deflecting through an angle of 18°24′. The chainage of the intersection of the tangents is 2140.00 m.

$$IT = 600 \tan \frac{18°24'}{2} = 97.20 \text{ m}$$

T may be located by measuring 97.20 m back from I

Therefore   chainage of tangent point $T = 2042.80$ m

$$\text{Length of curve} = 600 \times 18.4 \times \frac{\pi}{180} = 192.68 \text{ m}$$

Therefore
          chainage of tangent point $U = 2042.80 + 192.68$
                                    $= 2235.48$ m

Now $R/20 = 30$ m, and this interval would suffice for the peg spacings, but for convenience the chord length of 20 m will be adopted. Thus the first chord would be made 17.20 m followed by 8 chords each 20 m long, together with a final chord of 15.48 m length.

Chainage	Chord length (m)	Offset (m)
2042.80	0	—
2060.00	17.20	$\dfrac{17.20^2}{2 \times 600} = 0.25$
2080.00	20.00	$\dfrac{20(20.00 + 17.20)}{1200} = 0.62$
2100.00	20.00	$\dfrac{20.00^2}{600} = 0.67$
2120.00	20.00	$\dfrac{20.00^2}{600} = 0.67$
2140.00	20.00	$\dfrac{20.00^2}{600} = 0.67$
2160.00	20.00	$\dfrac{20.00^2}{600} = 0.67$
2180.00	20.00	$\dfrac{20.00^2}{600} = 0.67$
2200.00	20.00	$\dfrac{20.00^2}{600} = 0.67$
2220.00	20.00	$\dfrac{20.00^2}{600} = 0.67$
2235.48	15.48	$\dfrac{15.48(15.48 + 20.00)}{2 \times 600} = 0.46$

Setting out a radial   To establish a cross-section for levelling purposes, i.e. a radial line to a curve at say C in Fig. 11.8, use may be made of adjacent curve points B and D which are equidistant from C. The tangent at C is parallel to chord BD and the required radial line is at right angles to this tangent. Thus if the mid-point of BD is found by direct measurement between B and D the direction of the radial line is given by connecting the mid-point to C and extending as required.

<table>
<tr><td>(2) Setting out using one theodolite and a steel tape by the method of deflection angles</td><td></td></tr>
</table>

(2) Setting out using one theodolite and a steel tape by the method of deflection angles

Again it is recommended that the chords do not exceed $R/20$ in length, so that the length of the chord may be approximated to the length of the corresponding arc.

Field procedure:

(a) Set up theodolite at T, having calculated angles etc., for their particular chord lengths, set horizontal circle to read zero, clamp the upper and lower plates together, direct the telescope to I, obtaining coincidence with the lower clamp and tangent screw.

(b) Unclamp the upper plate (but ensure that the lower plate remains clamped and stationary throughout the operation), and set horizontal circle to read deflection angle α, thus directing the telescope to A. Clamp the plates together, hold the tape at T, so that chord TA of correct length will be obtained. As in the previous method with tapes, the first chord length TA will generally be some residual length calculated to bring the chainage of A to the next whole number (say 14 m). An arrow is thus held at the 14 m mark on the tape, which is now swung taut about the zero end at T until exact coincidence is obtained between the instrument cross-hairs and the arrow. This fixes A.

(c) Set horizontal circle reading to β and then reclamp. The tape is held at A and the arrow held at the 20 m mark, the taut tape is then swung about A until there is coincidence with the cross-hairs.

(d) Repeat for all points on the curve, and coincidence should be obtained at U located by measuring the tangent distance IU along the tangent from the intersection point. The final deflection angle $I\hat{T}U$ will of course equal $\theta/2$, and this could be read immediately on setting up at T to serve as a check that the tangent lengths have been correctly set out.

It may be mentioned here that α, β etc. are also known as tangential angles.

Calculation: Referring to Fig. 11.7

Fig. 11.7

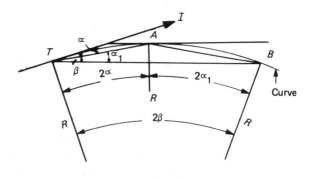

$$I\hat{T}A = \alpha$$

Hence the angle subtended at centre of curve by $TA = 2\alpha$

$$\text{arc } TA = 2R\alpha \quad (\alpha \text{ in radians})$$

$$= 2R\alpha \times \frac{\pi}{180} \quad (\alpha \text{ in degrees})$$

$$= \text{chord } TA \quad \left(\text{if chord } \not> \frac{R}{20}\right)$$

Therefore
$$\alpha = \frac{TA \times 180}{2\pi R} \text{ deg}$$

$$= \frac{TA \times 180 \times 60}{2\pi R} \text{ min}$$

$$= 1718.9 \times \frac{TA}{R} \text{ min}$$

or, if
$$TA = O_1 = \text{length of first chord}$$

$$\alpha = 1718.9 \times \frac{O_1}{R} \text{ min}$$

If α_1 equals the angle between the tangent at A and chord AB, then the angle subtended by AB at the centre of the curve is $2\alpha_1$, and since an angle subtended by a chord at the circumference is half that subtended by the chord at the centre, $A\hat{T}B = \alpha_1$.

Therefore
$$\text{arc } AB = 2R\alpha_1 \times \frac{\pi}{180}$$

and
$$\alpha_1 = 1718.9 \times \frac{AB}{R} \text{ min}$$

or, if
$$AB = O_2$$

$$\alpha_1 = 1718.89 \cdot \frac{O_2}{R} \text{ min}$$

Hence
$$I\hat{T}B = \beta = \alpha + \alpha_1$$

i.e. the angle set out from T while the chord AB is being set out.

Note that AB is the chord, not TB. It should also be noted that α_1 is the angle which would have to be set off from the tangent at A if we were setting out B with the instrument at A.

It will be seen that by repeating the above calculations, a series of deflection angles α, $\alpha + \alpha_1$, $\alpha + \alpha_1 + \alpha_2$ ..., etc., will be obtained corresponding to arcs TA, AB, BC, etc. The actual calculations are, of course, simplified because all the chords except the first and the last are equal, i.e. $\alpha_1 = \alpha_2 = \alpha_3 =$, etc.

Example 11.3 Tabulate data needed to set out by theodolite and tape a circular curve of radius 600 m to connect two straights having a deflection angle 18°24′, the chainage of the intersection point being 2140.00 m.

As shown in Example 11.2, the chainages of tangent points T and U are 2042.8 m and 2235.48 m respectively.

Chainage (m)	Chord (m)	Deflection angle	Total deflection angle	Total deflection angle set on 20″ inst.
2042.80	0	0	0	0
2060.00	17.20	$1718.9 \times \dfrac{17.20}{600} = 49.27'$	$00°\,49.27'$	$00°\,49'\,20''$
2080.00	20.00	$1718.9 \times \dfrac{20.00}{600} = 57.30'$	$01°\,46.57'$	$01°\,46'\,40''$
2100.00	20.00	$1718.9 \times \dfrac{20.00}{600} = 57.30'$	$02°\,43.87'$	$02°\,44'\,00''$
2120.00	20.00	$1718.9 \times \dfrac{20.00}{600} = 57.30'$	$03°\,41.17'$	$03°\,41'\,20''$
2140.00	20.00	$1718.9 \times \dfrac{20.00}{600} = 57.30'$	$04°\,38.47'$	$04°\,38'\,20''$
2160.00	20.00	$1718.9 \times \dfrac{20.00}{600} = 57.30'$	$05°\,35.77'$	$05°\,35'\,40''$
2180.00	20.00	$1718.9 \times \dfrac{20.00}{600} = 57.30'$	$06°\,33.07'$	$06°\,33'\,00''$
2200.00	20.00	$1718.9 \times \dfrac{20.00}{600} = 57.30'$	$07°\,30.37'$	$07°\,30'\,20''$
2220.00	20.00	$1718.9 \times \dfrac{20.00}{600} = 57.30'$	$08°\,27.67'$	$08°\,27'\,40''$
2235.48	15.48	$1718.9 \times \dfrac{15.48}{600} = 44.35'$	$09°\,12.02'$	$9°\,12'\,00'' = \dfrac{\theta}{2}$

When the curve deflects to the left It will be appreciated that the above procedure only applies directly when the curve deflects to the right of the tangent when proceeding in the direction of the chainage (as in Fig. 11.5) since only then will the telescope be traversing in a clockwise direction, and the theodolite reading the actual deflection angles α, $\alpha + \alpha_1$, etc., during the setting-out operation. Where the curve deflects to the left when approaching in the direction of the chainage, the deflection angles to be set out are $(360 - \alpha)$, $(360 - (\alpha + \alpha_1))$, $(360 - (\alpha + \alpha_1 + \alpha_2))$, etc., where α, α_1, α_2, etc., are computed as outlined above. This problem does not of course occur with a modern electronic theodolite where the direction of the circle readings can be reversed with a switch, nor with a theodolite whose horizontal circle is graduated both clockwise and anti-clockwise.

(3) Setting out using two theodolites

This method is suitable when the ground between the tangent points T and U is of such a character that taping proves difficult, e.g. very steep slopes, or if the curve is wholly or partly over water or marsh. Two theodolites are used, one being set at each tangent point and the main disadvantage of the method is that it requires two surveyors and two instruments as well as the assistants who locate the pegs.

Referring to Fig. 11.8 it will be seen that if the theodolite at T is set so that the line of sight deflects through angle α_E and the theodolite at U set so that its line of sight deflects through $[360° - (\theta/2) + \alpha_E]$ clockwise from tangent UI, the two lines of sight will intersect at point E. To locate E, therefore, the assistant is directed until the signal is seen at the intersections of the cross-hairs of both theodolites. Good liaison between the three groups is, of course, essential at this stage. This procedure is repeated for the location of the other points, and as an example the curve already tabulated for setting out with one theodolite and tape is now given for setting out with two theodolites.

In Fig. 11.8, angle subtended at T by $UE = E\hat{T}U$

$$= \frac{\theta}{2} - \alpha_E = E\hat{U}I$$

where α_E is the total deflection angle for setting out point E with the instrument at T.

Thus either a clockwise angle of

$$\left(360° - \frac{\theta}{2} + \alpha_E\right)$$

Fig. 11.8

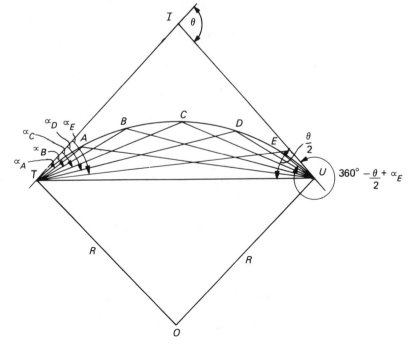

Table 11.1

Point	Chainage	Chord	Deflection angle	Deflection angle from T	Deflection angle from U
T	2042.80	0	0	0	350° 48′ 00″
	2060.00	17.20	49.27′	00° 49′ 20″	351° 37′ 20″
	2080.00	20.00	57.30′	01° 46′ 40″	352° 34′ 40″
	2100.00	20.00	57.30′	02° 44′ 00″	353° 32′ 00″
	2120.00	20.00	57.30′	03° 41′ 20″	354° 29′ 20″
	2140.00	20.00	57.30′	04° 38′ 20″	355° 26′ 20″
	2160.00	20.00	57.30′	05° 35′ 40″	356° 23′ 40″
	2180.00	20.00	57.30′	06° 33′ 00″	357° 21′ 00″
	2200.00	20.00	57.30′	07° 30′ 20″	358° 18′ 20″
	2220.00	20.00	57.30′	08° 27′ 40″	359° 15′ 40″
U	2235.48	15.48	44.35′	09° 12′ 00″	360° 00′ 00″

or, if the design of the theodolite allows this, an anticlockwise angle of $(\theta/2 - \alpha_E)$ will be required to locate E from U. Note that in Table 11.1 both sets of deflection angles show increasing values with increase in chainage with clockwise reading instruments.

(4) Setting out from remote control stations

To set out the curve from remote control stations the co-ordinates of the stations and some points on the curve, typically the tangent and intersection points, must be known. The co-ordinate information for the curve can be used to find the whole circle bearings of the first straight using the equation,

$$\text{WCB}_{TI} = \tan^{-1} \frac{(E_I - E_T)}{(N_I - N_T)}$$

Hence the WCB of the line from T to any point on the curve can be established by adding the deflection angles α, $\alpha + \alpha_1$, $\alpha + \alpha_1 + \alpha_2$ etc. Thus the WCB to point B, Fig. 11.7, would be:

$$\text{WCB}_{TB} = \text{WCB}_{TI} + \alpha + \alpha_1$$

The line joining T to the point on the curve, A in Fig. 11.9, is a chord and its length is given by the equation,

$$l = 2R \sin \theta_1$$

Hence the co-ordinates of the point on the curve are obtained from the equation,

$$E_A = E_T + l \sin \text{WCB}_{TA}$$
$$N_A = N_T + l \cos \text{WCB}_{TA}$$

When WCB are used, as above, the sign of the cos and sin values will alter automatically dependent upon the quadrant in which the line TA is based, giving the co-ordinates of A directly.

If we know the co-ordinates of two control stations X and Y, Fig. 11.9, an EDM can be set up at one station, say X. The bearing of line XY is given by the expression

$$WCB_{XY} = \tan^{-1} \frac{(E_Y - E_X)}{(N_Y - N_X)}$$

If the telescope is sighted at Y and the horizontal scale set to the value WCB_{XY} then a north pointing will be when the horizontal circle reads zero. A point on the curve such as A can be set out by sighting the instrument along horizontal circle reading WCB_{XA}, where

$$WCB_{XA} = \tan^{-1} \frac{(E_A - E_X)}{(N_A - N_X)}$$

and setting off distance XA, where

$$XA = \sqrt{(E_A - E_X)^2 + (N_A - N_X)^2}$$

Example 11.4 A 300 m radius circular curve is to be set out from remote control stations X (2134.091 mE, 1769.173 mN) and Y (2725.172 mE, 1696.142 mN). If the first tangent point T is at chainage 1109.27 m and has co-ordinates 2014.257 mE, 1542.168 mN and the intersection point I has co-ordinates 2115.372 mE, 1593.188 mN, calculate the bearing and distance from X required to set out point A which is on the curve at chainage 1180 m.

Fig. 11.9

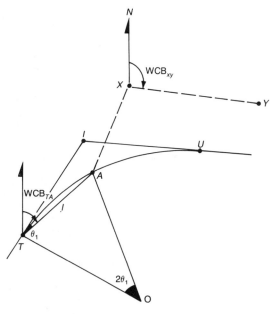

$$\text{WCB}_{TI} = \tan^{-1} \frac{2115.372 - 2014.257}{1593.188 - 1542.168}$$

$$= 63°13'32''$$

$$\text{Length of circular arc } TA = 1180 - 1109.27$$
$$= 70.73 \text{ m}$$

The circumference of a 300 m radius curve has a length of $2\pi \times 300$ and subtends an angle of $360°$ at the centre.

Thus the angle subtended at the centre by TA

$$= \frac{70.73 \times 360}{2\pi \times 300} \text{ degrees}$$

$$= 13°30'30''$$

and the corresponding deflection angle

$$= 13°30'30''/2$$
$$= 6°45'15''$$

$$\text{WCB}_{TA} = 63°13'32'' + 6°45'15''$$
$$= 69°58'47''$$

$$\text{Length } TA = 2 \times 300 \times \sin 6°45'15''$$
$$= 70.566 \text{ m}$$

$$E_A = 2014.257 + 70.566 \sin 69°58'47''$$
$$= 2080.559 \text{ m}$$

$$N_A = 1542.168 + 70.566 \cos 69°58'47''$$
$$= 1566.327 \text{ m}$$

Establish the theodolite at X and reference to Y

$$\text{WCB}_{XY} = \tan^{-1} \frac{2725.172 - 2134.091}{1696.142 - 1769.173}$$

$$= -82°57'23'' \text{ (on a pocket calculator)}$$

Now line XY lies in the SE quadrant and hence

$$\text{WCB}_{XY} = 180° - 82°57'23''$$
$$= 97°02'37''$$

This angle should be set on the theodolite when it is pointing at Y. Sighting onto A:

$$\text{WCB}_{XA} = \tan^{-1} \frac{2080.559 - 2134.091}{1566.327 - 1769.173}$$

$$= 14°47'1'' \text{ (on a pocket calculator)}$$

Now line XA lies in the SW quadrant and hence

$$\text{WCB}_{XA} = 180° + 14°47'1''$$
$$= \mathbf{194°47'1''} \text{ to be set on the theodolite.}$$

$$\text{length } XA = \sqrt{(2080.559 - 2134.091)^2 + (1566.327 - 1769.173)^2}$$
$$= \mathbf{209.791 \ m}$$

Possible
difficulties in
setting out simple
curves

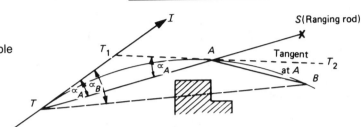

Fig. 11.10

(1) If entire curve
cannot be set out
from one tangent
point (Fig. 11.10)

Deflection angle from T to $A = \alpha_A$.
Deflection angle from T to B would be α_B.

(a) Set out as far as possible in normal manner, say to peg A, and
position a ranging rod at S on TA produced.
(b) Move instrument to A, set horizontal circle reading to zero, and
direct telescope on to ranging rod S, obtaining coincidence with
lower clamp and tangent screw (or, sight back on to tangent point
T and then transit the telescope to locate S).
(c) Free upper plate and set reading of α_B. The telescope is now
directed along AB and B is located in the normal manner, and
the remainder of the curve set out.

If T_1AT_2 is a tangent at A, the triangle TT_1A must be isosceles

and $\qquad\qquad T_1\hat{T}A = T_1\hat{A}T = \alpha_A$

But $\qquad\qquad T_1\hat{A}T = S\hat{A}T_2$

Also $\qquad\qquad T_2\hat{A}B = A\hat{T}B$

Therefore $\qquad\quad T_1\hat{T}B = S\hat{A}B = \alpha_B$

Thus, although the instrument has been moved to peg A, the same
total deflection angles apply to all points subsequent to A as if the
instrument were still set up at the tangent point, T.

Therefore, should another move be required in the general case where
the instrument is moved from peg N of total deflection angle δN to
peg X of total deflection angle δX,

(i) sight back to peg N with the horizontal circle reading set at δN,
(ii) transit and set off the next peg $(X+1)$ with total deflection angle
$\delta(X+1)$.

(2) Intersection
point inaccessible
(Fig. 11.11)

To locate T and U on the ground.

(a) Measure angles $T\hat{X}Y$ and $X\hat{Y}U$ by theodolite. Measure XY by tape,
XY being any pair of suitable points on the tangents.
(b) Deduce $I\hat{X}Y = 180 - T\hat{X}Y$, and $I\hat{Y}X = 180 - X\hat{Y}U$, and thence
evaluate $\theta = I\hat{X}Y + I\hat{Y}X$, and $X\hat{I}Y = 180 - \theta$.

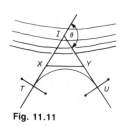

Fig. 11.11

(c) Evaluate $IX = XY(\sin I\hat{Y}X/\sin X\hat{I}Y)$, and $IY = XY(\sin I\hat{X}Y/\sin X\hat{I}Y)$.

(d) Knowing θ (from (b)) and the desired radius of the curve, determine $IT = R \tan \theta/2 = IU$.

(e) Determine $XT = IT - IX$, and $YU = IU - IY$, and measure these distances back from X and Y respectively to obtain the tangent points.

Transition curves

The forces acting on a vehicle change when it moves from the tangent on to the circular curve.

The centrifugal force P acting on the vehicle as it traverses the curve and the centrifugal ratio are given by the expressions $P = (WV^2)/(gR)$ and $P/W = (V^2)/(gR)$ respectively where W = weight of vehicle, V = velocity, R = radius of curve and g = acceleration due to gravity.

It will be seen from both formulae that the centrifugal force and ratio increase as R decreases, and thus, since the radius of the straight is infinity, the centrifugal force would increase instantaneously from zero to its maximum value (assuming no change in V), as the vehicle moved from straight to curve. Passengers in the vehicle would thus experience a lateral shock as the tangent was passed. To avoid this a curve of variable radius is inserted between the straight and the circular curve in order that the centrifugal force may build up in a gradual and uniform manner. This curve is called a *transition* curve.

The radius of a transition curve varies from infinity at its tangent point with the straight, to a minimum value at its tangent point with the circular curve. It will readily be seen that in any case where a single circular curve is to be replaced, two transition curves are required, joined together by the circular curve. In some cases the circular curve joining the two transitions is of zero length so that the single circular curve is replaced by two transition curves having one common tangent point.

Fig. 11.12

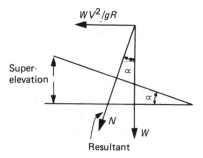

Super-elevation

A consideration of Fig. 11.12 will show that whereas a car or train going round a flat curve is subject to a side thrust equal to the centrifugal force, by lifting the outer edge of the road or rails, i.e. by applying cant or super-elevation to the curve, the resultant N can be made to lie along the normal to the road surface or rails for a given speed.

From the triangle of forces

$$\tan \alpha = \frac{WV^2}{gR} \bigg/ W = \frac{V^2}{gR}$$

The amount of cant or super-elevation is thus a constant for given values of V and R, and is given by

$$\text{Super-elevation} = G \sin \alpha = G \tan \alpha \text{ for small angles}$$

where $G =$ gauge of railway track or width of road.

The transition curve allows this super-elevation to be introduced in a gradual manner so that it varies from zero on the straight to its maximum value at the tangent point where the transition curve meets the circular curve.

Before proceeding with the design of transition curves, it is appropriate to say a word here about the actual amounts of super-elevation used in practice. The formula for $\tan \alpha$ contains three variables so that values of two of them will have to be used in the formula to enable the third term to be evaluated. Generally V is known, in railways being the maximum probable speed at which trains will use the track, and in roads the design average speed for the road, while the radius R is generally fixed by the land available. Thus $\tan \alpha$ and hence the cant can be worked out.

In *railways*, a maximum of, say, 150 mm lift on the outer rail is allowed, this figure being determined by considering the stability of lightly loaded vans moving slowly with high cross-wind. If the theoretical cant is more than this, then either

(*a*) a larger radius curve is used, or if this is impossible
(*b*) the value of 150 mm is used and velocity limited.

In *roads*, standards are fixed by the Department of Transport, having regard to the radius of the road and its design speed. Reference should be made to Department Standard TD 9/81 for the relevant data. Allowance may be made for friction between vehicle and road surface as follows. In Fig. 11.12 considering an equilibriant acting downwards along the road surface having a value of

$$\frac{WV^2}{gR} \cos \alpha - W \sin \alpha$$

and writing the coefficient of friction as μ

Then $\quad \mu = \dfrac{\dfrac{WV^2}{gR} \cos \alpha - W \sin \alpha}{\dfrac{WV^2}{gR} \sin \alpha + W \cos \alpha} = \dfrac{\dfrac{V^2}{gR} - \tan \alpha}{\dfrac{V^2}{gR} \tan \alpha + 1}$

Normally super-elevation can be determined from the expression

$$\text{Super-elevation} = 1 \text{ in } 314 \, \frac{R}{V^2}$$

where V is the design speed in km/h and R is the radius in metres. Super-elevation should not be steeper than 7 per cent (1 in 14.5, say) nor should it be flatter than 1 in 48 for effective drainage.

Derivation of the transition curve equations

Summarizing the points which have been made so far, we have

(a) P increases uniformly with distance l from the beginning of the transition curve, i.e. $P \propto l$.

(b) At the point where the radius of the transition curve $= r$

$$P = \frac{WV^2}{gr}$$

Hence $\qquad P \propto \dfrac{1}{r}$ for constant velocity

and so $\qquad l \propto \dfrac{1}{r}$ since $P \propto l$

(c) The amount of super-elevation provided increases uniformly with distance from beginning of curve and so is directly proportional to l.

Thus $lr = K$ (where K is some constant) and if the transition curve has a length L and the radius of the circular curve entered is R (see Fig. 11.13),

$$LR = K$$

or $\qquad lr = LR = K$

Fig. 11.13

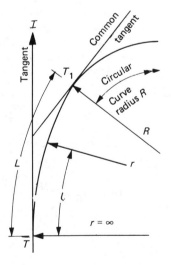

TT_1 is the transition curve along which the radius varies from infinity at T to R at T_1. The tangent at T_1 is common to the circular curve and transition curve.

Length of transition curve (Fig. 11.13)

The length of transition curve may be taken

(a) as an arbitrary value from past experience, say 50 m,
(b) such that the super-elevation is applied at a uniform rate, say 0.1 m in 100 m,
(c) such that the rate of change of radial acceleration equals a certain chosen value.

In this last case let L = total length of transition curve, R = radius of circular curve entered and V = uniform velocity of vehicle.

The radial acceleration is zero just as the vehicle is about to leave the tangent straight at T, whilst at the circular curve tangent point it is $V^2/R = f$.

Therefore, if time taken to travel along transition curve be t seconds,

$$t = \frac{L}{V}$$

Rate of change of radial acceleration $= \dfrac{f}{t} = \alpha$

then

$$\alpha = \frac{V^2}{R}\bigg/\frac{L}{V} = V^3/LR$$

and

$$L = V^3/R\alpha$$

Note: A rate of change of radial acceleration of $\frac{1}{3}$ m/s^3 is a 'comfort limit' above which side-throw will be noticed.

Consider points X and Y which are near together on the transition curve (Fig. 11.14), X being l from T,

$$dl = rd\phi = \frac{K}{l}\,d\phi$$

Therefore

$$d\phi = \frac{l}{K}\,dl$$

On integrating

$$\phi = \frac{l^2}{2K} + A$$

A is a constant of integration and is zero since $\phi = 0$ when $l = 0$.

Then

$$\phi = \frac{l^2}{2K} = \frac{l^2}{2LR} = \frac{l^2}{2lr}$$

The expression $\phi = l^2/2LR$ is the intrinsic equation of the *ideal transition spiral*.

Now we can write

$$dy = dl \cos\left(\phi + \frac{d\phi}{2}\right) \quad \text{and} \quad dx = dl \sin\left(\phi + \frac{d\phi}{2}\right)$$

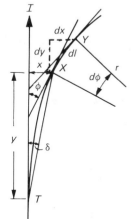

Fig. 11.14

Since $d\phi$ is small we can write $dy = dl \cos \phi$ and $dx = dl \sin \phi$

Then

$$dy = \left(1 - \frac{\phi^2}{2!} + \frac{\phi^4}{4!} - \ldots\right) dl$$

$$= \left\{1 - \left(\frac{l^2}{2K}\right)^2 \frac{1}{2!} + \left(\frac{l^2}{2K}\right)^4 \frac{1}{4!} - \ldots\right\} dl$$

$$y = l - \frac{l^5}{40K^2} + \frac{l^9}{3456K^4}$$

$$= l \text{ when we say that } K \text{ is large}$$

and

$$dx = \left(\phi - \frac{\phi^3}{3!} + \frac{\phi^5}{5!} - \ldots\right) dl$$

$$= \left\{\frac{l^2}{2K} - \left(\frac{l^2}{2K}\right)^3 \frac{1}{6} + \left(\frac{l^2}{2K}\right)^5 \frac{1}{120} - \ldots\right\} dl$$

Hence

$$x = \frac{l^3}{6K} - \frac{l^7}{336K^3} + \frac{l^{11}}{42\,240K^5}$$

$$= \frac{l^3}{6K} \text{ nearly}$$

In neither integration will there be a constant of integration, since

$$\phi = 0, \text{ when } 1 = 0$$

Thus
$$x = \frac{l^3}{6K} = \frac{l^3}{6LR}$$

which is the equation for the *cubic spiral*.

Writing $y = l$, then $x = \dfrac{y^3}{6LR}$, which is the *cubic parabola*.

Since $\phi = \dfrac{l^2}{2LR}$ we can also write

$$y = l\left(1 - \frac{\phi^2}{10} + \frac{\phi^4}{216} \ldots\right)$$

$$x = \frac{l^3}{6K}\left(1 - \frac{\phi^2}{14} + \frac{\phi^4}{440}\right)$$

Relation between δ
and ϕ

$$\tan \delta = \frac{x}{y} = \frac{l^2}{6K} \frac{\left(1 - \dfrac{\phi^2}{14} + \dfrac{\phi^3}{440}\right)}{\left(1 - \dfrac{\phi^2}{10} + \dfrac{\phi^4}{216}\right)}$$

$$= \frac{\phi}{3} \left(1 + \frac{\phi^2}{35}\right) \text{ say}$$

Therefore for small angles we can take $\delta = \phi/3$ radians and this expression is important in the setting out of the curve.

The maximum value of ϕ, which is denoted by ϕ_1, is equal to $L^2/2LR = L/2R$. It is the angle between the common tangent at T_1 and the tangent TI (see Fig. 11.15(b)).

Shift

Where transition curves are introduced between the tangents and a circular curve of radius R (see Fig. 11.15(a)), the circular curve is 'shifted' inwards from its original position by an amount $BP = S$ (the shift) so that the curves can meet tangentially. This is equivalent to having a circular curve of radius $(R+S)$ connecting the tangents replaced by two transition curves and a circular curve of radius R, although the tangent points are not the same, being B and T respectively in Fig. 11.15(a).

Referring to Fig. 11.15(b), TBN is tangential to the circular curve of radius $(R+S)$ and RT_1 is the common tangent at T_1

$$N\hat{B}O = R\hat{T}_1O$$
$$= 90°$$

i.e. R, B, O and T_1 are concyclic.

Fig. 11.15

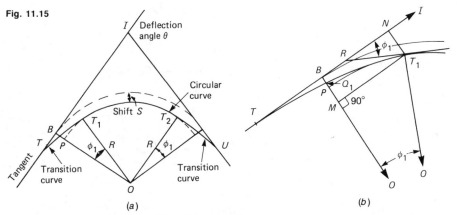

(a)

(b)

Therefore ϕ_1 is subtended at O by arc PT_1. Draw MT_1 perpendicular to OB as shown.

$$BM = NT_1$$
$$= \text{maximum offset on transition curve}$$

also shift $S = BP$
$$= BM - PM$$
$$= NT_1 - (PO - MO)$$
$$= \frac{L^3}{6LR} - (R - R\cos\phi_1)$$
$$= \frac{L^3}{6LR} - \left\{R - R\left(1 - \frac{\phi_1^2}{2!} + \frac{\phi_1^4}{4!} - \dots\right)\right\}$$

Ignoring higher powers than the second

$$S = \frac{L^3}{6LR} - \frac{R\phi_1{}^2}{2}$$

$$= \frac{L^3}{6LR} - \frac{R}{2} \left(\frac{L^2}{2LR}\right)^2 = \frac{L^2 \cdot}{6R} - \frac{L^2}{8R}$$

$$= \frac{L^2}{24R}$$

also
$$Q_1 T_1 \simeq PT_1$$
$$\simeq R \cdot \phi_1$$
$$= \frac{RL^2}{2LR}$$
$$= \frac{L}{2}$$

Therefore Q_1 is the mid-point of the transition curve and since the deviation of this from the tangent is small, then

$$TQ_1 \simeq TB \left(= \frac{L}{2} \right)$$

Setting out transition curves from the tangent point

To locate the tangent point T:

(1) Calculate the shift S from the expression

$$S = \frac{L^2}{24R}$$

(2) Calculate
$$IB = (R+S) \tan \frac{\theta}{2}$$

(3) Since
$$BT = \frac{L}{2}$$

Then
$$IT = (R+S) \tan \frac{\phi}{2} + \frac{L}{2}$$

Measure this length back from I and so fix point T.

The next step depends on whether it is intended to set out the transition with tapes using the cubic spiral or cubic parabola, or by the theodolite using the cubic spiral. Often both methods are employed on one job, the tape being used for preliminary work and the theodolite for final location. Usually the chords adopted are about half the length of those used for setting out the circular curve.

(4) *Either* calculate offsets from

$$x = \frac{l^3}{6LR} \quad \text{or} \quad x = \frac{y^3}{6LR}$$

Each peg is located by swinging a chord length from the preceding peg. (*Note*: l is the progressive chainage from T).

Or calculate the deflection angles δ for particular distances l from T using the fact that $\delta = \phi/3$ (*see* Fig. 11.14), and so since

$$\phi = \frac{l^2}{2RL}$$

then

$$\delta = \frac{l^2}{6RL} \text{ radians}$$

$$= \frac{180}{\pi} \frac{l^2}{6RL} \text{ degrees}$$

$$= \frac{1800}{\pi} \frac{l^2}{RL} \text{ min}$$

The final deflection angle which locates T_1 is $\dfrac{\phi_1}{3}$ when $l = L$.

Since

$$\phi_1 = \frac{L^2}{2RL} \text{ radians}$$

then

$$\delta_{max} = L/6R \text{ radians}$$

$$= \frac{1800L}{\pi R} \text{ min}$$

Setting out the circular curve (5) Calculate the deflection angles for the circular curve from

$$\delta_{circular\ arc} = 1718.9 \times \frac{c}{R} \text{ min}$$

where c is the chord length for the circular arc.

The angle subtended by the circular arc $T_1 T_2$ at the centre of that arc is $(\theta - 2\phi_1)$ as shown in Fig. 11.15(a).

Hence

$$\text{length of arc } T_1 T_2 = R(\theta - 2\phi_1) \frac{\pi}{180}$$

(6) Set up the theodolite at T_1, and sight back on T; then, transit the telescope and locate tangent $T_1 T_3$ by setting off an angle $\frac{2}{3}\phi_1$ (Fig. 11.16). Set out the circular curve by the deflection angle method, from this tangent.

Fig. 11.16

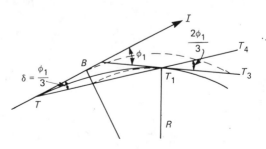

Alternatively when T_1 has been located as in (4) set a ranging rod at T_4 on TT_1 produced and sight T_4 after setting up at T_1.

Setting out the other transition   (7) Point U is located from I using the relationship $IU = IT$. The transition is then set out from tangent point U and tangent UI by either of the methods given in (4).

Example 11.5   Three straights AB, BC and CD have whole circle bearings of 30°, 90° and 45° respectively. AB is to be connected to CD by a continuous reverse curve formed of two circular curves of equal radius together with four transition curves. BC, which has a length of 800 m, is to be the common tangent to the two inner transition curves. Determine the radius of the circular curves if the maximum speed is to be restricted to 80 km/h and a rate of change of radial acceleration of 0.3 m/s^3 obtains. Give (i) the offset, and (ii) the deflection angle, with respect to BC, to locate the intersection of the third transition curve with its circular curve (Fig. 11.17).

Fig. 11.17

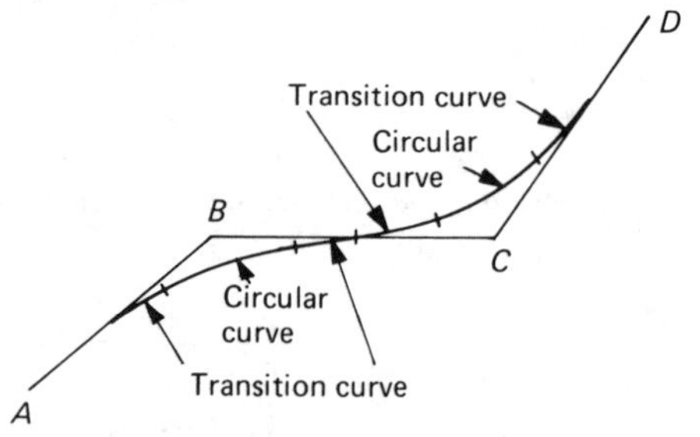

$$\alpha = 0.3 = \frac{V^3}{LR}$$

$$V = \frac{80 \times 1000}{3600} = \frac{1000}{45} \text{ m/s}$$

$$L = \frac{1000 \times 1000 \times 1000}{45 \times 45 \times 13.5R}$$

$$= \frac{36\ 580 \text{ m}}{R}$$

$$\text{Shift} = \frac{L^2}{24R}$$

$$= \frac{55\ 760\ 000}{R^3}$$

Since angles of 60° and 45° obtain for the pairs of straights AB, BC and BC, CD respectively

$$800 = \left(R + \frac{55\ 760\ 000}{R^3}\right) \tan 30° + \frac{36\ 580}{R}$$

$$+ \left(R + \frac{55\ 760\ 000}{R^3}\right) \tan 22.5°$$

Whence $R^4 - 806.7R^3 + 36\ 580R^2 + 55\ 760\ 000 = 0$

Therefore $\quad\quad\quad\quad R = \textbf{758.5 m}$ to the nearest 0.5 m

and $\quad\quad\quad\quad L = \dfrac{36\ 580}{R} = 48.23$ m

Fig. 11.18

Offset to locate circular curve (*see* Fig. 11.18)

$$= \frac{L^3}{6LR} = \frac{48.23^3}{6 \times 36\ 580}$$

$$= \textbf{0.51 m}$$

$$Deflection\ angle = \frac{1800L}{\pi R} = \frac{1800 \times 48.23}{\pi \times 758.5}$$

$$= 36.42'$$

$$= \textbf{00°36'20''}$$

i.e. 359°23'40" on instrument graduated clockwise.

Example 11.6 *Cubic spiral transition curves and circular curve* (Fig. 11.19). It is required to join two straights having a total deflection angle 18°36' right by a circular curve of 450 m radius, having cubic spiral transition curves at each end. The design velocity is 70 km/h and the rate of change of radial acceleration along the transition curve is not to exceed 0.3 m/s^3. Chainage of I is 2524.20 m.

Fig. 11.19

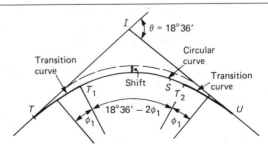

(i) *Determine length of required transition curve*

Design velocity = 70 km/h

$$= \frac{700}{36} \text{ m/s}$$

α = Rate of change of radial acceleration = 0.3 m/s^3

$$\alpha L = \frac{V^3}{R}$$

Therefore

$$L = \frac{\left(\dfrac{700}{36}\right)^3}{450 \times 0.3}$$

$$= 54.44 \text{ m}$$

(ii) *Calculate shift*

$$\text{Shift} = \frac{L^2}{24R} = \frac{54.44^2}{24 \times 450}$$

$$= 0.27 \text{ m}$$

(iii) *Tangent length*

$$IT = (R+S) \tan \frac{\theta}{2} + \frac{L}{2}$$

$$= 450.27 \tan 9°18' + \frac{54.44}{2} \text{ m}$$

$$= 100.97 \text{ m}$$

Hence   chainage of T = 2524.20 − 100.97
$$= 2423.23 \text{ m}$$

and   chainage of T_1 = 2423.23 + 54.44
$$= 2477.67 \text{ m}$$

(iv) *Cubic spiral transition curve with 10 m chord lengths.*
Deflection angle to locate a point on the transition distant l metres
from T

$$= \frac{1800}{\pi RL} \cdot l^2 \text{ min}$$

$$= 0.02339 \, l^2 \text{ min}$$

Chord (m)	l (m)	Chainage	Deflection angle	Angle set on 20″ instrument
0	0	2423.23	0	0
6.77	6.77	2430.00	0.02339 × 6.77^2 = 01.07′	00°01′00″
10.00	16.77	2440.00	× 16.77^2 = 06.58′	00°06′40″
10.00	26.77	2450.00	× 26.77^2 = 16.76′	00°16′40″
10.00	36.77	2460.00	× 36.77^2 = 31.62′	00°31′40″
10.00	46.77	2470.00	× 46.77^2 = 51.17′	00°51′20″
7.67	54.44	2477.67	× 54.44^2 = 69.31′	01°09′20″

(v) *Circular arc* $T_1 T_2$

$$\phi_1 = 3\delta_1$$
$$= 3 \times 69.31'$$
$$= 3°27.93'$$

Angle subtended by the circular arc

$$= 18°36' - 2\phi_1$$
$$= 11.669°$$

Length of circular arc

$$= \frac{\pi R}{180} (\theta - 2\phi_1)$$

$$= \frac{450}{180} \times \pi \times 11.669$$

$$= 91.65 \text{ m}$$

Hence chainage at end of circular curve

$$= 2477.67 + 91.65$$
$$= 2569.32 \text{ m}$$

Chord (m)	Chainage (m)	Deflection angle	Total deflection angle from tangent at T_1	Angle set on 20″ instrument
0	2477.67	0	0	0
22.33	2500.00	$1718.9 \times \dfrac{22.33}{450}$ $= 85.29'$	01°25.29′	01°25′20″
20.00	2520.00	$1718.9 \times \dfrac{20}{450}$ $= 76.40'$	02°41.69′	02°41′40″
20.00	2540.00	$1718.9 \times \dfrac{20}{450}$ $= 76.40'$	03°58.09′	03°58′00″
20.00	2560.00	$1718.9 \times \dfrac{20}{450}$ $= 76.40'$	05°14.49′	05°14′20″
9.32	2569.32	$1718.9 \times \dfrac{9.32}{450}$ $= 35.60'$	05°50.09′	05°50′00″

Chord length for circular arc is to be less than $R/20$, i.e. say 20 m normally.

A length of 2.33 m could be adopted for the first, followed by one of 20 m length, providing that the shortest focusing distance of a available instrument allows this. Alternatively a combination of say 12.83 m and 10.00 m could have been selected.

(vi) *The second transition curve.*

Therefore chainage of tangent point U

$$= 2569.32 + L$$
$$= 2623.76 \text{ m}$$

The final portion of the work is to set out the second transition curve from U. For running chainages, pegs will be placed as shown in the following table.

(*Note*: When locating this transition curve it is a *left*-hand curve from UI and if the theodolite reads clockwise and the telescope is swung anti-clockwise the angles will be as given.)

Chainage (m)	l metres from U	Chord (m)	Deflection angle $0.02339l^2$	Angle set on 20° instrument	Offsets from UI $x = \dfrac{l^3}{6LR}$ (m)
2569.32	54.44	0.68	$0.02339 \times 54.44^2 = 69.31'$	358°50′40″	1.10
2570.00	53.76	10.00	$0.02339 \times 53.76^2 = 67.60'$	358°52′20″	1.06
2580.00	43.76	10.00	$0.02339 \times 43.76^2 = 44.79'$	359°15′20″	0.57
2590.00	33.76	10.00	$0.02339 \times 33.76^2 = 26.66'$	359°33′20″	0.26
2600.00	23.76	10.00	$0.02339 \times 23.76^2 = 13.20'$	359°46′40″	0.09
2610.00	13.76	13.76	$0.02339 \times 13.76^2 = 4.43'$	359°55′40″	0.02
(U) 2623.76	0	0	0	360°00′00″	0

As mentioned earlier, the curve can consist of two similar transition curves TT_1 and T_1U (Fig. 11.20) only, T_1T_2 in Fig. 11.14 having zero length. The reader will now appreciate that $\phi_1 = \theta/2 = L/2R$ rad from the symmetry of the arrangement. Use should be made of the expressions

$$X = \frac{L^3}{6K}\left(1 - \frac{\phi_1^2}{14} + \frac{\phi_1^4}{440}\right)$$

and

$$Y = L\left(1 - \frac{\phi_1^2}{10} + \frac{\phi_1^4}{216}\right)$$

which relate offset X at distance Y from T when locating T_1. The terms involving ϕ_1^4 might be neglected as small, but those involving

Fig. 11.20

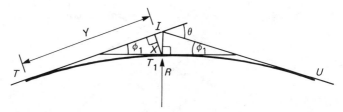

$\phi_1{}^2$ are usually of some importance in this situation: note also that $IT = Y + X \tan \phi_1$. The curves can be set out by deflection angles δ as explained previously.

Setting out transition curves from remote control

The logic of setting out a transition curve from remote control is the same as the process described earlier for circular curves. Knowing the co-ordinates of the tangent point to the straight and the intersection point the whole circle bearing of the straight may be calculated, WCB_{TI}. The length of the chord from the tangent to any point on the curve is given directly by $\sqrt{(x^2 + y^2)}$ and the deflection angle is given by $\tan^{-1}(x/y)$. For accuracy at least second order terms should be used in the calculation of x and y.

Thus
$$x = \frac{l^3}{6LR} - \frac{l^7}{336(LR)^3}$$

$$y = l - \frac{l^5}{40(LR)^2}$$

To a point B on the curve

$$\text{WCB}_{XB} = \frac{(E_I - E_T)}{(N_I - N_T)} + \tan^{-1} \frac{x}{y}$$

$$l_{XB} = \sqrt{(x^2 + y^2)}$$

$$E_B = E_T + l_{XB} \sin \text{WCB}_{XB}$$
$$N_B = N_T + l_{XB} \cos \text{WCB}_{XB}$$

Having established the co-ordinates for points on the curve the data is calculated and set out from the remote control stations as described earlier for circular curves.

Vertical curves

Curves will be required at the intersection of gradients, such curves being known as *vertical* curves. Two cases occur, gradients meeting at *summits*, and gradients meeting at *sags*, as shown in Fig. 11.21.

The gradients themselves are conveniently expressed as percentages, thus a 1 in 20 gradient is a 5 per cent gradient and, depending on its direction, is positive or negative. The following convention is used in this section:

Gradients rising to the right — positive
Gradients falling to the right — negative
Left-hand gradient — p per cent,

Fig. 11.21

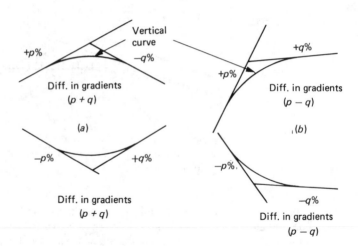

Diff. in gradients $(p + q)$

(a)

Diff. in gradients $(p + q)$

Diff. in gradients $(p - q)$

(b)

Diff. in gradients $(p - q)$

i.e. 1 in $100/p$ gradient, being inclined at $p/100$ rad to the horizontal

Right-hand gradient — q per cent,

i.e. 1 in $100/q$ gradient, being inclined at $q/100$ rad to the horizontal

Algebraic difference of gradients — $(p-q)$ per cent

It will be seen that every 100 m forward, the level along the gradient changes by p or q m.

Shape of curve

Where the ratio of length of curve to radius is less than 1 to 10, there is no practical difference between the shapes of a circle, a parabola and an ellipse, and since this condition can be shown to apply in the cases normally met with, the parabola will be used.

Length of curve

Factors affecting the length of a vertical curve are: (a) centrifugal effect, (b) visibility.

At sags, and at summits formed by flat gradients, centrifugal effect is the chief factor, but at summits where the algebraic change of gradient is large, visibility is the ruling factor.

Curve design

In the following analysis the conditions shown in Fig. 11.22 and 11.23 are used, but in any particular case from Fig. 11.21 the appropriate signs would have to be applied to p and q.

For flat gradients it is sufficiently accurate to treat the length along the tangents, the length along the curve, and chord AC as equal to length $2l$ so as to be in accordance with the survey information.

The properties of the parabola give:

(a) The vertical through the intersection of the tangents B is a diameter and bisects AC.

(b) $BD = DE$. D is the vertex of the parabola.

Fig. 11.22

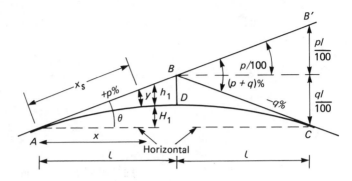

Fig. 11.23

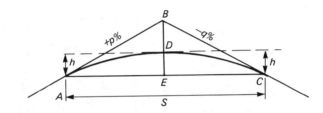

(c) Offsets from the tangent AB are proportional to the square (Fig. 11.22) of the distance from A. (These offsets should be at right angles to the tangent, but in the case of the flat gradients usually involved it is sufficiently accurate to take them vertically.)

(d) Offsets from the tangent at D are proportional to the square of the distance from D.

In Fig. 11.22

$$y = Cx_s^2$$
$$= c(x \sec \theta)^2$$

Therefore

$$y = Kx^2 \text{ where } K = c^2 \sec^2 \theta$$

$$\frac{dy}{dx} = 2Kx$$

$$\frac{d^2y}{dx^2} = 2K = \text{constant}$$

i.e. the parabola gives an even rate of change of gradient.

When
$$x = 2l$$

$$y = B'C = \frac{pl}{100} + \frac{ql}{100}$$

Hence
$$(p+q)\frac{l}{100} = K4l^2$$

$$K = \frac{p+q}{400l}$$

$$y = \frac{p+q}{400l} x^2$$

when $x = l$
$$y = BD = \frac{p+q}{400l} l^2$$

$$= (p+q) \frac{l}{400}$$

Height of B above $A = \dfrac{pl}{100}$ m

Height of C above $A = \dfrac{pl}{100} - \dfrac{ql}{100}$ m

Then the angle between AC and the horizontal (Fig. 11.23)

$$= \frac{l}{100} \frac{p-q}{2l} = \frac{p-q}{200} \text{ radians}$$

Thus
$$BE = \frac{pl}{100} - \frac{p-q}{200} \cdot l$$

$$= \frac{p+q}{200} l = 2BD$$

Therefore
$$BD = DE$$

As mentioned previously, the appropriate signs have to be given to p and q in a particular case since the offset $B'C$ at $x = 2l$ depends upon the angle $B'BC$. When the gradients are as in Fig. 11.21(b), the reader will appreciate that $y = [(p-q)/(400l)]x^2$ since the offset at $x = 2l$ will be $[(pl)/(100) - (ql)/(100)]$.

In setting out the curve it is necessary to compute the offsets at various chainages, which are naturally referred to the horizontal, and apply these to the known levels on the gradients; thus the final levels for the vertical curve may be determined. The procedure is shown in the examples given and, as mentioned previously for flat gradients, it is accurate enough to treat the length along the tangents as equal to any of the length of the curve, chord AC, and the horizontal projection of AC.

Highest point

Taking A as the datum point, the highest point of the curve will be at a height of $[(xp/100) - h_1]$ above A, where h_1 is the offset from the tangent AB at this point, whose distance from A is x.

Now
$$H_1 = \frac{xp}{100} - \frac{(p+q)}{400l} x^2$$

and for maximum value of H_1

$$\frac{dH_1}{dx} = 0$$

i.e. $$\frac{p}{100} - \frac{2(p+q)}{400l} \, x = 0, \text{ or } \frac{p}{100} - 2Kx = 0,$$

Hence $$x = \left(\frac{2p}{p+q}\right) l$$

Centrifugal effect

It has already been pointed out that the parabolas in vertical curves can be approximated to circular curves.

Let R = radius

$$\frac{d^2y}{dx^2} = 2K = \frac{1}{R}$$

Therefore $$\frac{1}{R} = 2 \cdot \frac{p+q}{400l}$$

and $$R = \frac{200l}{p+q} = \frac{v^2}{f}$$

where f = allowable centrifugal acceleration for velocity v.

Therefore length of curve

$$= (p+q) \, \frac{v^2}{100f}$$

$$= K' \, (p+q)$$

At summits on roads where speeds of 100 km/h are contemplated, the requirements of visibility, i.e. the sight line, can lead to longer curves than are given by the formula above.

Sight distances

Let two points on the curve at a height h from the ground be intervisible, and let the distance between them be S. The sight line is taken to pass tangentially through D parallel to AC (Fig. 11.23) and the sight distance thus represents the length of road over which an observer whose eye level is h above the road surface can just see an object h above the surface on the other side of the crest to the observer. A value of 1.05 m is usually taken as the eye level height above the road surface for an observer sitting in a motor car. Sight distances are laid down in the interests of road safety and the choice of any particular distance depends on the nature of the road and the speed of the traffic using it. K' values are listed in Department of Transport Standard TD 9/81 for different road conditions.

There are three cases to consider:

(a) Sight distance equal to length of curve. Then in Fig. 11.23,

$$S = 2l, \text{ and } DE = h = \left(\frac{p+q}{400}\right) l$$

Given h, p and q, then l may be determined and the offsets for the vertical curve computed from

Fig. 11.24

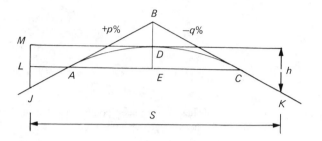

$$y = \left(\frac{p+q}{400l}\right) x^2$$

(b) Sight distance longer than curve AC (Fig. 11.24)

$$h = JL + ML$$
$$= JL + DE$$
$$= JL + BD$$

The angle which the tangent at D makes with the horizontal

$$= \frac{p}{100} - 2Kl$$

$$= \frac{p-q}{200} \text{ rad.}$$

Therefore AC and MD are parallel.

Thus $\qquad L\hat{A}J = \dfrac{p}{100} - \dfrac{p-q}{200}$

$$= \frac{p+q}{200} \text{ radians}$$

But $\qquad AL = \dfrac{S}{2} - l$

Hence $\qquad JL = \left(\dfrac{p+q}{200}\right)\left(\dfrac{S}{2} - l\right)$

Now $\qquad h = JL + BD$

$$= \left(\frac{p+q}{200}\right)\left(\frac{S}{2} - l\right) + \frac{l}{400}(p+q)$$

$$= \left(\frac{S-l}{400}\right)(p+q)$$

A relationship can be derived between h and K' for this case as follows. From above,

$$S = l + \frac{400\,h}{p+q}$$

and for $S > 2l$

$$\frac{400\,h}{p+q} > l$$

i.e.

$$\frac{400\,h}{p+q} > \frac{K'}{2}\,(p+q)$$

or $800\,h > K'\,(p+q)^2$

Example 11.7 If $S = 4l$, $p = 1$ per cent, $q = -1$ per cent, and $h = 1.05$ m, calculate l and the equation for offsets to the tangent.

---

then

$$1.05 = \left(\frac{4l-l}{400}\right)(1+1)$$

$$6l = 1.05 \times 400$$
$$l = \textbf{70 m}$$

Therefore equation for offsets from tangent measuring from A is

$$y = \left(\frac{p+q}{400l}\right)x^2$$

$$= \frac{(1+1)}{400 \cdot 70}\cdot x^2$$

$$= \textbf{0.0000714}x^2$$

(c) Sight distance less than length of curve (Fig. 11.25). The equation for offsets from the tangent FDG with the origin at D is

$$y_1 = Kx^2$$

Hence

$$h = K\left(\frac{S}{2}\right)^2 \quad \text{since } y_1 = h \text{ when } x = \frac{S}{2}$$

and

$$K = \frac{4h}{S^2}$$

At point J, $\quad x_1 = l$

so that

$$JA = \frac{4h}{S^2}\cdot l^2$$

But

$$JA = DE = BD = \frac{l}{400}\,(p+q)$$

Fig. 11.25

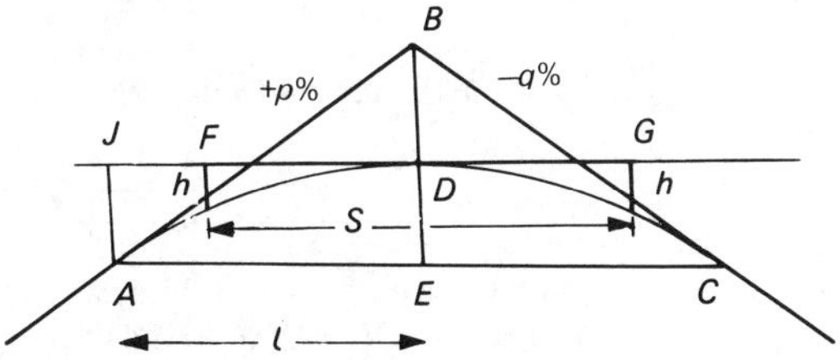

hence $\qquad \dfrac{4h}{s^2} \cdot l^2 = \dfrac{l}{400} \cdot (p+q)$

and $\qquad\qquad l = \dfrac{(p+q)}{1600} \cdot \dfrac{s^2}{h}$

Example 11.8 If the sight distance equals half the total length of the curve, $p = 2\%$ and $q = -2\%$ and the observer's eye level, $h = 1.05$ m, calculate the length of the curve.

$$s = l$$

then $\qquad\qquad l = \dfrac{4}{1600} \cdot \dfrac{l^2}{1.05}$

and $\qquad\qquad l = \mathbf{420\ m}$

Example 11.9 Two gradients of 1 in 50 and 1 in 75 meet at a summit (RL 30.35 m, chainage 2758 m). Design a vertical curve having $K' = 50$, such that two points 1.05 m above road level are intervisible. Determine the sighting distance between those points and give offsets at 20 m intervals. What is the rate of change of gradient?

Since $\qquad\qquad K' = 50$
Then $\qquad\qquad 2l = 50\,(p+q)$
$\qquad\qquad\qquad = 50\,(2+1.333)$
$\qquad\qquad\qquad = 166.67$ m
$\qquad\qquad\quad l = 83.33$ m (Fig. 11.26)

Fig. 11.26

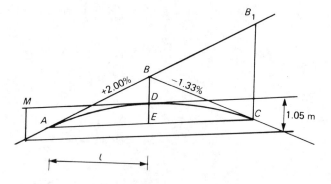

Now for the offsets from ABB_1

$$y = \left[\dfrac{p+q}{400\ l} \right] x^2 = 0.00010\ x^2$$

and writing $x = l = 83.33$ m
we get $y = BD = DE = 0.649$ m

Since we require intervisibility over 1.05 m the sighting distance $s > 2l$

Now
$$\frac{S-l}{400}(p+q) = h$$

$$\frac{S-83.33}{400} \times \frac{10}{3} = 1.05$$

$$S = \mathbf{209.33\ m}$$

The gradient levels are determined along ABB_1 from the formula,

$$\text{Grade level} = \text{level of } A + \frac{xp}{100}$$

and, as above, offsets from the tangent are given by the expression $y = 0.00010\ x^2$.

Setting out table

Point	A					B
Chainage	2674.67	2680.00	2700.00	2720.00	2740.00	2758.00
x	0	5.33	25.33	45.33	65.33	83.33
Grade level	28.68	28.79	29.19	29.59	29.99	30.35
Offset y	0.00	0.00	0.06	0.20	0.43	0.69
Curve level	28.68	28.79	29.13	29.39	29.56	29.66

Point	B					C
Chainage	2760.00	2780.00	2800.00	2820.00	2840.00	2841.33
x	85.33	105.33	125.33	145.33	165.33	166.69
Grade level	30.39	30.79	31.19	31.59	31.99	32.02
Offset y	0.73	1.11	1.57	2.11	2.73	2.77
Curve level	29.66	29.68	29.62	29.48	29.26	29.25

It should be noted that offsets could have been taken from tangent CB instead of AB with an appropriate adjustment to values in the formula.

Total change in gradient over 166.67 m = 3.33%

$$\text{Rate of change of gradient} = \frac{3.33}{166.67} = 0.02\%\ \text{per m}$$

Fig. 11.27

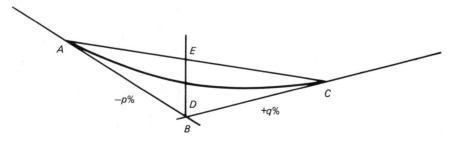

Example 11.10 Design a vertical sag curve 200 m long connecting a falling gradient of 1 in 50 with a rising gradient of 1 in 75. The RL of the intersection point of the gradients is 30.35 m AOD and the chainage is 2750 m (*see* Fig. 11.27).

$$AB = BC = l = 100 \text{ m}$$

$$A \text{ is now } 100 \times \frac{2}{100} = 2.00 \text{ m above } B$$

$$C \text{ is now } 100 \times \frac{1.33}{100} = 1.33 \text{ m above } B$$

Setting out table

Point	*A*						*B*
Chainage	2650	2660	2680	2700	2720	2740	2750
x (from A)	0	10	30	50	70	90	100
Grade level (level of A $-xp/100$)	32.35	32.15	31.75	31.35	30.95	30.55	30.35
Offset (m) $y = x^2/l^2\ BD$	0	0.01	0.07	0.21	0.41	0.67	0.83
Curve level	32.35	32.16	31.82	31.56	31.36	31.22	31.18

Point	*B*						*C*
Chainage	2750	2760	2780	2800	2820	2840	2850
x (from C)	100	90	70	50	30	10	0
Grade level (level of C $-xq/100$)	30.35	30.48	30.74	31.01	31.28	31.55	31.68
Offset (m) $y = x^2/l^2\ BD$	0.83	0.67	0.41	0.21	0.07	0.01	0
Curve level	31.18	31.15	31.15	31.22	31.35	31.56	31.68

Now
$$y = \frac{p+q}{400l} \cdot x^2$$

On substituting for p, q and l

$$y = \frac{1}{12\,000} x^2$$

Hence
$$BD = \frac{100^2}{12\,000}$$
$$= 0.83 \text{ m}$$

Therefore, formula for offsets from either tangent AB or CB is

$$y = 0.83 \left(\frac{x}{100}\right)^2$$

The lowest point on the curve is determined exactly as the highest point on the curve in the previous example. The rate of change of grade is the same as in the previous case.

(*Note*: Offsets are added to grade levels since the curve is above the gradients. Offset 50 m from A = offset 50 m from C, and so on.)

The second portion of the curve may be set off by offsets from AB and it is recommended that the reader completes the table, locates the lowest point, determines the rate of change of grade, and checks that the K' value for this curve is 60.

Example 11.11 An upgrade of 1 in 10 rising to the right is to be connected by a parabolic vertical curve to a second upgrade of 1 in 20 also rising to the right at a summit, where the reduced level and chainage are 105.76 m AOD and 1234.00 m respectively.

Determine the required length of curve (to the nearest ten metres) if two points each 1.05 m above the curve are to be intervisible over a distance of 200 m. Also locate that point on the curve whose reduced level is 103.00 m AOD and give the rate of change of grade for the curve (Fig. 11.28).

(*Salford*)

Fig. 11.28

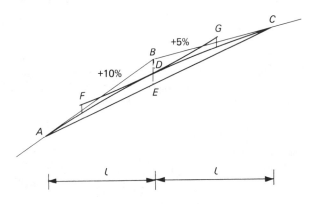

The reader is recommended to check that the sighting distance is shorter than the curve length.

Now
$$l = \frac{(p-q)S^2}{1600\,h}$$

(since p and q have the same sense)

$$= \frac{(10-5)}{1600} \times \frac{200^2}{1.05}$$

$$= 119 \text{ m}$$

Hence length of curve will be 238 m, say 240 m

$$DE = BD = \frac{5}{400} \times 120$$

$$= 1.50 \text{ m}$$

The setting-out data for the vertical curves designed previously have been given assuming the reduced level of the controlling feature is known to 0.01 m. Differences when offsets are calculated to 0.001 m and incorporated with grade levels taken to the same limits are shown below for this particular example.

Chainage	1114.00	1125.00	1150.00	1175.00	1200.00	1225.00	1234.00
Grade level	93.760	94.860	97.360	99.860	102.360	104.860	105.760
Offset	0	0.013	0.135	0.388	0.770	1.283	1.500
Curve level	93.760	94.847	97.225	99.472	101.590	103.577	104.260

If working to 0.01 m the following values would have been derived:

Curve level	93.76	94.85	97.23	99.47	101.59	103.58	104.26

A good quality level set up between A and B would allow the setting-out of these values expressed to 0.001 m, but for highway setting-out those to 0.01 m may be considered reasonable.

The second portion of the curve may be set off in the usual way and again the reader is recommended to complete the table when he should locate the required point at chainage 1218 m, say.

Graphical representation of vertical curve formulae

The relationship between sight distance, curve length and algebraic difference of gradients for slopes meeting at a summit is shown graphically in Fig. 11.29. The student will readily appreciate, from a comparison of the graph and the formulae, that where the curve length

Fig. 11.29 Relationship between sight distance, curve lengths and algebraic difference of gradients

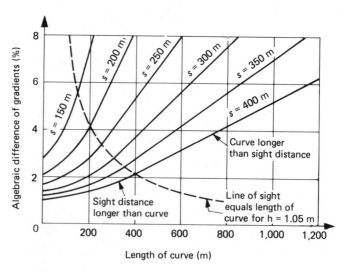

is longer than the sight distance, the relation between algrebraic difference of gradients and curve length is a linear one, so that the graph is easily extended to cover steeper slopes.

1 To locate the exact position of the tangent point T_2 of an existing 250 m radius circular curve in a built-up area, points a and d were selected on the straights close to the estimated positions of the two tangent points T_1 and T_2 respectively and a traverse $abcd$ was run between them (*see* below).

Station	Length (m)	Deflection angle
a		$9°54'R$
	89.0	
b		$19°36'R$
	115.5	
c		$30°12'R$
	101.5	
d		$5°18'R$

The angles at a and d were relative to the straights. Find the distance T_2d. (*London*)
Answer: 16.98 m.

2 Two straights AI and BI meet at I on the far side of a river. On the near side of the river, a point E was selected on the straight AI, and a point F on the straight BI, and the distance from E to F measured and found to be 85.00 m.

The angle $A\hat{E}F$ was found to be $165°36'$ and the angle $B\hat{F}E$, $168°44'$. If the radius of a circular curve joining the straights is 500 m, determine the distance along the straights from E and F to the tangent points.

Thence calculate the necessary data and explain in detail how to set out the curve,

(*a*) by tape and offsets only, and
(*b*) if a theodolite is available.
Answer: 75.55 m; 65.10 m.

3 A circular curve of radius 1000 m is to be connected to another circular curve of radius 600 m by a transition curve, so designed that the rate of change of radial acceleration is $\frac{1}{3}$ m/s^3 when the velocity is 100 km/h. Given that the chainage of the junction point with the curve of larger radius is 1224.00 m, determine the chainage of the other junction point. (*Hint*: $lr = K$.)
Answer: 1266.87 m.

4 A parabolic sag vertical curve connects a downgradient AB of 1 in 20 with a second downgradient BC of 1 in 50. Intersection point B had a chainage of 1459.00 m and its reduced level was 198.48 m above datum. In order to allow for clearance at a bridge, the reduced level of a point on the curve at chainage 1470.00 m is to be 198.56 m.

(*a*) Determine the total length of the vertical curve, and the rate of change of grade.

(*b*) The beams given out by the headlamps of a certain vehicle are parallel to the longitudinal axis of the vehicle. Calculate the sighting distance at night, given that the headlamps are 0.70 m above road level.

Answer: (*a*) 120 m; (*b*) 75 m.

5   In Fig. 11.30, the curve *ABC* represents the horizontal alignment of a slip road leaving a motorway at point *A* and forming a T-junction with a side road at *C*. The alignment consists of two reverse circular arcs *PQ, RS* connected by spiral transitions *QB, BR* at point *B*, so that the tangent at *B* cuts the motorway at an angle of 15°. Spiral transitions *AP, SC* connect the circular arcs *PQ, RS* with the straights T_1A, CT_2 respectively. The design speed for the slip road is 80 km/h and the rate of gain of radial acceleration is to be 0.45 m/sec³. Assuming that the alignments of the motorway and side road are both straight, calculate:

(*a*) the minimum clearance between the circular arcs, and
(*b*) the total length of the curve *ABC*.

Answer: (*a*) 1.306 m; (*b*) 510.32 m.

Fig. 11.30

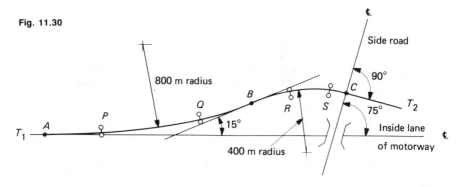

6   A straight portion of a railway runs due east and is followed by a further straight which runs N 15°E, the straights being jointed by a circular curve of radius 1800 m. In an improvement scheme it is decided to introduce at each end of a circular curve a transition (cubic parabola) 150 m long in such a manner that the route length remains the same.
    Calculate

(*a*) the distance of the new tangent points from the intersection point
(*b*) the amount by which the 'half way' point of the curve has to be moved.
Answer: (*a*) 1453.89 m; (*b*) 0.26 m

7   On a straight portion of a new road, an upwards gradient of 1 in 100 was connected to a downward gradient of 1 in 150 by a

vertical parabolic summit curve of length 150 m. A point P, at chainage 5910.0 m, on the first gradient, was found to have a reduced level of 45.12 m, and a point Q, at chainage 6210.0 m on the second gradient, of 44.95 m.

(a) Find the chainages and reduced levels of the tangent points to the curve.

(b) Tabulate the reduced levels of the points on the curve at intervals of 20 m from P and of its highest point.

Find the minimum sighting distance to the road surface for each of the following cases:

(c) the driver of a car whose eye is 1.05 m above the surface of the road;

(d) the driver of a lorry for whom the similar distance is 1.80 m.

(Take the sighting distance as the length of the tangent from the driver's eye to the road surface.) (*London*)
Answer: (a) Chainages 5944.8, 6094.8; (b) *RL* 45.92; (c) 163.8 m; (d) 253.8 m.

8 A curve connecting two straights is transitional throughout, no circular arc being present. The junction of the two transition curves is 5.00 m from the intersection point of the straights, and the tangent lengths are 148.00 m. Estimate, to the nearest degree, the total deflection angle between the straights.
Answer: 12°.

9 A sag vertical curve PQ is continuous with a summit vertical curve QR, the relevant tangents being PA, AQB and BR respectively, with PA and BR falling to the right. The gradients of PA, AQB and BR are 1 in 40, 1 in 50 and 1 in 30 respectively, and curve PQ has a length of 200 m. If the difference in level between the lowest point on curve PQ and the highest point on curve QR is 2.09 m, determine the length of curve PR to the nearest metre, and thence the reduced level of the curve point midway between P and R.

The reduced level of tangent point P is 80.00 m above datum.
(*Salford*)
Answer: 520 m, 80.40.

10 A compound reverse curve, which is wholly transitional, connects three straights AB, BC and CD whose whole circle bearings are 78°00′; 90°00′ and 80°00′ respectively. If the design speed is 70 km/h and the rate of change of radial acceleration is 0.3 m/s^3, determine,

(a) the length of common tangent BC

(b) the chainage of the tangent point on CD given that the corresponding point on AB has chainage 1736.91 m

(c) the deflection angles which locate the curve points at chainage 1820.00 m and 1940.00 m when a theodolite is set up at the junction point on BC with C as the reference pointing.

Answer: (*a*) 137.34 m   (*b*) 2010.99 m
(*c*) 178°35'19", 358°36'23"

11 A straight length of road 12 m wide is to be connected to a straight parallel alignment, the distance between the centre lines being 30 m. The connection is to be a symmetrical reverse curve, transitional throughout such that the rate of change of radial acceleration is limited to 0.3 m/s^3 at a design speed of 80 km/h. If the minimum radius of curvature is to be 500 m determine:

(*a*) the total length of the curve;
(*b*) the angle between the common tangent and a straight;
(*c*) the final deflection angle to locate the end of the first transition curve;
(*d*) the difference in level between outer and inner kerb lines to provide superelevation applicable to the design speed and hence the resisting force required for a vehicle traversing the curve at 33 m/s.   (*Salford*)
Answer: 600 m, 17°11'19.5", 2°51'53.2", 1.208 m.

12 A parabolic vertical curve has been designed to connect a gradient rising to the right at 1 in 40 to a gradient falling to the right at 1 in 30 such that two points 1.05 m and 0.3 m respectively above the curve are 160 m apart. The gradients intersect at chainage 2064.37 m and reduced level 147.380 m AOD, determine
(*a*) the length of the curve,
(*b*) the position and level of the highest point on the curve,
(*c*) the critical velocity of a vehicle to satisfy a centrifugal force of 0.2 m/s^2.
Tabulate the reduced levels of points on the curve at 20 m intervals of running chainage.
Answer: 302 m, 129.43 m, 145.223 m AOD, 115 km/h,
level at 2000 m = 145.033 m AOD.

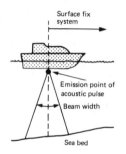

Surface fix
system

Emission point of
acoustic pulse

Beam width

Sea bed

12 Hydrographic surveying

Hydrographic surveying, so far as the civil engineer is concerned, conventionally covers the survey work for projects in, or adjoining, bays, harbours, lakes, or rivers, although nowadays it can well involve the construction of offshore production platforms and associated pipe lines. The types and purpose of the various branches of hydrographic surveying may be summarized as follows:

(1) Measurement of tides for sea coast work, e.g. construction of sea defence works, jetties, harbours, etc., for the establishment of a levelling datum, and for reducing sounding.
(2) Determination of bed depths, by soundings:
   (*a*) for navigation, i.e. for 'tow' out routes for production platforms, but also including the location of rocks, sand bars, navigation lights, buoys, etc.,
   (*b*) for the location of underwater works, volumes of underwater excavation etc.,
   (*c*) in connection with irrigation and land-drainage schemes.
(3) Determination of direction of current in connection with:
   (*a*) the location of sewer outfalls and similar works,
   (*b*) determination of areas subject to scour and silt,
   (*c*) for navigational purposes.
(4) Measurement of quantity of water, and flow of water — in connection with water schemes, power schemes, flood control, etc.

Normally, the civil engineer is not concerned with navigation, work in connection with this being carried out by such bodies as the Admiralty, US Coast and Geodetic Survey, etc., with the International Hydrographic Bureau acting as a co-ordinating body. The remaining aspects of hydrographic surveying are, however, of vital concern to the engineer, and will be dealt with here. It will be apparent from this brief outline that though some of the work is fundamental, much of it is of a specialized nature in practice, and is carried out by specialists in these fields.

If we exclude item (4) from the above list of various branches of hydrographic surveying (since it is concerned mainly with hydraulics), the fundamental task is the preparation of a plan or chart showing physical features above and below water, and involves:

(*a*) Vertical control. A chain of bench marks must be established near the shore line, and these serve for setting and checking tide gauges etc., to which the soundings are referred.

(b) Horizontal control. When making soundings of the depth of a river bed or a sea bed the location of the sounding vessel is made by reference to fixed control points on shore, and the accurate establishment of this shore framework is of the utmost importance.

(c) Determination of the bed profile by soundings and use of the fine wire sweep. At this stage, direct surveys of the bed might be carried out by divers working from suitably established control stations. It will be apparent that the co-ordinates of these control stations must be established from surface points.

(d) Location of all irregularities in shore-line, islands, rocks, etc., by normal surveying methods.

During the survey of a small area of sea or river bed, the vessel normally runs along a series of evenly-spaced parallel lines covering the area and, at set fixed intervals of time, records the water depth and the vessel's position. The spacing will depend upon the task in hand, and for civil engineering work near shore it could be 2 m to 5 m upwards.

Because the water depth varies, especially in tidal areas, the water level above a chosen datum has to be constantly monitored at a nearby shore base so that the depth measurements can be reduced to this datum.

Vertical control datums: tide measurement

According to the requirements of the particular survey, soundings will be reduced to one of two datum lines:

(a) The land-levelling datum, which in Great Britain is the *Ordnance datum*, this being the one generally used in civil engineering construction, since it enables the levels to be directly related to those of the adjoining shore installations (though for the work itself an arbitrary datum below the level of the lowest work may be adopted so that all levels are positive).

(b) The tidal datum, which is generally used for navigation purposes. The usual level adopted is that level of the water surface below which the tide rarely falls — lowest astronomical tide (LAT), and has the name *Chart datum*.

Tide gauges

Since we are not concerned primarily with the needs of navigation in this chapter, main attention will be given to vertical control using the land-levelling datum. This involves the location of tide gauges at intervals over the area to be surveyed, the number employed depending on the accuracy required. For example, if the tide levels at each end of the area under survey differ by a maximum of 0.5 m at any given time, and an accuracy of 0.1 m is required in the soundings, then 5 gauges should be used, and any particular set of soundings reduced by reference to the gauge nearest the sounding area.

The difference in tide level, or *tidal gradient*, may be determined by initially setting up gauges at each end of the survey area and comparing simultaneous readings over a complete tidal cycle.

The gauges themselves must allow the following measurements to be made simultaneously: (i) the level of the undisturbed sea, and (ii) the time at which this level occurs: they may be either visually-recording or self-recording. The simplest of the former type is the *staff gauge*, which is merely a vertical staff about 150 mm broad, with painted graduations covering sufficient length to deal with the highest and lowest known tides. It is not necessary for the zero to be set exactly at Ordnance datum. To calibrate the staff, a line of levels is taken, from the nearest bench mark, on to a graduation on the staff as fixed. In this way the actual RL of the staff zero may be determined and thence the correction which, when applied to the staff gauge readings, will reduce them to OD. Figure 12.1 shows a typical staff gauge whose zero graduation has been set at datum level.

Fig. 12.1
xy = sounding
$zy = xy - xz$
   = reduced sounding

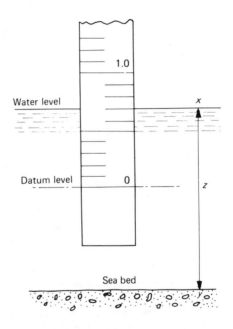

Such gauges are often difficult to read owing to wave action at the water surface, and *float gauges* can be used in which the staff is attached to a float which is enclosed inside a box open to the water at its lower end. The box acts as a stilling chamber. Otherwise, the float may be attached to a counterweight by means of a steel tape, suitably graduated and passing over a pulley.

These instruments naturally have to be read at fixed and frequent intervals so that when fairly lengthy records are required a self-recording gauge must be used. The principle of operation is shown in Fig. 12.2, time being recorded around the drum to produce a plot of related tide levels. It is good practice also to install a 'visual' staff gauge to check the behaviour of the automatic gauge at reasonably regular intervals.

Instruments are available which can be mounted on the sea or river bed or on the side of an underwater structure and will automatically

Fig. 12.2

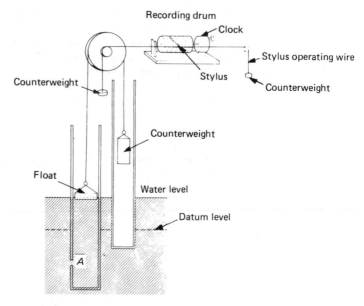

Inlet — A is not to be exposed at
low water.

monitor the action of waves and tides by measuring the change of water
pressure above the instrument.

Figure 12.3 shows the basics of the Valeport BTH700 Tide Gauge
which has been designed to measure water heights above a fixed datum
and to show the value on a gauging/display unit established locally.
Changes in water level are recorded by an immersed precision strain
gauge pressure transducer. The transducer is connected to the unit by
means of a signal cable which also contains a reference tube thereby
correcting the transducer output for changes in barometric pressure.

The transducer and cable are located with the aid of a stainless steel
slide wire attachment which comprises a top bracing plate and fixing
clamps: clips indicating metre intervals are fitted on the signal cable.

Fig. 12.3 System
configuration for a tide
gauge using pressure
transducers

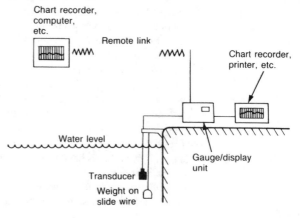

A sinker has to be provided to maintain tautness in the slide wire.

The gauging/display unit provides power to the transducer and also converts the output into tide information to a resolution of 1 mm. The unit is damped so that wave action is meaned and accordingly the stilling chamber shown in Fig. 12.2 is not required. Additional circuit boards can be housed in the gauging/display unit so that the tide data can be stored digitally or on charts. In addition the system can be extended by radio link or cable to a recording unit remote from the site and a series of such units along a coast line can be interlinked.

For accurate data, site calibration is needed due to changes in the density of sea water and temperature. Achievable accuracy of the order of 20 mm is claimed by the manufacturer over an operating range of 10 m. A typical chart is shown on Fig. 12.4.

Fig. 12.4 Tide chart (*Courtesy*: Valeport Marine Scientific Ltd)

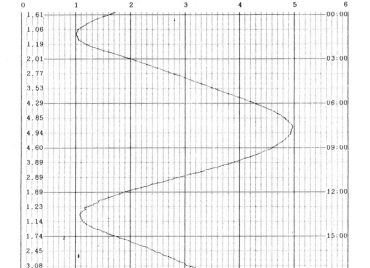

DARTMOUTH HARBOUR COMM.   DATE   : 05/02/88
RIVER   DART   TIDE GAUGE   TIME ZONE : GMT

Valeport Marine Scientific Ltd. Townstal Industrial Estate Dartmouth Devon TQ6 9LX UK Tel: + 44 (0)8043 4031 Fax: + 44 (0)8043 4320 Telex 42669 VALEFI E

Horizontal control: the shore framework

The precise nature of shore control depends on the method used to locate the soundings, which in turn depends to some extent on whether sea coast, river, or estuary, is being surveyed and upon survey techniques.

Fig. 12.5

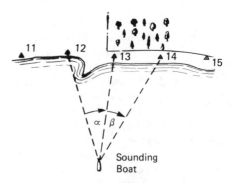

Sounding
Boat

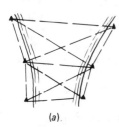

(a).

(b)

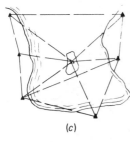

(c)

Fig. 12.6

Taking as an example the case in which soundings are to be taken along a stretch of coast and are to be located by sextant observations of pairs of angles subtended at the boat by shore stations, a series of beacons 11, 12, etc. must be located (Fig. 12.5). Where available, of course, salient features such as lighthouses or churches can be used, in conjunction with such beacons, and location of the control points is by closed traverse or by triangulation. Should the area have been surveyed already, it may be possible to obtain co-ordinates of salient features and trig. stations etc., but this previous survey must, of course, be of at least the same accuracy as the hydrographic survey being carried out.

For estuaries, rivers and inlets, networks of the type shown in Fig. 12.6 are employed. In addition to providing controls for the soundings, the beacons (or other control points) form the framework for the remainder of the shore survey. This may be done by any of the usual methods; i.e. closed traverses run between the control points, plane tabling, or tacheometry. In picking up the shore line, high and low water lines are located: the former may be obtained from the position of deposited material and the latter from tidal observations.

Sounding: determination of bed profile

This is one of the main operations in hydrographic surveying and corresponds to levelling in land surveying. Three methods of depth measurement are described below. With each method it is necessary to record not only position as described later but also the time of the particular observation so that the water level above datum, given by the nearest tide gauge, can serve as reference to relate the bed level to the selected datum as implied in Fig. 12.1.

Sounding or depth measurement may be carried out by the following methods:

(1) Direct: (a) Sounding rods; (b) Sounding leads on graduated lines.
(2) Indirect: Echo sounders.

1. Direct methods
(a) Sounding rods

Where the currents are not strong, graduated wooden poles may be used to measure the bed depth. This method is limited to depths of about 5 m. In strong currents it is difficult to maintain verticality of long sounding rods.

(*b*) Sounding lines

For depths from 5 m to some 30 m a lead line — a leaden weight attached to either a stretched and graduated hemp line or to a metal chain — can be used. Such a line may be incorporated in a *sounding machine*, in which a flexible wire is used, the amount paid out being measured by a friction-driven roller and shown on dials.

Care must be taken in swift-flowing water, or whilst the boat is in motion, since the sounding line will not hang vertically, thereby causing overmeasurement. Tables of correction to be applied to the sounding line are available.

Suspended weights may be used to transfer a surface point to a control point at bed level, analogous to the use of a plumb bob at ground level, but obviously drag can again be of great influence: typical accuracy of location is of the order of ± 1.8 m in depths of 30 m. Alternatively, taut wires taken from three bed points to a moored surface craft can be employed, the knowledge of their lengths and the distances apart of the three points being used to fix one of these points with respect to the surface craft.

2. Indirect
methods
Echo sounding

This method is somewhat analogous to the methods of distance measurement discussed in Chapter 5, depth being measured by timing the interval between the transmission of a pulse of sound energy from the boat and its reception after reflection at the seabed. Transducers, which transmit and receive the acoustic signals, can be hung over the side, but for permanent installation on larger craft the transmitting and receiving transducers are mounted on opposite sides of the keel.

The record of depth is made by a stylus on a moving band of dry paper as shown in Fig. 12.7 for the Kelvin Hughes MS48 Echo

Fig. 12.7 The Kelvin Hughes MS48 Echo Sounder (*Courtesy*: Kelvin Hughes)

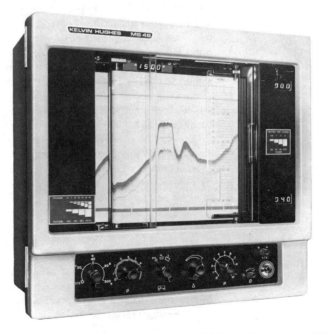

Sounder: the draught of the vessel can be compensated for, so that transmission is effectively from water level.

The maximum depth attainable is stated to exceed 1000 m but it has basic depth scales of 0 to 20 m, 0 to 40 m, 0 to 80 m and 0 to 200 m. The scale divisions for the depth range in use, controlled by the left-hand switch, are indicated on the right between the upper and lower limits of that range, i.e. 0 to 40 m in Fig. 12.7. With phasing of scales, effected by the next switch, the range can be extended. It thus covers the range of civil engineering hydrographic surveying requirements so far as sounding as such is concerned.

Special points in echo sounding

The accuracy of measurement depends upon matching the recorder's time scale with the velocity of the acoustic pulse. This is approximately 1500 m/s in sea water but it varies with salinity and temperature which, in turn, vary with the depth, weather and time. One of the many expressions to calculate acoustic velocity in sea water is

$$V = 1410 + 4.21\,T - 0.037\,T^2 + 1.14\,S$$

Where T = surface temperature in degrees celsius
S = salinity in parts of sodium chloride per 1000.

To check on instrument delays and on the determination of the velocity, the echo sounder can be calibrated directly. A target (usually a bar — hence bar check) is lowered into the water to a series of known depths and the echo sounder is adjusted to read those depths: this is a practical proposition in shallow water only.

The impulses do not form a true beam, but, as implied by the illustration at the beginning of the chapter, they have a conical shape, the main strength of the impulse being in an acute-angled cone at the centre, and this has a number of repercussions. Echo sounders record minimum depth, so projections above mean bed level falling into the beam will be recorded, whilst potholes can be missed altogether, and false depth readings will be obtained when operating over a steeply sloping sea bed. Furthermore, surface rolling of the boat will result in measurement of slant lengths rather than vertical depths, unless stabilizing equipment is provided. It will be realized that features of height less than half the pulse length will merge into the sea bed detail.

Care must be taken when operating close to jetties and quay walls, since the return signal from the wall may blank out that from the bottom. In addition pockets of air bubbles which may occur under the bottom of a moving boat can cause reflections, so the transducer assembly needs to be carefully located.

Acoustic techniques may be adopted in the transfer of surface positions to sea-bed positions, and these include 'range-range' and 'range-bearing' systems. The former uses a number of transponders (acoustic beacons) at known distances apart on the bed, which are interrogated in turn by acoustic pulses from a transducer unit at the surface. The times of travel of signals are measured and 'slant' distances to each are evaluated or presented directly by the transducer unit.

In the range-bearing method, a single transponder lies at bed level and three or four transducers, at known separations on the surface, are used to determine slant distances and thence the position of the transponder. Bearings are also established for position verification.

Sampling and sweeping

These are two operations often associated with sounding, sampling being the most likely to be of interest to the civil engineer. If indicative samples of the bed bottom are required a sounding lead can be used. This has a hollowed out base filled with tallow which is dropped onto the sea bed collecting a sample in the base. For larger samples special grabs may be utilized. Valeport manufacture a variety of small grabs such as the SK180 which is designed to work up to 20 m and to sample an area of 0.1 m^2 with minimum disturbance. It can be fitted with a plate to prevent excessive penetration into very soft material.

Sweeping is carried out to verify the presence, or otherwise, of underwater obstructions or other features not located on a vertical echo sounding. This is necessary because the echo sounding process produces a single profile of the bed and obstructions could be missed unless the spacing of the sounding lines was less than the beam width at bed level. Rather than reduce the spacing of the lines it is quicker and cheaper to perform further examination of the bed. The traditional method was a fine wire drag but it is now more common to use side scan sonar.

Essentially this is an oblique looking echo sounder which can record all return echoes from the sea bed rather than record only the first signal, thereby presenting a picture of its surface. A common system is for a 'fish' to be towed just above the sea bed behind the survey vessel with sets of transducers pointed either side of the ship's track, whilst

Fig. 12.8

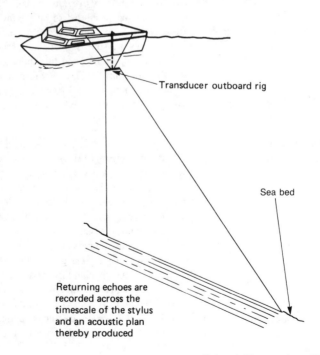

Transducer outboard rig

Sea bed

Returning echoes are recorded across the timescale of the stylus and an acoustic plan thereby produced

an alternative is shown in Fig. 12.8. Experts in interpretation of side-scan sonar records can identify not only pinnacles and wrecks, etc. but also, because they have varying reflective properties, the different surface materials on the sea bed.

Location of soundings: position fixing

Most of the methods of position fixing for hydrographic work are variations on the methods described for land survey but these need some modification due to the different requirements. The main factors to consider when designing, or selecting, a position fixing system are as follows:

(1) Speed of measurement. The survey vessel is moving and so the fix must be instantaneous.
(2) Shore control. Virtually all methods require some shore control. The number and position of shore beacons is an important factor since much work can be involved in setting them up and establishing their position by land survey.
(3) Speed of plotting. Position must usually be determined and plotted immediately to assist the helmsman to steer the vessel.
(4) Rate of fix. The interval between one fix and the next must be matched with the vessel's speed and the plotting rate.
(5) Manpower, logistics and communications. Some of the simpler methods require no more than two people on the survey vessel, whereas the more advanced systems for work offshore require only one person to operate the equipment on the survey vessel but permanently manned shore transmitters.
(6) Accuracy. Accuracy offshore tends to be lower than that onshore because of the problems of surveying from a moving vessel. However, the requirements are generally lower.
(7) Cost. In general, the simpler the observing system, the less the cost of the equipment and the operation.
(8) Weather and time of day. Some of the electronic systems are capable of continuous all-weather operation but this may not be true for the simple visual systems.

The sextant

At this stage it is appropriate to describe the construction and use of the sextant since several of the simpler fixing methods require its use.

There are in fact, two versions of this instrument: (i) the nautical sextant, and (ii) the box sextant. The former instrument consists essentially of a $60°$ arc (i.e. $\frac{1}{6}$th of a circle), the periphery of which is graduated in degrees. An index arm, pivoted at the centre of the arc, carries a wholly-silvered index glass at the pivot end and at the other a vernier which moves along the graduated arc. A half-silvered glass A, known as the horizon glass, is attached to the sector as shown in Fig. 12.9, i.e. the plane of this glass is parallel to the radius through the zero of the graduated scale. When the index arm is set at zero the index glass should be parallel to the horizon glass, and adjusting screws are provided on the index glass to enable this condition to be complied with. A pinhole sight or a telescope is fitted opposite to the horizon

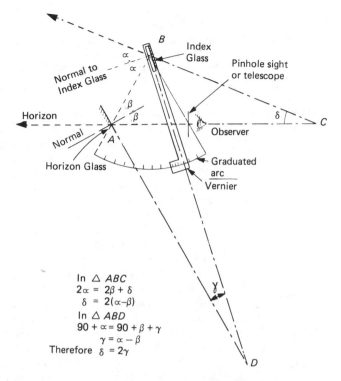

Fig. 12.9 Optical principles of sextant

In △ ABC
$2\alpha = 2\beta + \delta$
$\delta = 2(\alpha-\beta)$
In △ ABD
$90 + \alpha = 90 + \beta + \gamma$
$\gamma = \alpha - \beta$
Therefore $\delta = 2\gamma$

glass, and to measure vertical angles the instrument is held upright and a sight taken on to the horizon. The index arm is now swung so that the image of the sun or star, seen in the mirror half of the horizon glass, cuts the horizon. The angle between the two glasses is half the angle of elevation of the object sighted, and to give direct reading the scale is graduated so that $\frac{1}{2}°$ reads as 1°. Thus the scale reads to a total angle of 120°, with perhaps an extra degree or two at each or one end.

When measuring horizontal angles between two targets, the instrument must be held as horizontal as possible, horizontality being assessed with reference to the horizon or shore line. One target is now sighted through the horizon glass and the other, viewed through the index glass, is made to coincide with the first by rotating the index arm. The horizontal angle can now be read off the scale.

If the targets are elevated such that the horizontal angle cannot be measured directly, the angle in the plane of the observer and the two targets with respect to the observer then have to be measured and a correction applied. This method is very laborious and not recommended.

The box sextant is a compact version of the nautical instrument and is primarily intended for use in land surveying as a pocket instrument; it is much inferior to the theodolite in angular measurement. Both types of sextant have the merit of portability and although readings can be taken to 10 seconds of arc the accuracy is more likely to be of the order of 30 seconds.

Position fixing

The salient features of some of the methods of position fixing will now be considered. As a broad classification, these will be divided into (i) those in which the sounding vessel is used in channels or close to banks etc., and (ii) those in which the vessel is used in open and wide waters.

1. Soundings in channels

With rivers and channels which are narrow enough for a line or steel wire to be slung across, soundings taken with the sounding vessel secured probably constitute the most accurate method. The cable supports on each bank must be accurately located; the cable is graduated, so that the boat can be pulled or manoeuvred to the appropriate station and the sounding made. This technique is usually adopted when stream flows are being measured with the current meter, which is dealt with later in the chapter. The boat used should be heavy enough to damp out wave action; maximum width of water for this method may be taken at 300 m. (By running between an anchored boat and the shore the method can be used for off-shore work, but it is not so convenient as other methods.)

When soundings are being made within 60 m of a jetty or river bank, instead of using the taut wire a special subtense board may be erected on the bank and the boat brought in at right angles to the shore with the sextant, set at a fixed angle of 3°, sighted on the board. The general principle is shown in Fig. 12.10. The subtense board is set up so that the lower sighting mark, which is viewed directly through the horizon glass of the sextant, is at the observer's eye-level as he stands in the boat. The height of the graduations above the sighting mark are such that they subtend 3° at known distances from the board so that, with the sextant set at 3°, as each graduation (viewed through the index glass) becomes coincident with the sighting mark (viewed directly) the

Fig. 12.10

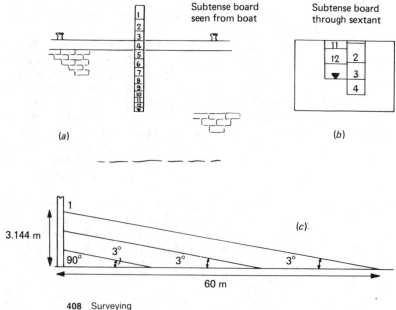

Subtense board seen from boat

Subtense board through sextant

(a)

(b)

(c)

3.144 m

90°   3°   3°   3°

60 m

observer knows his distance from the board. This fixing is rapid enough for use with the echo sounder; when in tidal waters the run must be made rapidly so that no correction for change in water level is necessary.

The heights of the graduations above zero for a 3° subtense bar can be readily calculated, the fix at 60 m being given by mark 1 which is 3.144 m above that zero.

The bar may be hinged at line No. 6.

2. Soundings in open water

In the general case, where the survey vessel is not secured, there are two problems; the first of which is to keep the boat on a known course, so as to obtain systematic sounding, known as *conning* the vessel, and the second to locate the soundings so that they may be charted.

Fig. 12.11

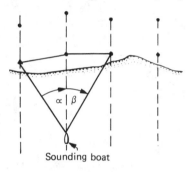

Sounding boat

(i) Conning the survey vessel

This task is, of course, mainly one of seamanship. One of the most common methods is to fix markers (poles, beacons etc.) on shore, as shown in Fig. 12.11 a method suitable for work in rivers and open-sea areas up to 5 km off shore where working on the larger scales, e.g. 1/2500. The vessel is run down each track, the steersman lining himself in with the successive pairs of shore stations which have been fixed in advance and which must be easily identifiable. Tracks generally run normal to the depth curves, but crosslines can be run at 45° or 90° to the tracks for checking purposes. An alternative in common use is to steer on the compass, but this is suitable only for smaller scales of working, say 1:10 000 or less.

(ii) Fixing by sextant observations: plotting the soundings

The above methods serve to guide the steersman. The survey vessel is accurately and continuously fixed by other methods, some of which are given below.

For up to about 5 km offshore, the position of a survey vessel can be located by the observation of resection angles with a sextant. For one fix, two angles are required: in Fig. 12.5, angles α and β between stations 12 and 13 and 13 and 14 are sufficient to locate the survey vessel. If the markers in (i) above have been brought into the horizontal control, then they themselves can be used as appropriate; measurement of either α or β will suffice for a fix near shore. All control points have to be co-ordinated and should be placed such that there will be three stations always visible from the sounding points which will not subtend angles of less than 30° or more than 115°. The two angles should be observed from the same point at the same time but this, of course, is impossible. A compromise has to be adopted: either two sextants are used with the observers standing side by side, or only one sextant is used and one angle taken immediately after the other.

According to some sources, a well-conditioned fix by sextant can realize accuracies of the order of ±1.5 m when plotting at 1:1250 scale.

There are three ways to determine the vessel's position from the observed angles.

Fig. 12.12

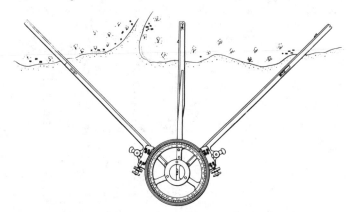

(a) The station pointer This can be used to plot the position. It has three arms, the outer two being movable such that α and β can be set off on the graduated circle to approximately the same accuracy as that with which they were measured. After rotating the three 'straight edges' to lie over the positions of the shore stations on the plan, the required point is obtained by pricking through at their intersection at the centre of the graduated circle (*see* Fig. 12.12).

(b) Resection This method is identical with the last except that it is required to plot the fix to greater accuracy than is possible with a station pointer.

Fig. 12.13

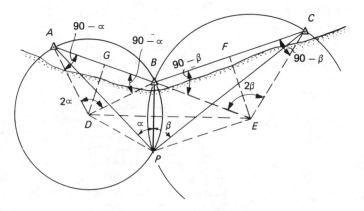

The problem is, therefore, referring to Fig. 12.13, to locate P, given the co-ordinates of positions A, B and C, and angles $A\hat{P}B$, $C\hat{P}B$, measured by sextant as α and β.

D and E are the centres of the circumscribing circles through points ABP and CBP respectively; AB and BC are known or can be calculated from the co-ordinates.

D and E are located by either setting off angles $(90-\alpha)$ from A and B and $(90-\beta)$ from B and C respectively, or by calculation. In the latter case

$$A\hat{D}B = 180 - 2(90-\alpha) = 2\alpha = 2A\hat{P}B$$

If DG be the perpendicular bisector of AB then

$$A\hat{D}G = G\hat{D}B = \alpha \text{ since triangles } AGD \text{ and } BGD \text{ are congruent.}$$

Thus

$$AD = AG \text{ cosec } \alpha = \tfrac{1}{2} AB \text{ cosec } \alpha$$

and

$$GD = AG \text{ cot } \alpha = \tfrac{1}{2} AB \text{ cot } \alpha$$

Thus D (and similarly E) can be located. The intersection of the two circles gives P since α and β are simultaneously subtended by AB and BC respectively.

In the Δ's ABD, CBE, the lengths of AB, BC and the angles are known, so that it is possible to calculate the co-ordinates of D and E. Then, in ΔDPE, the length of side DE and the angles are known, so that the co-ordinates of P may be calculated. The following alternative method is however often preferable.

The angle $A\hat{B}C$ being known or computed from the co-ordinates, P may be located by determining all angles in triangles ABP and BCP and then calculating sides AP, BP and CP.

Example 12.1 A, B and C are three shore stations on a coastline (Fig. 12.14), and P is a sounding point at sea. $AB = 400$ m, $BC = 381$ m, $A\hat{B}C = 122°30'$, $A\hat{P}B = 48°36'$, and $B\hat{P}C = 45°24'$. A and C are respectively east and west of BP, B and P are respectively north and south of AC. Calculate the distances AP, BP, and CP.

Fig. 12.14

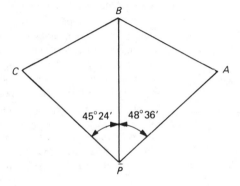

$$B\hat{C}P + B\hat{A}P = 360° - 48°36' - 45°24' - 122°30'$$
$$= 143°30'$$
Therefore $B\hat{C}P = 143°30' - B\hat{A}P$

Now $\dfrac{BP}{\sin B\hat{C}P} = \dfrac{381}{\sin 45°24'}$ and $\dfrac{BP}{\sin B\hat{A}P} = \dfrac{400}{\sin 48°36'}$

so that $BP = 381 \dfrac{\sin B\hat{C}P}{\sin 45°24'} = 400 \dfrac{\sin B\hat{A}P}{\sin 48°36'}$

and $\sin B\hat{C}P = \dfrac{400}{381} \dfrac{\sin 45°24'}{\sin 48°36'} \sin B\hat{A}P$

$$= \sin(143°30' - B\hat{A}P)$$
whence $B\hat{A}P = 72°02'$
and $B\hat{C}P = 71°28'$
Therefore $C\hat{B}P = 180° - 45°24' - 71°28' = 63°08'$
$A\hat{B}P = 180° - 48°36' - 72°02' = 59°22'$

$\dfrac{AP}{\sin 59°22'} = \dfrac{400}{\sin 48°36'}$ $\therefore AP = \textbf{459 m}$

$\dfrac{BP}{\sin 72°02'} = \dfrac{400}{\sin 48°36'}$ $\therefore BP = \textbf{507 m}$

$\dfrac{CP}{\sin 63°08'} = \dfrac{381}{\sin 45°24'}$ $\therefore CP = \textbf{477 m}$

P may now be located by striking arcs equal in length to these values.

The co-ordinates of P can of course be calculated as mentioned in Chapter 6. A programmable calculator is ideal for this sort of work since, once programmed, the work reduces to no more than the input of two angles. However, if the vessel is being kept on course by the resection fixes, the co-ordinates will then have to be plotted. The principles governing the relation between the beacon positions and the vessel's position are given in that chapter in so far as they affect the strength of fix.

(c) Using circle plotting sheets This method eliminates the use of the station pointer and enables positions to be plotted immediately as on a graph. It is a graphical development of the previous method. One measured angle at the survey vessel defines its position to be somewhere on a circle. The second angle fixes the position of the vessel at the intersection of two circles so, if on a sheet, a series of circles are plotted for each pair of shore beacons, the circles representing a range of subtended angles, the vessel's position can be plotted quickly and easily, directly from the observed angles.

Considering one pair of control stations, A and B (Fig. 12.15), we require to know AD and the location of D for various values of α. Join A and B, and construct the perpendicular through the mid-point E. The length AB is known, and as previously

Fig. 12.15

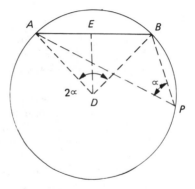

$$ED = AE \cot \alpha \text{ which gives the position of } D$$
and
$$AD = AE \operatorname{cosec} \alpha$$
$$= \text{ radius of the circle on whose}$$
$$\text{circumference } AB \text{ subtends } \alpha$$

A range of values is chosen for α, and the appropriate curves plotted. The procedure is then repeated for values of β subtended by the other control points, and a graph is obtained, as in Fig. 12.16.

Referring to Fig. 12.16 three control points, the middle one of which is a lighthouse, have been chosen and a series of circular arcs passing through pairs of points for different values of α and β are given. Three

Fig. 12.16 Circle plotting sheet showing sounding lines

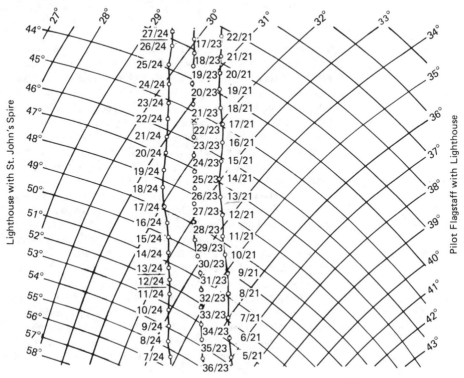

runs have been indicated and individual soundings numbered; for example, the angles subtended at 22/23 were 45°24′ and 31°25′ respectively and the sounding position is obtained by interpolating between the curves given by whole degrees.

The reverse situation can be used when locating say a raft-mounted shell and auger rig during a site investigation for a proposed bridge. Values of α and β could be predetermined for a particular borehole and these would be established by direct observation to position the rig.

Note also that a simplified method of resection involving the measurement of only one angle is possible close to the shore. As shown in Fig. 12.11, when shore beacons have been established to define lines along which the survey vessel can be steered, the measurement of one angle only, α or β, is required to define its position.

(iii) Fixing by theodolite angles

Two theodolites set up at known poisitons on the shore can be used to fix the position of a survey vessel by intersection but the main drawback with this method is the amount of co-ordination required. The angles must be read at the same instant, so the surveyors manning the theodolites must be constantly tracking a target on the survey vessel and then, on a command from the boat, recording the circle reading. This command should coincide with a depth measurement. Also, this method cannot easily be used to navigate the boat since the data is read and recorded at two points on the shore so that immediate plotting on the vessel is not possible.

(iv) Fixing by EDM

A theodolite and infra-red EDM instrument set up at a shore station can be used to fix a vessel by the polar method of bearing and distance from a reference station on shore. This approach is logistically easier than intersection using two theodolites.

As an example, if we consider the Geodimeter 220 instrument discussed in Chapter 5, its fast tracking facility and wide beam assist in keeping contact with the moving prism system. With the distance being read every 0.4 seconds the boat could move at 4 m/s without any measurement problems. Moreover, the one-way Unicom speech communication with a range of up to 1600 m allows some measure of co-ordination of soundings with the distance measurement. Communication can be extended by radio link. The boat may be kept on course as indicated in Fig. 12.11 and horizontal and vertical distances to the prism(s) can be determined by the Geodimeter. These, together with times, can be stored in a data logger such as the Geodat 126. Assuming that an echo sounder is being carried on the boat, depth and time readings can be similarly logged for later comparison.

If instantaneous comparison is required, then it is possible to transmit data from the shore station to a computer on board ship by radio link. Position and time can be stored in the computer with on-line connection to the echo sounder. Consequently the data can be plotted or displayed graphically as the survey progresses to keep the vessel on line and ensure that important data is not missed.

The main problem with this method is maintaining the orientation of the EDM on shore with the reflector offshore so that a return signal is constantly received. Obviously the calmer the water the easier the task in this respect. However, Geotronics have introduced the Geodimeter 140T (Fig. 12.17) which incorporates a three-dimensional tracking system especially for this task.

The instrument is based upon the Geodimeter 140, with the auto-tracking device mounted on top of the measuring unit. Once the EDM has been fixed on the prism system the instrument will automatically follow the movement of the boat. The circles use electrodynamic high frequency fields to give an accuracy of ±3 seconds of arc, as described in Chapter 5. The EDM component is a gallium arsenide diode giving an infra-red carrier wave with a nominal range of 2.5 km to one prism and 5.5 km to a bank of eight prisms. The 'dynamic' accuracy depends upon the velocity of the boat and is said to be ±0.3 m at a speed of 4 m/s. A mode selector switch enables the slope distance, horizontal distance or vertical distance to be displayed at will and a function switch allows the horizontal or vertical circle readings to be displayed. The recording Geodat will accept the readings for over 1400 points of detail together with their coding data. A video camera can be connected to monitor the pointings.

Fig. 12.17 Geodimeter 140T (*Courtesy*: Geotronics (UK) Ltd)

(v) Fixing over long distances

The infra-red systems mentioned in (iv) cover most civil engineering requirements but for the longer ranges recourse has to be made to microwave systems and radio waves. In the former case Tellumat's Hydroflex system will measure up to 100 km, a master unit on the

Fig. 12.18 Principle of
Hi-Fix System

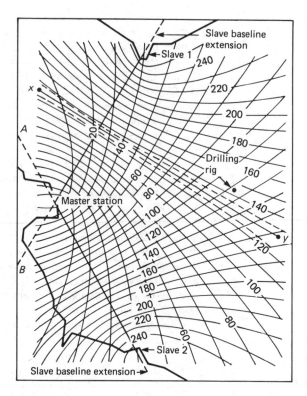

vessel measuring to two known slave stations. However, since 'line of sight' conditions have to be established, work over the horizon has to be performed with radio waves. Decca Hi-Fix allows location offshore by means of two 'families' of hyperbolae established by three transmitting stations, a master and two slaves.

The full lines on Fig. 12.18 indicate equal phase difference for signals received from the transmitters situated at the focal points. On the perpendicular bisector xy the signals from the two stations are in phase since these stations are equidistant therefrom, and zero phase difference would be given in measurement. Phase differences of 360° then apply in turn as one moves to adjacent in-phase hyperbolae: this implies the traversing of one complete lane, equal in length on the base line to half a wavelength. At the dotted lines the signals are exactly out of phase. By means of a suitably designed phase meter the number of lanes crossed can be counted from a 'zeroing' at the start point, and location within a family arranged following phase measurement. A similar situation arises for the master and slave 2 which produce a second family, and a fix can be made on a marked lattice chart. It is likely that these techniques will be superseded as Global Positioning systems, discussed in Chapter 7, will provide a simpler solution to long distance fixes.

Sea-bed detail

In so far as the civil engineer is concerned, three broad sub-divisions of underwater surveying can be identified: (*a*) surface location, (*b*) establishment of sea-bed control, and (c) determination of sea-bed detail. Some discussion of (a) and (b) has already been given.

The control points need to be permanent, i.e. stainless steel stakes driven well into the bed or drilled into rock. They should be capable of resisting displacement by dredges or anchors and to be unaffected by either erosion or deposition at bed level. Grids of wires, angle sections or pipes can also be used in this context, and accuracies of the order of ±0.1 m have been mentioned for detail fixing, but on large sites it is likely that transponders will be set out, either forming a grid system or acting as individual reference marks.

For the collection of detail measurements of distance, bearing and height are needed as on land. Distance is frequently measured by tape, corrections for pull, slope etc. being made as on land, but current drag can well be of importance. Reasonable visibility is demanded and maximum distances of 50 m tend to apply so that acoustic methods are demanded for longer lengths (see 'Calibration of an underwater acoustic distance measuring instrument', N. C. Kelland, *Proceedings of the Institution of Civil Engineers*, August 1975, and discussion, May 1976; and 'Underwater photogrammetry', E. J. Moore, *Photogrammetric Record*, October 1976).

The compass, possibly in conjunction with some form of horizontal protractor, can be used to establish magnetic bearings, and thence angles, but a theodolite has been developed for underwater observations (see 'The development and use of a practical underwater theodolite', R. Farrington-Wharton, *Proceedings of the Institution of Civil Engineers*, March 1970 and its subsequent discussion, March 1971) and in addition, the plane table with a vane-sighting alidade has been adapted for underwater surveys. Obviously, clear visibility is demanded and sighting distances tend to be short.

Bourdon tube pressure gauges are frequently used to measure water depths and thence to deduce relative heights. Accuracies of the order of ±2 per cent have been quoted when measuring depths of about 35 m: water densities need to be known to the same accuracy as that of the pressure-measuring device.

The measurement of currents and flows

As indicated at the beginning of this chapter, the engineering surveyor can be called on to measure the speed and the direction of sea and river currents and also to determine the discharge of a stream or river. The use of a current meter is the most common method for the measurement of currents but floats can also be used.

The current meter

One type of current meter is shown in Fig. 12.19. It consists essentially of an impeller, which is the only moving part and which is of neutral buoyancy, a reed switch, and counter-control unit. The bearings of the impeller, which has a helix of 0.27 m, are made from PTFE plastic and only need water as their lubricant: their design is such as to prevent entry of all but a minimal quantity of silt.

The impeller houses magnets which rotate about the reed switch, causing it to open and shut alternately with each revolution of the

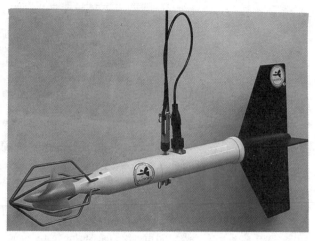

impeller as this turns in the current. A circuit is accordingly opened
and closed, a pulse then being produced and converted electronically
to register digitally at the surface readout unit.

On older units the velocity of flow is indirectly determined because
the control unit is allowed to operate over a measured time interval,
say 100 seconds, and the number of pulses recorded is divided by that
time interval. Reference to a calibration chart will give the velocity
over a range from 0.03 m/s to 6.0 m/s, to an accuracy of $\pm 1.5\%$,
or better, according to the manufacturer.

The Valeport BFM108 MK II eliminates the need to refer to
calibration charts because, in addition to velocity and direction, it
records pressure and temperature. If the meter is used in conjunction
with Valeport's BFM240 surface unit then the output is in engineering
units and can be hard-copied to a printer. Further developments include
solid state electromagnetic current meters which measure the velocity
by recording the disturbance to a magnetic field around the probe and
thus eliminate the need for an impeller.

To use a meter for discharge measurement, a station is chosen on
the stream or river, the suitability of the station being governed by
the following requirements:

(1) Channel should be regular in shape and straight up- and down-
     stream of the section;
(2) Channel bed should be free of obstructions;
(3) Flow should be as streamlined as possible, since eddy currents
     affect the meter.

A graduated rope or wire is pulled taut across the river at the station
(note that bridges may be used as stations) and divided into intercepts
from 1 m to 5 m depending on the width of the river and any perceived
local current effects. The intercepts do not have to be equal but they
must be recorded. At each of these points the current meter is lowered
to the water and the velocity measured at, say, 0.2, 0.4, 0.6 and 0.8
of the depth. It will be noticed that the velocity, which at the river
bed tends towards zero due to skin friction, rises to a maximum at

Fig. 12.20

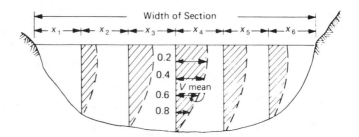

0.2 depth of about 1.2 times the mean velocity. By plotting these velocities for each vertical section, as shown in Fig. 12.20 the mean velocity, V_{mean}, in each case may be found:

$$V_{mean} = \frac{\text{area enclosed by curve (hatched)}}{\text{depth at that position}}$$

Note, by plotting this value of V_{mean} on the velocity/depth diagram for the vertical section concerned, that V_{mean} occurs in each case at *about* 0.6 × depth. Thus, if time allows only one velocity measurement at each vertical section, this measurement should be made at 0.6 × depth.

The value V_{mean} having been obtained for each vertical section, it is possible to obtain mean velocities for the water passing through the trapezoids into which the cross-section is divided, and hence by integration, using planimeter, Simpson's rule, or trapezoidal rule, the discharge is determined from:

$$Q = \Sigma a \cdot V_{mean}$$

The areas are in m^2, and the velocities in m/s, so that Q is in m^3/s. This is known as the area-velocity method.

Alternatively, 'contours' may be drawn on the cross-section, joining up points of equal velocity. The areas of these curves are measured by planimeter, and these are treated as equidistant cross-section areas of a solid, the volume of which gives the discharge. Yet another way is to use the 'spot height' analogy assuming the cross-section to be subdivided into a series of areas, the velocities of flow through these being the mean of the velocities at the corners.

A compass unit is contained within the particular meter shown in Fig. 12.19, near the tail fin, and the compass card can be clamped at the moment of reading in any one of 36 positions within the range of 0° to 360°. The direction of flow is read on the digital counter mechanism, the functions of velocity or direction measurement being selected by means of a switch at the readout unit to which the current meter is connected by cable. Streamlined sinker weights can be attached to assist in maintaining verticality: the quoted maximum operating depth is 300 m. It will be realized by the reader that the meter must be static since the speed of the current can only be measured by water flowing past the measuring device.

Floats

Although the current meter is by far the most widely used for the measurement of current direction and velocity (and normally it is the most accurate method), there are times when floats are used instead — notably when excessive velocities, depth and floating drift prohibit the use of the current meter.

Many types of float are available, ranging from small surface floats with flags on them to the double float which has a perforated cylinder or a canvas vane suspended at a known depth below the surface from a small floating buoy. An ordinary 2 m ranging pole can also be used, as it floats upright with about one-third showing. Flags of different colours are attached (small enough to avoid acting as sails) and the floats released at intervals.

The fixing of the positions of the floats may be done by simultaneous theodolite observations from the shore or, better, by frequent visits to each float in turn by a surveying boat, whose position is fixed by sextant observations on the shore-control points. The time of observation is noted in each case, so that the rate of drift can be calculated.

Surface floats give the velocity of the surface water only, and apart from the sensitivity of such light floats to wind, the choice of coefficient to convert $V_{surface}$ to V_{mean} ranges from about 0.7 to 0.95. The results are thus of doubtful accuracy.

The double float already mentioned can be used to give velocities at different depths (though allowance must be made for the effect of the surface float), the calculation of discharge thus being exactly as described for the current meter.

Float velocities are measured by releasing the floats at the appropriate point upstream and then timing them along the measured distance to a second station downstream.

A further alternative is the release of floats from points at bed level on the cross-section of the river or channel. Such floats will rise towards the surface with velocities dependent upon their buoyancy and size, and naturally will be carried downstream in the process under the influence of the water flowing down the river or channel. We have already indicated in Fig. 12.20 that the velocity of flow varies with depth and so, provided that the floats have a constant rate of rise, it will be apparent to the reader that the distances of displacement from the 'release' cross-section to the surfacing points will give measures of discharge per unit width at the corresponding points of release. Each displacement effectively represents an integration of a velocity profile with depth, and the accuracy of this method is related to simultaneous measurement of those displacements. One approach has been the use of a floating grid to trap solid floats released mechanically from bed level whilst another developed by Sargent and Rosser at the University of Lancaster (*New Civil Engineer*, 16 October 1980 and 13 November 1980) embodies the use of air bubbles as floats.

Chemical methods

These methods involve the introduction of a chemical into the stream, and there are many variations; one of these, the *salt-dilution* method, will be described briefly.

No measurements of area or distance are necessary, and the method is pre-eminently suitable for use in turbulent mountain streams. A salt solution of known concentration is added at a constant rate to the stream to be gauged and, by analysis, the subsequent dilution of the solution is determined. The samples are taken far enough below the entry point for complete mixing and uniform distribution to have taken place. If W is the weight of water discharged per second, W' is the weight of salt solution added per second, X is the percentage (by weight) of natural salt in the stream, X' is the percentage (by weight) of salt in concentrated solution and X'' is the percentage (by weight) of salt in the sample after mixing.

$$WX + W'X' = (W+W')X''$$

and so
$$W = W' \frac{(X' - X'')}{(X'' - X)}$$

X'' must be uniform at all points in the cross-section and $(X'' - X)$ must be found accurately: the salt used should be detectable in small quantities and should be stable in the water. When obtaining samples, bottles can be immersed in the stream at the relevant cross-section or a hand pump can be used for drawing off. Sodium dichromate is a suitable chemical according to Hutton and Spencer ('Gauging water flow by the salt-dilution method', S. Hutton and E. Spencer, *Proceedings of the Institution of Civil Engineers* August 1960), although at concentrations of 30 parts per million it is toxic to fish. Five parts per million have been used to achieve an accuracy of $\pm 1-2$ per cent in measurement of flow. Larger quantities of sodium chloride would be needed since it is present in natural waters and in this case X should not exceed $0.15\ X''$.

Exercises 12

1 Describe with the aid of sketches:

(a) how the profile of the bed of a tidal river, approximately 100 m wide and having a minimum depth of 5 m, may be determined;

(b) three methods of locating soundings off shore.

2 During a hydrographic survey three shore stations A, B and C were established such that $AB = 792$ m and $BC = 870$ m, the three stations lying in a straight line. Angles APB and BPC were measured simultaneously by sextant as $48°36'$ and $46°24'$, respectively, from a float P which was then due east of B. Determine the bearing of ABC, given that A lies southwards of B. *Answer*: $355°31'$.

3 Two stations A and B are 846 m apart. From theodolite stations P and Q on opposite sides of AB the following angles were observed:

$$A\hat{P}Q = 61°12'; \quad B\hat{Q}P = 53°28';$$
$$Q\hat{P}B = 44°11'; \quad P\hat{Q}A = 41°29'.$$

Calculate the distance between stations P and Q.

(London)

Answer: 713 m.

4   A, B and C are three stations on a coastline used to fix the position of a buoy, P, at sea, which lies on the opposite side of AC to B. $AB = 482$ m, and $BC = 344$ m. The seaward angle $ABC = 143°30'$, and angles APB and BPC are found to be $45°36'$ and $40°48'$ respectively. Find dimensions AP, BP and CP.
Answer: 671 m, 518 m and 453 m respectively.

5   A survey has to be carried out as a preliminary to the construction of a small harbour off a rocky coast. Describe how you would execute this work, the equipment you would need, and the data you would collect.

6   Describe a method of finding approximately the discharge of a river, stating the requirements of a site for the relevant measurements.

    Calculate the discharge of a river, given the following measurements made with a flow meter:

Distance across river from one bank (m)	0	10	20	30	40	50	60	70
Depth of bed (m)	0	0.7	1.2	1.5	1.8	1.5	0.9	0
Rate of flow at 0.6 depth (m/s)	0	0.15	0.24	0.30	0.36	0.33	0.24	0

Answer: 21 m^3/s approx.

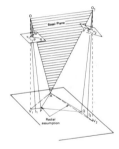

13 Photogrammetry

Photogrammetry can be defined as the method of determining the shapes, sizes and positions of objects using photographs and, therefore, it is an indirect method of measurement since photographic images are under scrutiny rather than the objects themselves. Some linear or angular measurements in the 'object space' need to be obtained or to be known for control purposes, but primarily the photography provides the information. As a method of measurement, photogrammetry has the disadvantage that it involves reducing the size of the object to the scale of a photograph, but when direct measurement is impracticable, impossible or uneconomical, photogrammetry might well offer the only method of dimensional analysis.

Photogrammetry has been developing for well over one hundred years and in fact the Frenchman Laussedat published a paper on the subject in 1854. The first photogrammetric measurements were made from terrestrial photographs taken from a photo-theodolite at ground stations. This method has been largely superseded by aerial photogrammetry which nowadays finds itself amongst a number of other remote sensing techniques for the production of maps.

Some basic principles of photogrammetry are given in this chapter together with a brief description of the techniques and instruments.

Aerial photogrammetry

Aerial photogrammetry involves the use of photographs taken in a systematic manner from the air. They are then controlled by land survey and measured by photogrammetric techniques. Accuracies achieved are comparable with those obtained by land survey and in many cases the work is carried out more economically. Very expensive equipment ·is required, and so the civil engineer will call upon the services of the specialist firms when plans for a proposed construction project are to be prepared. This means that he must have some understanding of the whole process, together with an appreciation of the specialist firms' problems or points of view, since he will be involved in the drafting of specifications and the adjudication of tenders. Furthermore the same air photographs may be used by him for a number of simple measurement techniques and for visual interpretation to provide much valuable information.

For the photographs to give a true plan certain conditions must be fulfilled, namely: (i) the ground on the photograph should be horizontal, (ii) the camera must not be tilted from the vertical when the exposure is made, (iii) the camera lens and photographic material should be as perfect as possible and there should be no atmospheric refraction. In addition, when flying at high altitude, the curvature of the earth is of some account.

These ideal conditions are rarely, if ever, satisfied. Normally the ground has some relief and small tilts are unavoidable. These produce distortions on the air photograph and associated scale variations. As a result the air photograph is not a true plan or map.

Fig. 13.1 The aerial photograph is a central perspective; the map is an orthogonal projection of the terrain, ground point A being positioned thereon at A_1

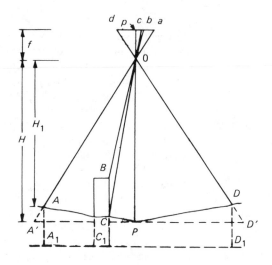

The vertical air photograph

Figure 13.1 represents a photograph taken over undulating ground, the camera being directed vertically downwards. Rays from points on the ground pass through the perspective centre O of the lens, and the images of those points appear on the negative, i.e. A on the ground appears at a. The point p is the photograph principal point, P is the ground principal point, and also they are the plumb points for a vertical photograph. The *principal point* p lies at the base of the perpendicular dropped from O on to the plate and is located by the intersection of lines joining the *collimation* or *fiducial* marks which are usually found at the corners or on each of the boundaries of the photograph: Op is known as the principal distance. If the ground were level, as shown by the dotted line through P, the scale for the photograph would be $f{:}H$ (triangles Oda and $OD'A'$ are similar, and so $da/D'A' = f/H$). But if a horizontal plane were drawn through A, a further pair of similar triangles could be obtained from which the scale would be $f{:}H_1$ and so on for all points not at datum. Thus the scale at a point on the photograph depends upon the height of that point above the chosen datum.

Example 13.1 Vertical photographs at a scale of 1:20 000 are to be taken of an area whose mean ground level is 500 m above mean sea level. If the camera has a focal length of (a) 210 mm, (b) 152 mm, find the flying height above mean sea level.

(a) $f = 210$ mm

$$\frac{1}{20\ 000} = \frac{0.210}{H - 500}$$

Therefore

$$H = \textbf{4700 m}$$

(b) $f = 152$ mm

$$\frac{1}{20\ 000} = \frac{0.152}{(H - 500)}$$

Therefore

$$H = \textbf{3540 m}$$

Fig. 13.2

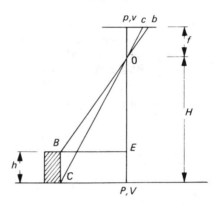

Consider the side of the high building BC in Fig. 13.2 and its consequent image cb on the negative. B is vertically above C, and in plan the two coincide, but on the photograph the side of the building cb would be observed as well as the roof, and this building would appear to be leaning outwards from the centre of the photograph. It may be shown that the distortion of the image is proportional to the distance from the photograph plumb point, v, which in this particular case coincides with the principal point, p.

Figure 13.2 shows the building BC, shown for the sake of clarity with its base on the datum plane, and with B at a height h above that plane. So long as pOP is truly vertical, then v, V (or p, P), B, C, c and b are all contained in one vertical plane, since B is vertically above C. This vertical plane intersects the plane of the negative, and since two planes intersect in one line only, v, c and b must be collinear and the displacement of b from c is radial from the plumb point v.

Consider the similar triangles vbO, EBO,

$$\frac{vb}{vO} = \frac{EB}{EO}, \quad \text{i.e.} \quad \frac{vb}{f} = \frac{EB}{H - h}$$

Also from similar triangles vcO, VCO,

$$\frac{vc}{vO} = \frac{VC}{VO}, \quad \text{i.e.} \quad \frac{vc}{f} = \frac{VC}{H} = \frac{EB}{H}$$

Photogrammetry **425**

Therefore

$$\frac{vb}{vc} = \frac{H}{H-h} = \frac{vb}{vb-bc}$$

and

$$\frac{vb}{bc} = \frac{H}{h}$$

Thus, the distortion due to height BC

$$= bc = \frac{h}{H} \cdot vb$$

This expression only holds for a truly vertical photograph, but height distortion is radial from the plumb point whether the photograph be vertical or tilted.

Example 13.2 In Fig. 13.2, for a certain photograph, $H = 1200$ m, $f = 152$ mm, vc is measured as 88.36 mm and vb as 90.78 mm, estimate the height of the building.

It will be seen that

$$bc = vb - vc$$
$$= 90.78 - 88.36$$
$$= 2.42 \text{ mm}$$

Now

$$bc = \frac{h}{H} \, vb$$

Thus

$$h = \frac{1200 \times 2.42}{90.78}$$

$$= \mathbf{31.99 \text{ m}}, \text{ (say 32 m)}$$

The reader should note that had C not been at datum level but (say) 95 m above we would now write

$$\frac{vb}{f} = \frac{EB}{H-(95+32)}$$

and

$$\frac{vc}{f} = \frac{VC}{H-95}$$

Thus for the same values of vb and vc we would obtain,

$$\frac{90.78}{152} = \frac{EB}{H-127} = \frac{VC}{H-127}$$

and

$$\frac{88.36}{152} = \frac{VC}{H-95}$$

whence

$$VC = \frac{98.78}{152} (H-127) = \frac{88.36}{152} (H-95)$$

and so $H = \mathbf{1295.4}$, say 1295 m

Small tilts from the vertical in unknown directions are unavoidable, and the average angle of tilt (from the vertical) generally will not be below $2°$ without gyroscopic aid, but it may be reduced to say $\frac{1}{2}°$ when using an automatic pilot. Like relief, tilt causes distortion.

Fig. 13.3

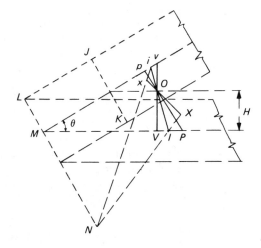

Figure 13.3 shows perspectively the negative plane tilted at an angle θ to the horizontal. O is the perspective centre of the camera lens, V and v are the ground and photograph plumb points, and are in the same vertical line with O. P and p are the ground and photograph principal points, pOP being perpendicular to the negative plane. iOI is the bisector of angle pOv and POV, both of which equal θ. The points I and i are termed the ground and photograph *isocentres*. Points such as i and I, v and V, x and X which are on any ray through O, are known as *homologous points*. It will be seen that iOI makes the same angle $(90 - (\theta/2))$ with both negative and ground planes. The vertical plane containing O, v, V, p and P is termed the *principal plane*, its intersection with the negative plane giving the negative, or plate, principal line pv.

The horizontal plane containing O gives the horizon trace JK on the negative, and any horizontal line parallel to this on the negative is termed a *plate parallel*. Also the images of lines parallel to VIP on the ground intersect on the horizon trace at a point termed the vanishing point. The scale at the principal point and at any point on the plate parallel through p is equal to

$$\frac{Op}{OP} \quad \text{i.e.} \quad \frac{f}{H \sec \theta}$$

Whilst the scale at v is

$$\frac{f \sec \theta}{H}$$

Accordingly the scale of a tilted photograph changes along the principal line (and therefore between plate parallels) at the rate $\sin \theta/H$, increasing from p to v. Note that the scale at i is f/H.

The image of a point X on the ground plane is given at x, so that the points I, X, x and i all lie in one plane. This plane, the ground plane and the plane of the negative intersect at one point only, which is denoted as N.

It has been stated earlier that the line iOI joining the isocentres makes the same angle with the ground plane as with the negative, with a result that the triangle iMI is isosceles.

Hence
$$iM = IM$$

Since MN is common and $i\hat{M}N = I\hat{M}N = 90°$ then triangles iMN and IMN are congruent and $M\hat{i}N = M\hat{I}N$. Thus in Fig. 13.3 the angle subtended at I by X and P on the ground equals the angle subtended at i by x and p on the negative, and by extension of the analysis it will be seen that in general angles subtended at the ground isocentre by points on the ground are equal to angles subtended at the photo isocentre by the corresponding images of these points.

Fig. 13.4

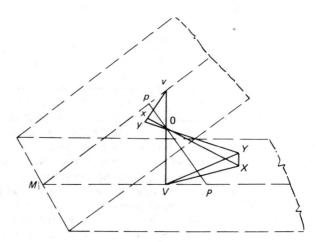

In Fig. 13.4 the displacement of the image of Y with respect to the image of X is shown to be radial from the plumb point on the tilted negative since Y is vertically above X, so that Y, X, V, O, v, x and y are all contained in a vertical plane which intersects the negative to give the straight line containing v, x and y.

Thus in the tilted photograph,

(1) distortions due to height are radial from the plumb point,
(2) distortions due to tilt are radial from the isocentre.

An assumption which can be adopted is that both distortions are radial from the principal point provided that the tilt angle is small — say less than 2° or 3° — and the ground height variations are small compared to the flying height. This assumption makes for considerable simplification, since the principal point can be located by means of

Fig. 13.5

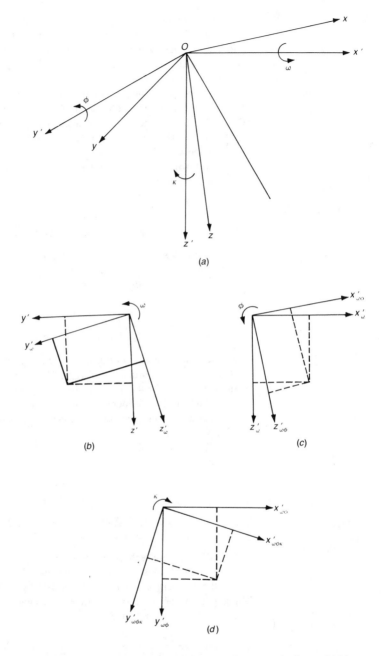

(a)

(b)

(c)

(d)

the fiducial marks, whereas the plumb point cannot be located without a knowledge of the plan positions and heights of at least three points.

Analysis of tilt   The combined effects of relief and tilt on the position of a point on a tilted photo are complex. So too are the scale variations produced.

Whilst an air photograph represents a perspective view of the ground this is not a true plan or map. Any methods of mapping using air photographs must seek to eliminate these scale variations. Tilt θ can be resolved into three rotational components acting about three mutually perpendicular axes. Figure 13.5(a) shows a ray from an object point to the perspective centre of a lens, which is the origin of mutually perpendicular axes, Ox', Oy' and Oz' and Ox, Oy and Oz. Assume that $x'Oy'$ forms a horizontal plane and that rotations ω, ϕ and κ be applied in turn (Fig. 13.5(b), (c), (d)) so that finally plane $x'Oy'$ lies in the plane xOy.

The following co-ordinate relationships can then be derived and expressed in matrix form (see Example 13.8)

$$
\begin{pmatrix} x' \\ y' \\ z' \end{pmatrix} = \begin{pmatrix} \cos\phi\cos\kappa \\ \cos\omega\sin\kappa+\sin\omega\sin\phi\cos\kappa \\ \sin\omega\sin\kappa-\cos\omega\sin\phi\cos\kappa \end{pmatrix}
$$

$$
\begin{pmatrix} -\cos\phi\sin\kappa & \sin\phi \\ \cos\omega\cos\kappa-\sin\omega\sin\phi\sin\kappa & -\sin\omega\cos\phi \\ \sin\omega\cos\kappa+\cos\omega\sin\phi\sin\kappa & \cos\omega\cos\phi \end{pmatrix} \begin{pmatrix} x'_{\omega\phi\kappa} \\ y'_{\omega\phi\kappa} \\ z'_{\omega\phi\kappa} \end{pmatrix}
$$

$$
= (M_{\omega\phi\kappa}) \begin{pmatrix} x'_{\omega\phi\kappa} \\ y'_{\omega\phi\kappa} \\ z'_{\omega\phi\kappa} \end{pmatrix} \tag{13.1}
$$

Thus the point, whose co-ordinates were x', y' and z' in the first system, now has co-ordinates of $\chi'_{\omega\phi\kappa}$, $y'_{\omega\phi\kappa}$ and $z'_{\omega\phi\kappa}$ in the rotated system. We can take this latter as the xyz system when we put $x'_{\omega\phi\kappa}$ $= x$ etc. and we can consider the system to now refer to a non-vertical photograph at exposure, the z axis lying in the optical axis of the camera, pOP in Fig. 13.3

For the near-vertical case,

$$
\begin{aligned}
x' &= x - \kappa y + \phi z \\
y' &= \kappa x + y - \omega z \\
z' &= -\phi x + \omega y + z
\end{aligned}
$$

and $\qquad \dfrac{x'}{z'} = \dfrac{x-\kappa y+\phi z}{-\phi x+\omega y+z}, \qquad \dfrac{y'}{z'} = \dfrac{\kappa x+y-\omega z}{-\phi x+\omega y+z}$

with x, y as photograph co-ordinates and x', y' as equivalent or corrected co-ordinates in the horizontal plane. The z'-axis is vertical, whilst the z-axis is perpendicular to the tilted photograph plane and if this photograph be rotated into the horizontal plane, and the fact that $z = f = z'$ be accepted, then

$$
x' = f\,\frac{x-\kappa y+\phi f}{-\phi x+\omega y+f} \qquad \text{and} \qquad y' = f\,\frac{\kappa x+y-\omega f}{-\phi x+\omega y+f}
$$

If the ground plane and photograph plane be taken to be horizontal and parallel (Fig. 13.6), then

$$
x_G = x'\,\frac{H}{f} \qquad \text{and} \qquad y_G = y'\,\frac{H}{f}
$$

Fig. 13.6

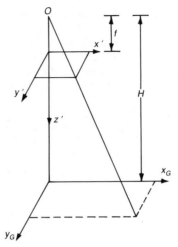

where x_G and y_G are ground co-ordinates of an object point: x' and y' are those of the corresponding image point.

If the photograph be slightly tilted at exposure to produce co-ordinates of x and y then,

$$x_G = H \frac{x - \kappa y + \phi f}{-\phi x + \omega y + f} = H \frac{(Ax + By + C)}{(Dx + Ey + 1)}$$

$$y_G = H \frac{\kappa x + y - \omega f}{-\phi x + \omega y + f} = H \frac{(-Bx + Ay + F)}{(Dx + Ey + 1)}$$

where $A = \dfrac{1}{f}, \quad B = -\dfrac{\kappa}{f}$ etc.

The six constants call for a knowledge of the corresponding co-ordinates, image and ground, for three points and thence the co-ordinates of other image points can be used to devise ground-point information. Also ω, ϕ and κ could be estimated.

For the general case (13.1) when ω, ϕ and κ are of appreciable magnitude, the reader can verify that eight unknowns are involved and that ground and photograph co-ordinates can be related as follows:

$$x_G = H \frac{A'x + B'y + C'}{D'x + E'y + 1}$$

$$y_G = H \frac{F'x + G'y + K'}{D'x + E'y + 1}$$

Four suitably positioned ground control points are now required to determine the eight constants. This type of approach can be applied to photography carried out to study traffic flows in a city environment or along slip roads, the camera being mounted on a tall building with its axis obliquely inclined to the ground plane. Control points can consist of manhole covers, lamp-post bases etc. if so wished, but no three points should be collinear.

Example 13.3 The image of a ground point has co-ordinates of $x = +62.48$ mm and $y = +64.19$ mm on a photograph taken at a height of 1250 m with a camera of focal length 152 mm. Assuming that rotations of $\omega = \phi = \kappa = +1°$ actually occurred at exposure determine the corrected co-ordinates and hence ground co-ordinates with respect to the camera position.

Since $x' = f\dfrac{x - \kappa y + \phi f}{-\phi x + \omega y + f}$

$$= 152 \left[\frac{62.48 - \dfrac{\pi}{180} \times 64.19 + \dfrac{\pi}{180} \times 152}{\dfrac{-\pi}{180} \times 62.48 + \dfrac{\pi}{180} \times 64.19 + 152} \right]$$

$$= 152 \left[\frac{62.48 - 1.12 + 2.65}{-1.09 + 1.12 + 152} \right]$$

$$= 64.00 \text{ mm}$$

Also $y' = f\dfrac{\kappa x + y - \omega f}{-\phi x + \omega y + f}$

$$= 152 \left[\frac{\dfrac{\pi}{180} \times 62.48 + 64.19 \dfrac{-\pi}{180} \times 152}{\dfrac{-\pi}{180} \times 62.48 + \dfrac{\pi}{180} \times 64.19 + 152} \right]$$

$$= 152 \left[\frac{1.09 + 64.19 - 2.65}{-1.09 + 1.12 + 152} \right]$$

$$= 62.62 \text{ mm}$$

Also $x_G = x'\dfrac{H}{f}$

$$= \frac{64.00 \times 1250}{152}$$

$$= \mathbf{526.32 \text{ m}}$$

and $y_G = y'\dfrac{H}{f}$

$$= \frac{62.62 \times 1250}{152}$$

$$= \mathbf{514.97 \text{ m}}$$

Air photography

The air photograph is the basic source of data and unless it satisfies certain conditions, accurate measurements cannot be made. The two essential conditions of good air photography for photogrammetric use are:

(1) each photograph is to be of the highest quality, both geometrically and pictorially, and

(2) each photograph is to be taken from a given position in space with the camera axis pointing as near as possible vertically downwards.

Lenses

For photogrammetric use, any lens should be virtually distortion free yet have a high resolving power, excellent light distribution and optimum colour correction of the visible and near-visible parts of the spectrum. In a modern lens, a distortion of less than 0.01 mm in the focal plane is normal.

Camera lenses may be classified as super-wide-angle, wide-angle, normal-angle and narrow-angle. Super-wide-angle lenses are those types having angular fields of the order of 120°, e.g. the Wild Super-Aviogon II, which has a focal length of 88.5 mm. Such lenses are intended for small-scale mapping and they have much larger ground coverage than wide-angle lenses at a given flying height. Wide-angle lenses have angular fields of the order of 90°, e.g. the Wild Universal Aviogon, which has a focal length of 152 mm. This type of lens has an extensive usage in topographic and large-scale mapping. Normal-angle lenses have fields of the order of 60°: the Zeiss Topar A is typical of this group. Its focal length is 305 mm and its applications include orthophoto mapping and surveys of built-up areas, since this type gives a reduction of 'dead spaces' when photographing such regions. Narrow-angle lenses, which are also of value in urban surveys, have fields of the order of 30° and focal lengths of 610 mm (e.g. Zeiss Telikon A) and they allow larger-scale photography at minimum flying height than the other types mentioned.

Films

In modern practice, film is normally used as the photographic medium, rather than plates. It is important that the base on which the light sensitive emulsion is coated be dimensionally stable so that the geometrical properties of the photography are maintained. Black and white, infra-red, colour negative, colour reversal and false colour films are all available, although the former is still the most commonly used.

Black and white panchromatic film is sensitive to wavelengths ranging from 300 nm to 700 nm, the resulting photographs consisting in the main of various shades of grey (tones) which depend upon the amount of light reflected by the subject to the camera. The more light reflected, the whiter or brighter the image appears, and so an asphalt-surfaced road can appear white if the sun's rays are reflected from that surface directly on to the film: if light is not directed thereon then a black image is filmed. Since natural features can reflect light from the sun or sky in many directions, generally only some rays will strike the film thereby resulting in grey tones. The individual details then become identified by changes or contrasts in these grey shades as well as by, say, shape, shadow and associated features. Panchromatic film is sensitive to blue light (in the shorter range of wavelengths quoted) and also to orange light, with a minimum sensitivity to green light around 550 nm. As light passes through the atmosphere from ground to camera it can encounter fine suspended particles in the form of dust or water and as a result scattering can occur, the shorter wavelengths being scattered to a greater degree than the longer lengths. In order that sharp images of the ground detail be produced, this unwanted light is removed by a yellow filter.

The reflectance of the infra-red range by green living vegetation (the chlorophyll effect) results in woods and meadows appearing more brightly in infra-red photographs, with the possible differentiation of various species. In contrast, areas of man-made structures, or polluted ground give a low reflection and appear dark. This factor was first used in camouflage detection since green-painted surfaces do not reflect this radiation to the same extent as green living vegetation. Now it has a number of civilian applications not only in the control of pollution but also for tracing buried pipelines, etc. Film sensitive to infra-red radiation of the order of 900 nm wavelength is used, but since this film too is sensitive to blue light a deep red filter is used to absorb visible radiation. On such film water, having a reduced reflectance, becomes sharply defined in dark shades in the photographs and, accordingly, shore lines and rivers become marked.

Multi-band cameras producing separate images of the same terrain are of value during interpretation studies as follows: blue waveband — penetration into shaded areas; green waveband — plant species identification; red waveband — general interpretation, underwater detail; infra-red waveband — vegetation differentiation, drainage patterns. By means of filters it is possible to produce colour and false colour images from these 'black and white' originals. Colour and false-colour photographs both show general topographic details and they are of particular value in interpretation studies, since the different wavelengths are depicted as colours, but the latter was developed with the detection of camouflage as a primary objective. Accordingly, the infra-red radiation is then represented such that an object which is green-coloured will appear magenta-coloured if it strongly reflects the radiation, but will appear blue-coloured if it absorbs that radiation: water appears blue-coloured for that reason.

Cameras

Figure 13.7 shows the Wild RC20, a film camera which takes photographs 230 mm square. Amongst its components are the mount, lens cone, drive unit, cassettes, control unit and view finder.

The mount is formed by a two-point base plate and box-type housing with anti-vibration dampers. Servo motors are fitted that can rotate the camera about three axes, thus levelling corrections against tip and tilt up to $\pm 5°$ can be introduced together with a drift correction up to $\pm 30°$, all operated by remote control from the viewfinder/navigation sight.

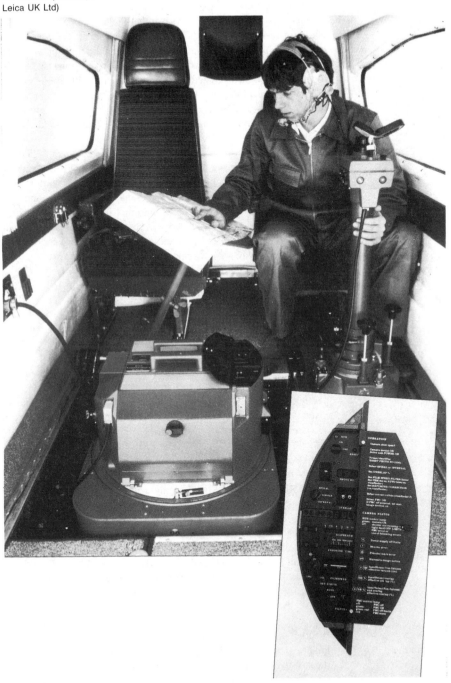

Fig. 13.7 Wild RC20 and navigation sight, with close up inset of the camera (*Courtesy*: Leica UK Ltd)

The camera can be fitted with the range of Wild lens cones, all of which are interchangeable in flight. The lens cones are fitted with shutters with continuously adjustable exposure times between 1/100 and 1/1000 second; the aperture is also continuously adjustable. A set of filters is provided for each lens so that the full range of black and white, infra-red and colour film can be handled.

The drive unit contains the means to advance the film and to flatten it against a pressure plate during exposure whilst take-up and feed spools for the film are incorporated into a separate removable cassette. The drive functions of the unit are controlled by a microprocessor which is one of six included in the electronic control. A second microprocessor sets up the operator requirements and the parameters for serial exposures, whilst the third controls an automatic exposure meter mounted within the lens cone and makes the shutter and aperture adjustments. Microprocessors four, five and six arrange respectively for the imaging of digital data in the film margin, forward motion compensation to make the photographic image sharp and the input of shutter speed for the projection of marginal information. The latter also sets the fiducial marks which are exposed exactly half way through the period in which the shutter is open. The forward motion compensator (FMC) is an important feature if modern, slow, fine grained film is used, otherwise blurred imagery can result.

Four optical fiducial marks are given in the margin frame of the photograph, one in each corner and there is also a fiducial mark at the centre of each side which may be either V-shaped or of optical form on request. In addition data such as time and date, height, number and identification are presented.

The navigation sight of the RC20 reveals to the operator the flight line, the nadir point, the frame size for the lens cone being used, the lateral limits of the format and auxiliary strips for maintenance of parallel flight lines. In addition a system of moving spiral lines is displayed which are synchronized with the speed of the image of the ground and thus allow overlap control. Accurate synchronization is also essential for the control of the FMC system. In addition, for a high degree of accuracy of overlap, an intervalometer is provided which allows integral time measurement.

The RC20 can be used in a single camera system or dual camera system, controlled from the same view finder/navigation sight. In the latter system the two can be run synchronously or separate with the same or different film.

Whilst similar in principle to ordinary cameras, it will be realized from the above description that in order to meet their objectives air cameras must possess many special design features; they are thus expensive precision instruments.

Weather conditions

Weather conditions have an obvious effect upon the photographic quality. The ideal day for air photography is a bright one, and below the required aircraft flying height it should be clear of cloud and haze. Preferably some thin cloud should lie above to reduce the direct rays of the sun, thereby preventing objects on the ground casting strong shadows which obscure detail: if the sun is shining it should be at a high altitude so that shadows are short. There should also be little or no air turbulence at the required flying height as this would adversely affect the stability of the slow-flying aircraft.

These conditions vary with the time of year and the locality. In the United Kingdom the best months for air photography are April to October. Delays due to unfavourable weather are usually inevitable and it is prudent to plan well ahead when using aerial photogrammetry for a mapping project.

Survey flying

For photogrammetric use, the photographs must be taken in a particular way. Not only must the area of ground be covered by a series of photographs which have no gaps between them but also it is essential for stereoscopic examination that every point on the ground appears on at least two photographs.

To achieve this the photographs are taken in a series of strips. As the aircraft flies along a straight line, photographs are taken at intervals such that there is a minimum overlap of 60 per cent between adjacent photographs. This *fore and aft overlap* ensures stereoscopic coverage along the strip with some margin for error. Two adjacent photographs taken in this way are called *a stereopair*.

Figure 13.8 illustrates three successive photographs taken from air

Fig. 13.8

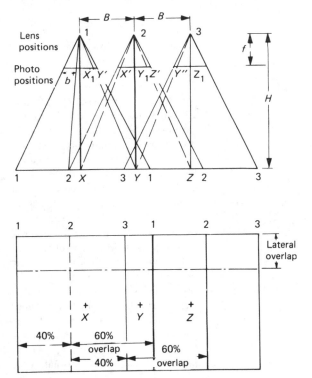

stations 1, 2 and 3, respectively. X, Y and Z are ground principal points, and X_1, Y_1 and Z_1 are photograph principal points. The distances between the air stations are termed air bases, B. It will be seen that there is a common overlap of 20 per cent between the three photographs, and that the ground principal point Y, which is vertically below the photograph principal point Y_1, will appear on the three photographs, appearing at Y' and Y'' on photographs 1 and 3 respectively. Thus by similar triangles,

$$\frac{f}{H} = \frac{X_1 Y'}{XY} = \frac{X_1 Y'}{B} = \frac{b}{B}$$

$$B = \frac{bH}{f}$$

b is also equal to the portion of the photograph not overlapped by its immediate neighbour. If the fore and aft overlap is 60%, then $b = (40/100) \times 230$ mm for a 230 mm square photograph. The number of photographs in a strip for any given scale of photography can now be derived since B can be calculated. At least one photograph should be added to that number to make certain of end coverage.

Subsequent strips are flown parallel to the initial one and at a spacing which gives a lateral overlap between strips of between 15 and 35%. This allows the strips to be connected together and leaves some margin for error in navigation. If w be the lateral portion of the photograph not overlapped and W is the corresponding distance on the ground,

$$\frac{w}{W} = \frac{f}{H}$$

$$W = w \cdot \frac{H}{f}$$

If the lateral overlap is 25%, then $w = (75/100) \times 230$ mm for a 230 mm square photograph. The corresponding ground dimension, W, can now be derived for any given scale of photography. W is also the distance between flight lines and the number of flight lines used to cover an area of given width can be worked out. Again, one extra can be allowed for complete coverage.

The net area of ground covered by one photo is $W \cdot B$ and it would seem that the number of photographs required to cover an area A would be $A \div (W \cdot B)$. This would, however, only give a preliminary estimate, since the number required will depend upon the arrangement by which the strips of photographs cover the area.

The flying of air photography requires great skill and practice. At the time of photography the camera axis must be within a degree or two of the vertical and the predetermined flying height must be maintained. Also the camera orientation must be correctly set in relation to the aircraft heading. To prevent the aircraft drifting off course under the action of cross winds, the aircraft must be directed so that it will be kept along the correct line, but in this case the photographs will not form a continuous strip unless the camera is turned so that the photographs are taken squarely on course: sighting devices, previously mentioned, will allow for the necessary corrections. Course may be set by compass or on to a distant landmark, but it is difficult to hold on to a course in lengths over, say, 20 to 25 km without intermediate marks. Systems are manufactured for tracking and controlling the flying of straight and parallel lines. One of these is the Litton Photogrammetric Integrated Control System, a major component being a navigation system which utilizes inertial platforms. Procedures and computer software have been developed with particular reference to aerial photography, and an Interface Unit allows the passage of control commands and annotation data to two cameras, such as the Zeiss RMK series.

Before carrying out any air photography a flight plan and specifications for the work must be prepared. Model specifications in respect of air photography are given in *Specification for Vertical Air Photography* (March 1980) published by the Land Surveyors Division of the Royal Institution of Chartered Surveyors.

Plotting and measurement

When the photographic process has been completed, measurements can be taken from the photographs by a number of different methods which are dependent upon the required end product.

For instance Fig. 13.9 shows an aerial photograph of a railway station and part of a line map derived therefrom. Heighting information in the form of spot heights or contours can also be provided on the map.

The most accurate methods of measurement are of a rigorous form and involve the formation of a stereomodel, which is an exact replica of the ground, at a reduced scale, from each stereopair of photographs. This stereomodel is formed when the two camera positions and camera tilts, as at the time of photography, have been re-established. The measuring device produces a three-dimensional model of the terrain from two photographs and dependent upon the design either ensures the accurate plotting and heighting of features within the models without calculations or, more typically, determines the three dimensional co-ordinates of the model points.

The stereomodel is viewed through a binocular viewing system consisting of a series of lenses, prisms and mirrors with in-built magnification. By this means the left eye sees one photograph at the same time as the right eye sees the other, and the slightly differing views fuse to form a model. Measurement of the model is based upon the *floating mark principle*. Within the viewing system for each eye there is a small measuring mark (dot) which appears to lie in the same focal plane as the photograph. When these marks fall on the same ground point, as imaged on the two photographs, the operator sees a single fused mark (*the floating mark*) which appears to rest on the surface of the stereomodel. If one mark is slightly displaced by adjustment of the height (Z) screw of the device, then the floating mark will appear above or below the model surface.

Three steps are to be followed when establishing the stereomodel, namely, *inner orientation, relative orientation* and *absolute orientation*.

Fig. 13.9(a) Aerial photograph of a Railway Station and surrounding land (*Courtesy:* British Transport Commission)

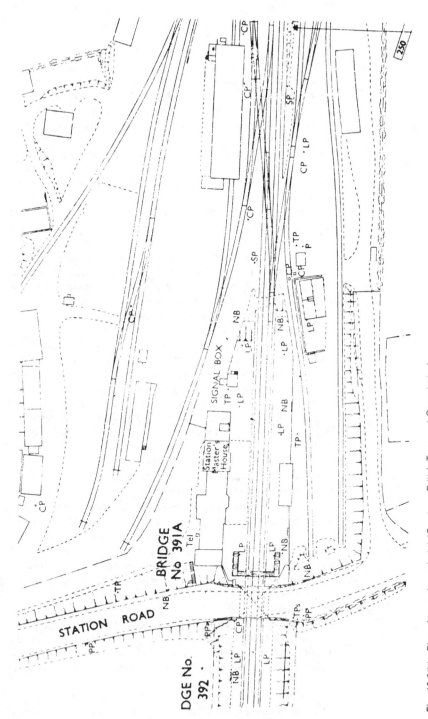

Fig. 13.9(b) Plan of area shown in (a) (*Courtesy:* British Transport Commission)

In the first step the internal geometry of the air camera used during photography is established in the plotting machine so that the projected bundle of rays will now be congruent with that which entered the camera at the time of photography. Relative orientation places the pair of photographs in their correct relative positions and, when completed, a stereomodel is formed by the intersection of corresponding rays. During the third stage, the stereomodel is rotated until it lies in its correct relationship to the datum plane above which heights are to be measured: it is also scaled and set in its correct azimuth. Theoretically the plan positions of two ground points and the height of three points need to be known. In practice, more points are fixed so as to provide redundant information which improves the accuracy and prevents mistakes being undetected. Ground control, established in horizontal (plan) and vertical (height) position by surveyors working in the field is essential for the absolute orientation stage. All such points can be selected (and surveyed) after the photography but if there is little firm detail in the area they will need to be premarked. Any points selected as control must be clearly identifiable on the photographs and the surveyor must prepare a clear witness diagram to enable the plotting machine operator to identify the point on the stereomodel. It will be apparent that the accuracy of photogrammetric work will depend upon that of the ground survey and consequently this should be carefully planned and carried out.

To minmize the amount of control fixed in the field *aerial triangulation* may be used. Aerial triangulation is a process for intensifying a control network from a number of ground control points which may be several overlaps apart. It is carried out from measurements made on the photograph along each strip and in the lateral overlaps, and it is then possible by computation to link together the individual models to form strips and blocks. Errors which occur in this process may be reduced to an acceptable level by adjusting the block to fit ground control points positioned every five or six models along each strip. Following this adjustment, the space co-ordinates of photo control on the intervening models are obtained.

A map produced by first-order plotter is comparable in every respect with one produced by land survey. At this stage, however, it must be borne in mind that there will always be some detail and contours which cannot be plotted from the photography owing to difficulties of identification, foliage etc. The map must be taken into the field for completion and checking by the surveyor.

Digitized systems. Over the last decade manufacturers have been moving away from analogue machines to digitized systems, and devices like the Wild A10 (Fig. 13.10) which is a first-order plotting machine that incorporates mechanical projection, have been taken out of production. Such machines will, however, remain in service for some time.

Zeiss, for instance, now offer the P series Planicomps, launched in 1987, which are data acquisition workstations designed to be integrated into the PHOCUS photogrammetric-cartographic software system. The

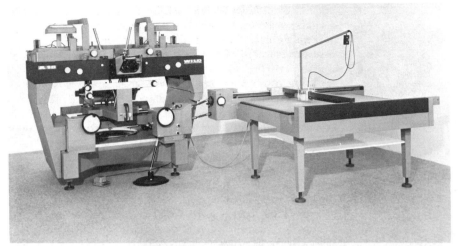

Fig. 13.10 Wild A10 Universal Plotter (*Courtesy*: Leica UK Ltd)

Fig. 13.11 Planicomp P2 analytical plotter (*Courtesy*: Carl Zeiss)

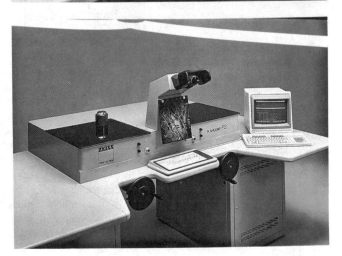

Fig. 13.12 Planicomp P3 analytical workstation (*Courtesy*: Carl Zeiss)

Photogrammetry **443**

prime object of PHOCUS is to establish, and manage, a cartographic data base object code to cater for discrete applications.

There are three Planicomps in the series namely, the Universal Analytical Plotter P1, Analytical Plotter P2 (Fig. 13.11) and Analytical workstation P3 (Fig. 13.12). PHOCUS is based upon the HP1000/RTE-A and MicroVAX computers. However, the three devices can be used with a PC running MS/DOS and in this configuration Zeiss provide interface software so that the machine can be operated from Microsoft Windows and the data output through computer aided design packages such as AutoCAD.

The essential elements of the PC/P3 Planicomps are a desktop viewer into which the stereo pair of photodispositives is placed: a usable photo area of 240 × 240 mm is maintained with up to 100% overlap. Photo alignment is achieved by servomotors which are controlled by microprocessor, operating the P-CAP base module software. The machine is operated by a mouse, the P-Cursor, on a digitizing tablet, although optional handwheels and foot disks can be supplied. In addition to the sliding ball for $X - Y$ movement the P-Cursor contains a knurled knob for Z direction and keys for fast movement; a resolution of 0.025 mm is claimed.

Fig. 13.13 Typical configuration of a modern plotting system

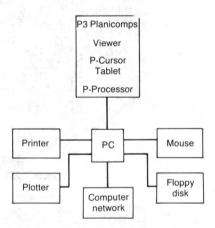

As well as operating the viewer and mouse the P-CAP software module provides the interface to other software operated by the PC or workstation, where the data is stored and processed. Full capabilities of the machine are dependent upon the software modules available in the computer but will generally include a data base which can be viewed, edited and plotted. A typical arrangement is shown in Fig. 13.13 and it will be noted that connection can be made to a plotter so that plans can be produced. However, one of the most important attributes of such systems is the application to digital mapping when strings of co-ordinates of all salient features are derived and stored. These can be used to produce, or to update an existing, plan and transferred to a CAD package where civil engineering project details can be added.

There are various ways of abstracting digital information from maps and photographs. Zeiss, for instance, market a high resolution black and white or colour scanner, PS1 PhotoScan. This can digitize a 230 mm wide air-photograph in 16 passes each 15.36 mm wide and consisting of 2048 pixels (dots making up the picture), the data being transferred to the memory of the machine in raster form, i.e. as a regular array grid of pixels. The high quality image can be inspected on a graphics terminal and for example used to recognize and correlate well-defined points like the control stations for the air survey.

Measurement techniques with simple equipment

For most civil engineering applications of aerial photogrammetry the rigorous methods of photo measurement as previously discussed will be needed to obtain the desired accuracy. Nevertheless there may be occasions where a lower order of accuracy is acceptable and there are a number of measurement techniques which employ simple and inexpensive instruments. Atlhough all are of a non-rigorous nature and make certain assumptions concerning the air photography or the ground, many of these simpler techniques (and a knowledge of their limitations) can be of value to the civil engineer.

Stereoscopic observation

For stereoscopic observation, pairs of photographs must be arranged so that the left eye observes the left-hand picture, and the right eye observes the right-hand picture only. The two images combine in the brain to give a single image of the common overlap, which gives an impression of relief in a manner similar to an observer himself looking over the terrain. The simplest and most common aid to produce a stereomodel is the stereoscope, of which there are two main types: the *pocket stereoscope* and the *mirror stereoscope*.

The pocket stereoscope is cheap, portable and very suitable for field use but it is difficult to use with air photographs which are 230 mm square and accordingly the mirror stereoscope, shown in Fig. 13.14,

Fig. 13.14 Folding mirror stereoscope and stereometer

Fig. 13.15

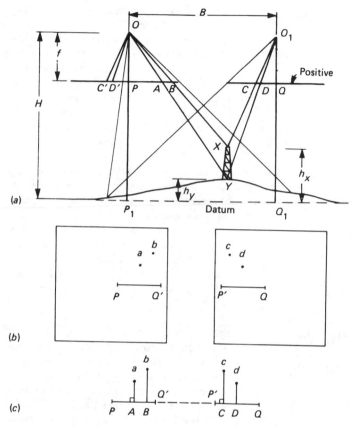

(a)

(b)

(c)

is the more popular. This provides for a physical separation of the two air photographs, even when they are 230 mm square, by employing double reflection of the rays off a system of mirrors (at 45° to the horizontal).

The pair of photographs must be orientated in a specific way under the stereoscope. The principal point of each photograph is found, and the principal point of the right-hand photograph identified and marked on the left-hand photograph, and the principal point of the left-hand photograph marked on the right-hand photograph. The photographs are separated with the overlap inwards, to give satsifactory fusion and positioned such that the two transferred principal points are collinear with the principal points as in Fig. 13.15(c). Viewing should always be parallel to this base and the photographs should be evenly illuminated, and secured to prevent movement.

Parallax heighting

Simple formulae, giving the heights of objects, may be developed if it be assumed that the stereopair of photographs have no tilt and are taken from the same flying height.

In Fig. 13.15(a), O and O_1 are successive air stations B apart, from which vertical photographs have been taken. Consider the top and bottom of the pylon XY with images (Fig. 13.15(b)) at b, c and a, d respectively. These images are projected in Fig. 13.15(c) to B, C and A, D on the line PQ' $P'Q$, which contains the principal points P and Q and their conjugate images Q' and P'.

Parallax is defined as the algebraic difference of the distances of two images of a ground point from their respective principal points, measured parallel to the air base. Therefore, referring to X, the parallax has a magnitude of $PB-(-CQ)$ when distances are measured positive to the right, whilst for Y the parallax is $PA-(-DQ)$. In Fig. 13.15(a), OC' and OD' have been drawn parallel to $O_1 C$ and $O_1 D$ respectively so that $C'B$ equals the parallax of $X(p_x)$ and $D'A$ equals the parallax of $Y(p_y)$.

Triangles $OD'A$ and $YO_1 O$ are similar, and so

$$\frac{OO_1}{D'A} = \frac{H-h_y}{f}$$

and since triangles $OC'B$ and $XO_1 O$ are similar,

$$\frac{OO_1}{C'B} = \frac{H-h_x}{f}$$

$$D'A = p_y = f\frac{B}{H-h_y} \quad \text{[parallax of } Y\text{]}$$

$$C'B = p_x = f\frac{B}{H-h_x} \quad \text{[parallax of } X\text{]}$$

In each case the parallax depends upon $(H-h_y)$ or $(H-h_x)$ and in general all points at a height of h_x and h_y above datum must have the same parallax on this pair of photographs. Differences in parallax values are caused by changes in distance from the air base, and these differences give rise to the perception of relief when the pair of photographs is viewed stereoscopically.

Let the difference in parallax between X and Y be dp,

$$dp = p_x - p_y = fB\left[\frac{1}{H-h_x} - \frac{1}{H-h_y}\right]$$

$$= fB\left[\frac{h_x - h_y}{(H-h_x)(H-h_y)}\right]$$

Since $\quad \dfrac{fB}{H-h_x} = p_x$

then $\qquad dp = p_x\dfrac{h_x - h_y}{H-h_y} = p_x\dfrac{dh}{H-h_y}$

or $\quad dh = \dfrac{dp}{p_x}(H-h_y) = $ difference in level between X and Y

If h_y be taken as small compared to H, then

$$h_x - h_y = dh = dp\frac{H}{p_x} \simeq \frac{H^2}{fB}dp$$

The difference in level between two points may thus be determined from the difference of parallax of the pairs of images measured in the direction of the air base.

The difference in parallax (dp) is normally measured with a *stereometer* or *parallax bar*. This consists of two glass windows, each with an identical mark on the centre, carried in frames attached to a bar. The left-hand window may be moved through a small range relative to this by means of a micrometer reading to 0.01 mm. The stereometer is used in association with a mirror stereoscope under which the photographs are set up as explained previously.

The right-hand mark in the centre of its run is laid over a point of detail on the right-hand photograph, and the left-hand mark adjusted over that point on the left photograph. When this graticule is clamped, both glasses should be flat so that the marks do not cast shadows. The marks should appear fused together when viewed through the stereoscope, and rotation of the micrometer, which alters the separation of the marks, causes the fused mark to appear to move vertically. It is made to touch the point selected, at ground level. The micrometer is then read, and the mean of several readings then obtained. Another point is now treated, and the difference of parallax of the two determined. The parallax bar must always be used parallel to the base line.

There are two types of stereometer and before using any instrument the type should be determined. Stereometers in which the separation between the measuring marks increases as the micrometer readings increase are called *inverse reading* and an increase in readings represents a decrease in parallax and a decrease in height. In a *direct reading* instrument an increase in micrometer readings causes a decrease in the separation between the measuring marks, and an increase in the parallax and the height.

Stereoscopic examination of photographs at the reconnaissance survey stage of extensive engineering projects such as, say, highway construction, allows the evaluation of various controlling factors, such as river crossings and swampy land, and of relative heights by stereometer. Possible routes can be located and marked on the photographs where they appear in stereoscopic correspondence with the topography, and comparison can then be made in the field.

Example 13.4 In a pair of overlapping photographs (mean photo base length 89.84 mm) the mean ground level is 70 m above datum. Two nearby points are observed and the following information obtained:

Point	Height above datum	Parallax bar reading
X	55 m	7.34 mm
Y		9.46 mm

If the flying height was 2200 m above datum and the focal length of the camera was 150 mm find the height of Y above datum. (Assume a direct reading stereometer.)

The air base B can be found from the expression $(b/f) \simeq (B/H_m)$ where H_m is the flying height with respect to mean ground-level and b is the mean photographic base.

$$\text{Thus} \frac{89.84}{150} \simeq \frac{B}{2130}$$

Therefore
$$B \simeq 1276 \text{ m}$$

Now
$$p_x = \frac{fB}{H - h_x} = 150 \times \frac{1276}{2145}$$

$$= 89.22 \text{ mm}$$

With a direct reading instrument

$$p_y = p_x + dp = 89.22 + (9.46 - 7.34)$$
$$= 91.34 \text{ mm}$$

Also
$$dp = p_y \left(\frac{dh_{xy}}{H - h_x} \right)$$

Therefore
$$dh_{xy} = \frac{2.12}{91.34} \times 2145$$

$$\simeq 50 \text{ m}$$

Thus Y is **105 m** above datum.

If that height was known to be so, we could directly write that a height change of 50 m was equivalent to a parallax change of 2.12 mm, i.e. 23.6 m/mm of parallax, and so if the reading for an intermediate point was 8.18 mm a heighting difference of $0.84 \times 23.6 = 20$ m above X would be implied.

Heights obtained in this way are termed *crude heights* as they are subject to error if the assumptions (that there is no tilt and no variation in flying height above datum) used in deriving the parallax formulae are not complied with. Only when the points under consideration are near to each other in the overlap will the answer derived from these formulae be reasonable. A false parallax is given by tilt and for accurate work the measurements should be corrected for that tilt. If the photographs are set up in the same relative positions they occupied in the air at exposure, then a point on the ground, its image on each photograph, and the two air stations, are in the same plane (the basal plane). Each ground point can be fixed by a basal plane which must include the air base, and which can be located by a rotation about that base. If the relative positions are not the same, the five points will not be coplanar, giving a 'want of correspondence' or y-parallax. This can be expressed as

$$dp_y = y_1 - y_2$$

where y_1 and y_2 are the distances from the x-axis or air base to a point in each of two adjoining photographs.

Methods are available for correcting the crude heights when several points of known height are located on the overlap. One of the best known, and most successful, of these is reviewed in papers by E. H. Thompson in *Photogrammetric Record*, October 1954 and October 1968 and by B. D. F. Methley in that journal dated April 1970.

The radial line assumption

It has been shown that on an air photograph distortions due to height are radial from the plumb point whilst tilt distortions are radial from the isocentre. Moreover we mentioned that an assumption often adopted is that both distortions are radial from the principal point provided that the tilt angle is small — say less than 2° or 3° — and the ground height variations are small compared to the flying height.

Maps may be prepared graphically using the assumption but from the point of view of the civil engineer its main value is that it can be used analytically to obtain the co-ordinates of ground points from measured photo co-ordinates taken after setting up the photographs as in Fig. 13.16(*a*). *P* and *Q* are the principal points of the two photographs and *P'* and *Q'* are their conjugate images. Let the co-ordinate origin be *P* and the *X*-axis coincide with the photo base

Fig. 13.16

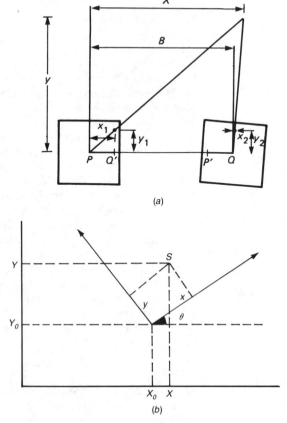

(a)

(b)

direction,
then
$$\frac{X}{x_1} = \frac{Y}{y_1} \quad \text{and} \quad \frac{(X-B)}{x_2} = \frac{Y}{y_2}$$

$$X = \frac{Bx_1 y_2}{x_1 y_2 - y_1 x_2} \quad Y = \frac{By_1 y_2}{x_1 y_2 - y_1 x_2}$$

If b is the mean value of PQ' and $P'Q$ we can write

$$x = \frac{bx_1 y_2}{x_1 y_2 - y_1 x_2} \quad \text{and} \quad y = \frac{by_1 y_2}{x_1 y_2 - y_1 x_2}$$

These latter two expressions give preliminary co-ordinates at approximately photo scale which can be used directly to derive ground co-ordinates. Since these may not be related to the ground insofar as directions of axes are concerned, the transformation equations

$$X = ax + by + X_0 \quad \text{and} \quad Y = -bx + ay + Y_0$$

can be adopted, where x and y relate to the photographic measurements and X and Y to the ground measurements: X_0 and Y_0 allow for displacement of the origins of the two systems, as in Fig. 13.16(b). It will be seen that

$$X = x \cos \theta - y \sin \theta + X_0$$
$$Y = x \sin \theta + y \cos \theta + Y_0$$

Writing $\cos \theta = a$ and $\sin \theta = -b$ leads to the transformation equations above. These equations can be used to place a local survey into the National Grid (Chapters 6 and 7) or other fixed system.

Example 13.5 The image co-ordinates of three points A, B and C and of the principal points P and Q on two overlapping photographs were determined as follows:

	Left photo		Right photo	
	x (mm)	y (mm)	x (mm)	y (mm)
P	0.0	0.0	−76.2	0.0
Q	+76.0	0.0	0.0	0.0
A	+10.6	+60.5	−66.0	+59.0
B	+11.2	−6.3	−64.5	−6.7
C	+14.5	+34.3	−61.5	+33.7

If the ground co-ordinates of A and B are 79 000 mE, 92 940 mN and 78 910 mE, 92 760 mN respectively, estimate those of C. (*Salford*)

From the data it will be seen that b = photo base = 76.1 mm and so for A,

$$x = \frac{76.1 \times 59.0 \times 10.6}{10.6 \times 59.0 + 66.0 \times 60.5} = 10.3 \text{ mm}$$

$$y = \frac{76.1 \times 59.0 \times 60.5}{10.6 \times 59.0 + 60.5 \times 66.0} = 58.8 \text{ mm}$$

In the same way the preliminary co-ordinates for B and C can be derived as $x = 11.9$ mm, $y = -6.7$ mm for B, and $x = 14.3$ mm, $y = 33.9$ mm for C.

Conversion of the co-ordinates now gives:

$$79\ 000 = 10.3a + 58.8b + X_0 \text{ for } A$$
$$92\ 940 = -10.3b + 58.8a + Y_0$$
$$78\ 910 = 11.9a - 6.7b + X_0 \text{ for } B$$
$$92\ 760 = -11.9b - 6.7a + Y_0$$

Hence $a = 2.71$, $b = 1.56$, $X_0 = 78\ 881$ and $Y_0 = 92\ 797$

Thus we can establish the co-ordinates for C as

$$X = 2.71 \times 14.3 + 1.56 \times 33.9 + 78\ 881$$
$$= \mathbf{78\ 973\ mE}$$
$$Y = -1.56 \times 14.3 + 2.71 \times 33.9 + 92\ 797$$
$$= \mathbf{92\ 867\ mN}$$

This approach has been used to observe traffic flows and speeds on main roads and motorways. Ground control is required and use may be made of 'street furniture' in this respect.

Perspective methods   If we assume that the area of ground covered by the air photograph, or a part of it, is a plane surface (not necessarily horizontal), the normal projective relationships hold from ground to photograph and from photograph to map. Many simple methods, based on perspective

Fig. 13.17

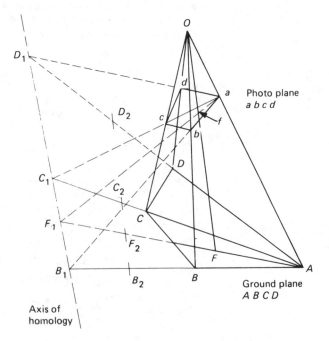

transformation and using single air photographs, are then available for locating and plotting planimetric detail.

One such method to locate the map position F of an image point f uses the 'paper-strip' method which makes use of the principle of the cross-ratio. Figure 13.17 shows four control points A, B, C and D imaged at a, b, c and d on the photograph. The two planes intersect in a line (the axis of homology) and lines on those two planes connecting pairs of points must intersect on that line, i.e. ab and AB at B_1.

$$\frac{C_1 B_1}{C_1 F_1} \;\bigg/\; \frac{D_1 B_1}{D_1 F_1}$$

is known as a cross-ratio and it can be readily shown (by use of the sine rule with respect to vertex A) to have the same magnitude as

$$\frac{C_2 B_2}{C_2 F_2} \;\bigg/\; \frac{D_2 B_2}{D_2 F_2}$$

given by any other transversal cutting the corresponding lines in the ground plane, or an equivalent transversal cutting the lines in the photo plane.

The intersection of rays emanating from one image, say a to b, c, d and f, is marked on a strip of paper and identified as in Fig. 13.18(a). A similar procedure is adopted at d. These strips can now be positioned over the corresponding lines drawn through the control points on the map, so that the lines previously marked lie over the respective lines to B, C and D or A, B and C. The position of F can be fixed using rays Af_1 and Df'.

Fig. 13.18

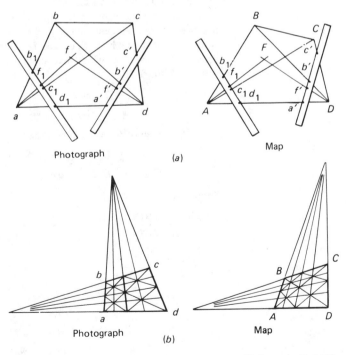

Photograph   (a)   Map

Photograph   (b)   Map

Fig. 13.19

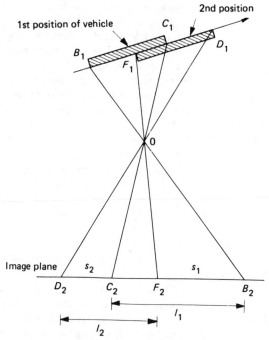

If many points are to be located, then grids can be drawn on both map and photograph by producing corresponding lines on them as in Fig. 13.18(b). Detail can now be transferred from photograph to map by reference to the network.

The cross-ratio can be of value in various other situations. For instance, vehicle velocities can be measured if the vehicle be imaged twice on a photograph. If $B_1 C_1$ and $D_1 F_1$ are the first and second positions of the vehicle, with corresponding images at $B_2 C_2$ and $D_2 F_2$ in Fig. 13.19, then

$$\frac{C_1 B_1}{C_1 F_1} \bigg/ \frac{D_1 B_1}{D_1 F_1} = \frac{C_2 B_2}{C_2 F_2} \bigg/ \frac{D_2 B_2}{D_2 F_2}$$

Let the length of the vehicle be L and let $B_1 F_1 = C_1 D_1 = s$

$$\frac{C_1 B_1}{C_1 F_1} \bigg/ \frac{D_1 B_1}{D_1 F_1} = \frac{L}{L-s} \bigg/ \frac{L+s}{L} = \frac{L^2}{L^2 - s^2}$$

Similarly,

$$\frac{C_2 B_2}{C_2 F_2} \bigg/ \frac{D_2 B_2}{D_2 F_2} = \frac{l_1}{l_1 - s_1} \bigg/ \frac{l_1 + s_2}{l_2} = \frac{l_1 l_2}{l_1^2 + l_1(s_2 - s_1) - s_1 s_2}$$

Since $l_2 + s_1 = l_1 + s_2$

$$\frac{L^2}{L^2 - s^2} = \frac{l_1 l_2}{l_1 l_2 - s_1 s_2}$$

$$s^2 = \frac{s_1 s_2}{l_1 l_2} L^2$$

s is then determined since all other quantities are known, and hence velocity s/t. The relative directions of vehicle and image plane are not important, but the inverval (t) between exposures must be found accurately.

Map revision

As mentioned previously, the influences of tilt, ground relief and flying height can be eliminated during the production of the map or plan. Updating of maps can be effected by means of relatively simple instruments like the SB215 Zoom Stereosketch manufactured by Cartographic Engineering Ltd. Essentially two photographs are positioned for stereoscopic fusion through a system of binoculars and mirrors in a manner somewhat comparable with Fig. 13.14. The existing map, covering the area of the photographs, is mounted on a table at a level below the photographs. It can also be viewed through the binoculars and by means of a beam splitter the image of the map and photographs are adjusted until they are superimposed: the scale of the map can be adjusted by use of a zoom lens until it exactly matches that of the three-dimensional ground image of the stereo photographs. Revisions can thus be easily identified and transferred to the map. It is preferable to use the central portion of the photographs and to arrange for three local 'control' points to be selected in the map near to, surrounding, and if possible at the same level as, the revision.

Mosaics

If a set of overlapping vertical photographs is joined together on a backing board, a rough map, or aerial view, of the area is obtained which has no uniformity of scale. This is called an uncontrolled mosaic, and its assembly calls for much skill since relief and tilt distortions which occur in each photograph inevitably cause a mismatching of detail between adjacent photographs. These must be disguised as far as possible, and so only the centre portions of the photographs, where distortions are least, are used. The mosaic is assembled usually from the centre outwards and, if required, features such as roads, rivers, etc. can be emphasized, and lettering added. The mosaic is then re-photographed so that the final appearance is that of a single photograph.

However, if the photographs be rectified and brought to a common scale the mosaic is said to be controlled, resulting in a superior mosaic. Such photographs are produced by a plotting instrument known as a rectifier, for example the Zeiss SEG6. This instrument treats single photographs, preferably taken over flat terrain, in such a manner that distortions are eliminated. Although photographs taken by wide-angle lens can be used, in hilly country distortions due to relief require the adoption of normal-angled or narrow-angled lenses with only the central portion of the photograph being utilized to minimize distortion effects. As with the uncontrolled mosaic, features can be emphasized and lettering added, together with a grid system in the margin, so that the mosaic becomes effectively a photomap.

The mosaic can be of great assistance at the reconnaissance and planning stage as it contains all that which could be seen directly from the aircraft at the time of photography. The photographs show such points as the type and density of any development, land use, existing access and that required for a highway scheme, alterations in topography etc. It reveals information in far greater detail than can be plotted on a map.

Terrestrial photogrammetry

Terrestrial photogrammetry employs photographs taken from points on the earth's surface for measurement purposes and its principles are very similar to those of aerial photogrammetry. However, the photography is taken from fixed ground stations and so it is usually possible to determine the absolute orientation of the camera at the time of exposure thereby simplifying the subsequent measurement process.

Terrestrial photogrammetry can be carried out with a photo-theodolite, one such type being shown in Fig. 13.20. The optical axis of the camera is arranged parallel to the line of collimation of the theodolite telescope and hence the theodolite allows (*a*) the determination of the orientation of the camera axis, (α and β, Fig. 13.21) and (*b*) the angular measurements needed to fix the camera stations

Fig. 13.20 Wild P32 terrestrial camera mounted on T16 theodolite (*Courtesy*: Leica UK Ltd)

Fig. 13.21

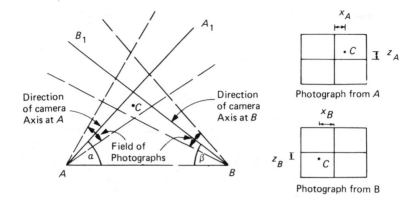

Direction of camera Axis at A

Direction of camera Axis at B

Field of Photographs

Photograph from A

Photograph from B

during the control survey (*A* and *B*, Fig. 13.21). The image plane of the camera is represented by the surface of a glass plate upon which are engraved four fiducial marks which appear on each photograph and allow the position of the principal point to be located as explained for aerial cameras. These marks allow two axes to be constructed on the photographs as shown in Fig. 13.21. The presence of such marks can lead to the camera being designated a 'metric' camera, although another definition is 'a camera designed specifically for photogrammetric purposes'.

To fix one point in plan requires that photographs be taken from at least two camera stations whose positions must also be known and very important points should lie on photographs from three stations. When obtaining these photographs the directions of the camera axes may either converge from the stations or be parallel.

Figure 13.21 refers to the former case and it shows that the camera was positioned at each station in turn so that the required point *C* was included in the detail shown in each photograph. It is most convenient if the plates be vertical at the time of photography and this will be assumed in the subsequent discussions. Angles α and β, which denote the directions of the camera axes, must be derived at the relevant station for each photograph, the line of sight or the theodolite being initially directed along either *AB* or *BA* respectively for this. After processing, the co-ordinates of the image of point *C* can be readily determined with respect to the principal horizontal and vertical axes connecting the respective fiducial marks.

In Fig. 13.22, *a* and *b* represent the two camera stations *A* and *B* plotted at the required map scale on a sheet of drawing material. The lines aa_1 and bb_1 indicating the directions of the camera axes are drawn in at the derived angles α and β. The positions of the plates are then plotted at a convenient distance representing the focal length of the camera at right angles to aa_1 and bb_1; co-ordinates x_a and x_b being plotted at the same scale.

Heighting may also be carried out from the photographs.

Having found the plan position C_1 (Fig. 13.23) of a certain point, it is required to determine the elevation *H*, of that point with respect to the horizontal plane containing the camera axis. If *C* is the position

Fig. 13.22

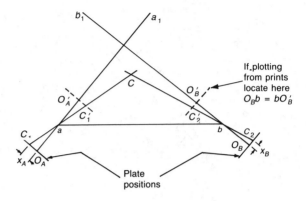

If plotting
from prints
locate here
$O_B b = b O_B'$

Plate
positions

Fig. 13.23

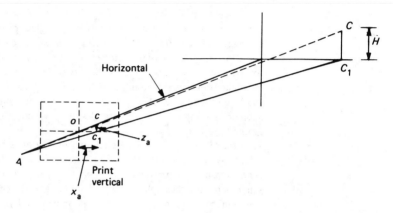

Horizontal

Print
vertical

of the point in elevation, then, from similar triangles,

$$\frac{H}{z_a} = \frac{AC_1}{Ac_1} = \frac{AC_1}{\sqrt{x_a^2 + f^2}} \quad \text{since } Ao = f$$

Therefore

$$H = \frac{z_a}{\sqrt{x_a^2 + f^2}} \cdot AC_1$$

AC_1 will be known either from the plot or by calculation as indicated in the following example. Note that in Fig. 13.22 the angles between the directions of the camera axes at a and b and the lines drawn to c are $\tan^{-1}(x_A/f)$ and $\tan^{-1}(x_B/f)$ respectively.

Example 13.6 A phototheodolite having a focal length of 200 mm (Figs. 13.24(a) and (b)) was used at two stations A and B having co-ordinates (0, 0) and (0, 152.0) m, station B being 9.1 m higher than A. In both cases station C, a tower 47.5 m high to the north, is on the vertical centre line of the photograph. In that from A it measures 26.00 mm and its base is 6.50 mm above centre, while in that from B it measures 24.00 mm and its base is 1.00 mm above the centre.

Determine the co-ordinates of C, the level of the tower base

---

relative to *A*, and the directions of the sights from *A* and *B*. There is some doubt as to the horizontality of the sight from *B*. Check for any error: and then determine the co-ordinates and level above *A* of a point *D* located 16.20 mm right and 17.60 mm up in the photograph from *B*, if it is in line with the tower as seen from *A*. Would it be visible from *A*?

(*London*)

Fig. 13.24

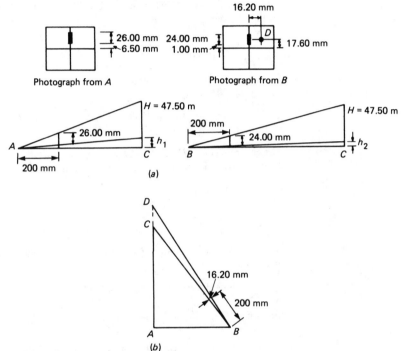

26.00 mm
6.50 mm

24.00 mm
1.00 mm

16.20 mm

17.60 mm

Photograph from *A*

Photograph from *B*

H = 47.50 m

26.00 mm

h_1

200 mm

200 mm

24.00 mm

H = 47.50 m

h_2

A

C

B

C

(a)

D

C

16.20 mm

200 mm

A

B

(b)

By similar triangles,

$$\frac{200}{32.50} = \frac{AC}{H+h_1} \quad \text{and} \quad \frac{200}{25.00} = \frac{BC}{H+h_2}$$

also

$$\frac{200}{6.50} = \frac{AC}{h_1} \quad \text{and} \quad \frac{200}{1.00} = \frac{BC}{h_2}$$

whence $\quad AC = 365.4$ m $\qquad\qquad BC = 395.8$ m

$\qquad\qquad h_1 = \textbf{11.88 m} \quad \text{and} \qquad h_2 = \textbf{1.98 m}$

now $\qquad CB^2 = CA^2 + AB^2 - 2CA \cdot AB \cos C\hat{A}B$

$\quad 395.8^2 = 365.4^2 + 152.0^2 - 2 \times 365.4 \times 152.0 \cos C\hat{A}B$

Therefore $\qquad\qquad\qquad$ **C$\hat{A}$B is 90°**

Co-ordinates of *C* are then $\qquad$ **(0 E, 365.4 N).**

Since $$C\hat{A}B = 90°$$

then $$\tan CBA = \frac{CA}{AB} = \frac{365.4}{152.0}$$

$$C\hat{B}A = \mathbf{67°25'}$$

From the photograph from B the tower base is apparently 1.98 m above B, whereas it should be $(11.88 - 9.10)$ or 2.78 m higher. Thus the sight of B was inclined upwards, so that an error of 0.80 m in level is given in a distance of 395.8 m, i.e. 1 in 495, or 0.40 mm difference on the photograph.

Now in Fig. 13.24(b) take

$$\tan CBD = \frac{16.20}{200} \quad \text{(since the above inclination is small)}$$

Therefore $C\hat{B}D = 4°35'$

and $$AD = AB \tan(4°35' + 67°25') = 152.0 \tan 72°00'$$
$$= 467.8 \text{ m}$$

The co-ordinates of D are (0, 467.8). Apparent difference in level of D with respect to the horizontal axis of the instrument at B

$$= BD \times \frac{z_D}{\sqrt{f^2 + x_D^2}}$$

where $BD = \sqrt{467.8^2 + 152.0^2} = 491.9$ m

Now $z_D = 17.60$ mm (measured) or 18.00 mm (corrected) and since

$$x_D = 16.20 \text{ mm and } f = 200 \text{ mm}$$

therefore $BD \dfrac{z_D}{\sqrt{f^2 + x_D^2}} = 44.1$ m, when $z_D = 18.00$ mm

Then the level of D is 44.1 m above B, i.e. **53. 2 m above A,** and so D will not be seen from A, since the tower is 47.5 m high and its base is 11.9 m above the horizontal axis at A.

When the plate is not vertical the image co-ordinates of points such as D will not be the same as those produced when the plate is vertical.

Stereo photogrammetry

In this method, pairs of photographs are taken from stations at each end of a line PQ (Fig. 13.25(a)) with the phototheodolite set up normally so that the image planes are in the same vertical plane and parallel to the line joining the stations. Thus the camera axes when set up at each station in turn must be made parallel by setting them at right angles to the line: this is achieved by the appropriate orientation of the theodolite.

Alternatively for close range work devices known as stereometric cameras can be obtained. These essentially consist of two similar cameras, mounted on a bar at known separation, such that their optical axes are parallel and perpendicular to the bar. Control points are needed so that the photography can be checked: their bearings from PQ are measured by theodolite and their plan positions should be known during the survey. Three or four such points are taken, one being near P and the others being well separated at the limits of the area to be plotted.

Photo measurement

The plan position R' of a ground point R is shown in Fig. 13.25(a). Its co-ordinates X_p and Y_p with respect to P can be readily established, since by similar triangles

$$\frac{X_p}{x_p} = \frac{Y_p}{f} = \frac{X_q}{x_q}$$

Fig. 13.25

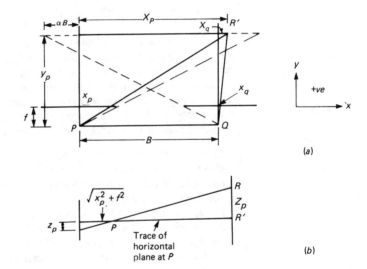

(a)

(b)

Now $X_p - X_q = B$, which is measured independently

Therefore

$$X_p - X_p \frac{x_q}{x_p} = B$$

and so

$$X_p = B \frac{x_p}{(x_p - x_q)}, \quad X_q = \frac{Bx_q}{(x_p - x_q)}$$

and

$$Y_p = \frac{Bf}{(x_p - x_q)}$$

It will be noted that points at different distances (Y) from the base line must produce different values ($x_p - x_q$) since Bf is constant. Note the similarity with parallax as defined earlier in the chapter.

Put

$$x_p - x_q = p$$

Then

$$Y = \frac{Bf}{p} \quad \text{and} \quad X_p = \frac{Bx_p}{p}$$

Similarly, from Fig. 13.25(b)

$$\frac{Z_p}{z_p} = \frac{PR'}{\sqrt{(x_p{}^2 + f^2)}} = \frac{Y_p}{f}$$

Therefore $$Z_p = Y_p \frac{z_p}{f} \quad \text{and} \quad Z_q = Y_p \frac{z_q}{f}$$

$(Z_p - Z_q)$ indicates the difference in level between the horizontal axes of the instrument at P and Q. Note also that the whole of the terrain between P and Q will not be photographed. αB in Fig. 13.25(a) can be readily determined by similar triangles since $(\alpha B + B)/Y = (w/2)/f$, where w is the width of the negative or positive.

Errors in Y due to errors dB, df and dp can be readily determined as

$$dY = \frac{Y}{B} dB = \frac{Y}{f} df = \frac{-Y^2}{Bf} dp$$

This last equation allows the base-distance ratio B/Y to be established for a given distance Y and the required accuracy dY in respect of that distance: the value actually given to dp can be taken as the accuracy of stereoscopic measurement. In this respect it can be mentioned that p becomes larger as R approaches PQ and a 'minimum-range' condition of $4B$ is suggested by Wild for their P32 to ensure that stereoscopic viewing of pairs of photographs can be achieved.

Example 13.7 Two photographs are taken with a phototheodolite from stations P and Q, 100 m apart, the camera axes being 90° to PQ in each case. A point R appears on the photograph from P as 12.78 mm to the right of the vertical hair, and 10.48 mm above the horizontal hair, while on the photograph from Q it appears as 24.88 mm to the left of the vertical hair, and 9.05 mm above the horizontal hair. Q is east of P.

Calculate the co-ordinates of R from P, and the difference in level of the two camera stations if the focal length of the camera is 165 mm and the camera axes are at the same height above the respective stations.

Refer to Fig. 13.25(a), noting that x is measured positive to right,

$$x_p = 12.78 \text{ mm} \quad z_p = 10.48 \text{ mm}$$

$$x_q = -24.88 \text{ mm} \quad z_q = 9.05 \text{ mm}$$

$$X_p = \frac{Bx_p}{(x_p - x_q)} \quad \text{and} \quad Y_p = \frac{Bf}{(x_p - x_q)}$$

$$x_p - x_q = 12.78 - (-24.88) = 37.66 \text{ mm}$$

Therefore $$X_p = 100 \times \frac{12.78}{37.66} = 33.9 \text{ m}$$

$$Y_p = \frac{165}{37.66} \times 100 = 438.1 \text{ m}$$

Hence co-ordinates of R are **(33.9, 438.1)** m w.r.t. P

$$Z_p = Y_p \frac{z_p}{f} = 438.1 \times \frac{10.48}{165} = \mathbf{27.8\ m}$$

$$Z_q = Y_p \frac{z_q}{f} = 438.1 \times \frac{9.05}{165} = \mathbf{24.0\ m}$$

and so, since these values refer to R, which is of course fixed in position, P must be lower than Q by 3.8 m.

Plotting machines are available which allow continuous line plotting of detail and contours from terrestrial photographs. Many of the plotting machines available for mapping from air photographs cannot be used as they are restricted to mapping in a plane which is almost parallel to that of the photography.

Close range photogrammetry

Photogrammetry is a non-contact measuring technique and as such can be used to measure objects which are inaccessible for direct measurement or of a delicate nature. Also it can provide measurements of non-rigid and moving objects provided that a synchronized pair of cameras is available to freeze the condition at the instant of exposure. Further a permanent record in the form of controlled photography is provided and these photographs can be measured at any time in the future if the need arises.

Close range photogrammetry employs the techniques and instrumentation of both aerial and terrestrial photogrammetry discussed earlier. It is generally limited to a maximum distance of 300 m and it has many applications in civil engineering as mentioned later in this chapter.

As an example consider the following situation which may arise say in a traffic study.

Fig. 13.26

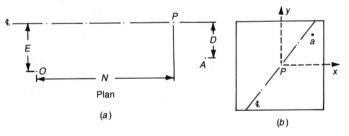

Plan

(a)

(b)

Example 13.8 In Fig. 13.26 O is a camera station at the top of a tall building. The camera ($f = 100.0$ mm) is pointed obliquely downwards with the ground principal point P on the centre line of the road. Figure 13.26(b) represents diagrammatically a positive photograph showing the principal point p, the image a of an arbitrary point A on the road surface and the axes of a rectangular co-ordinate system having the x-axis along the plate parallel through p.

If $O = 25.0$ m above the road surface (which may be assumed to be a horizontal plane), $E = 16.6$ m, $N = 40.0$ m and the photo co-ordinates of a are $x = 35.5$ mm, $y = 31.2$ mm, calculate the length of the offset from A to the road centre line (D in Fig. 13.26(a)). (*London*)

From Fig. 13.26(a) $O_1p = \sqrt{40.0^2 + 16.6^2}$

$$= 43.3 \text{ m}$$

In Fig. 13.27(a), $\tan \theta = \dfrac{43.3}{25.0} = 1.732$

Therefore $\theta = 60°00'$

Fig. 13.27

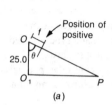

(a)

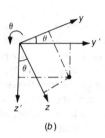

(b)

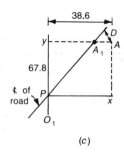

(c)

Examination of Fig. 13.26(b) in conjunction with Fig. 13.27(a) shows that the positive has been rotated about the x-axis. Its position, analogous with the case shown in Fig. 13.5(c), is represented in Fig. 13.27(b) in which axis y' is in the horizontal plane with axis x'. Thus the co-ordinates of the points are related as follows:

$$x' = x$$
$$y' = y \cos \theta + z \sin \theta$$
$$z' = -y \sin \theta + z \cos \theta$$

Therefore $\dfrac{x'}{z'} = \dfrac{x}{-y \sin \theta + z \cos \theta}$

and $\dfrac{y'}{z'} = \dfrac{y \cos \theta + z \sin \theta}{-y \sin \theta + z \cos \theta}$

But $z = z' = f$

therefore $x' = f \left[\dfrac{x}{-y \sin \theta + f \cos \theta} \right]$

and $y' = f \left[\dfrac{y \cos \theta + f \sin \theta}{-y \sin \theta + f \cos \theta} \right]$

in which $x = 35.5$ mm, $y = 31.2$ mm and $f = 100.0$ mm

464 Surveying

Hence
$$x' = 100 \frac{35.5}{-31.2 \sin 60° + 100 \cos 60°}$$

$$= 154.3 \text{ mm}$$

and
$$y' = 100 \frac{31.2 \cos 60° + 100 \sin 60°}{-31.2 \sin 60° + 100 \cos 60°}$$

$$= 444.3 \text{ mm}$$

These are the co-ordinates which would have been realized theoretically on a horizontally positioned positive at O for point A.

Whence with respect to O

$$x_a = x' \frac{H}{f} = \frac{154.3}{100} \times 25.0$$

$$= 38.6 \text{ m}$$

and
$$y_a = y' \frac{H}{f} = \frac{444.3}{100} \times 25.0$$

$$= 111.1 \text{ m}$$

Thus in Fig. 13.27(c)

$$Py = 111.1 - 43.3 = 67.8 \text{ m}$$

and since
$$\tan yPA_1 = \frac{16.6}{40.0}$$

$$yA_1 = 67.8 \times \frac{16.6}{40.0}$$

$$= 28.1 \text{ m}$$

and
$$A_1 A = 38.6 - 28.1$$
$$= 10.5 \text{ m}$$

Hence perpendicular distance $D = A_1 A \cos yPA_1$

$$= 10.5 \times \frac{40.0}{43.3}$$

$$= \mathbf{9.7 \text{ m}}$$

Leica R5 camera
The hand-held Leica R5 camera is a recent introduction into this particular field of surveying. It is a metric camera, having calibrated focus stops together with a reseau or register glass, which automatically projects 35 crosses in a 5 × 7 grid on each exposure. It is worthy of mention that the use of a reseau was pioneered in the UK for analytical applications of aerial photogrammetry. The Williamson F49 camera had 529 crosses at 10 mm intervals on the reseau, each being known in position to 0.001 mm.

The Leica camera is designed to be used with the Elcovision 10 interpretation and plotting software. A precision digitizer is used to scan the photograph and the crosses, the ensuing information being stored and processed by an IBM PS/2 computer. Photographs are used in pairs or with check dimensions to construct a three-dimensional image of the survey. The results can then be output as co-ordinates or plan and elevation plots. Since the only instrument required on site is the camera the method is fast and ideal for recording events like traffic accidents, fire damage for insurance claims and on the smaller scale physiological measurements for use in orthopaedics and surgery. In the construction industry it is popular with architects for surveying the facades of existing buildings that are to be modified.

Other developments in close range photogrammetry include the use of non-conventional image systems such as X-rays to produce photographs and the use of non-metric cameras, i.e. cameras which have not been specifically designed for photogrammetric use. They may reveal appreciable lens distortion in addition to not having fiducial marks, but nevertheless, when used with an appropriate analytical measurement system, a similar order of accuracy to that expected from metric cameras can be achieved. The reader is referred to *Developments in Close Range Photogrammetry*, K. B. Atkinson (Editor), Applied Science Publishers, for a comprehensive review of cameras, plotters and techniques.

Applications of photogrammetry

The civil engineer is often concerned with mapping at scales between 1 in 500 and 1 in 2500, and photogrammetry can be readily applied to such tasks when considering highway and railway projects, pipeline locations, industrial layouts and catchment problems. He also has an important role nowadays in monitoring the behaviour of completed structures.

Applications of aerial photogrammetry

With modern cameras, film and digital plotters which are of maximum precision, aerial photogrammetry can produce plans at a scale of 1:500 with contours at intervals of 0.5 m. In fact mean square errors in plan and height position of some ±0.1 m are obtainable at that scale. The use of the traditional ground-levelling methods discussed in Chapter 3 should be considered when spot heighting along existing roads and highways. But the contouring produced photogrammetrically is generally superior to that provided by ground survey. No interpolation between spot heights is involved and the contours are traced out directly, so that consequently they fit the terrain well.

The factors to be studied when choosing between aerial survey and conventional ground survey for such mapping include (i) the size, shape and character of the area, (ii) its position, (iii) the amount of detail to be plotted, and (iv) the supply of permanent ground marks.

Ground survey is likely to be more economical for sites of area 10 to 20 hectares as the distance increases from the 'headquarters' of the aerial survey company. The same is true for narrow bands of mapping which may arise in highway projects, when the area increases to 50 hectares or so. But as the bands widen, thereby increasing the area covered for a given length, and the amount of detail increases, air surveys tend to become more economical. Except for surveys of small sites, the mapping will usually be carried out quicker by air survey. Once the air photography is taken there is little delay in an air survey due to the weather. Provision of permanent ground marks for setting out purposes does not appreciably alter the cost of ground survey, but it can do so to a greater extent in air surveys as the character of the ground-control survey is changed. In addition, dense urban areas and wooded areas can give rise to obscured points of detail in air surveys: the need for appreciable ground verification then arises, as it does for the location of services such as sewers on the map. On the other hand an air survey is able to map areas where ground survey would be impossible, difficult or unpleasant. Surveys of busy railway junctions and airports present few problems when air survey is used.

Topographic mapping

Accurate and up-to-date maps and plans are essential at the design stage of a project, and the planning and design of highway construction will be discussed in a general manner in this context. It can be divided into the location stage and the design stage, the former applying to all work up to, and including, the adoption of the alignment. Usually the photogrammetric work is kept separate for the two stages, but in remote areas it is likely to be a more economical proposition to take two sets of photography for one positioning of the aircraft.

The intention of the location stage is to have the horizontal alignment of the proposed road shown on 1:2500 maps: side-slope intersections can also be marked. To achieve this the location stage is subdivided into (i) preliminary reconnaissance, and (ii) preliminary design. The purpose of (i) is to obtain feasible alternative routes, and of (ii) to select the most suitable of those alternatives after studying all legal and technical aspects. In the UK, maps at scales of 1:2500, 1:10 000 and 1:10 560 cover an appreciable part of the whole and these can be studied at the reconnaisance stage. Photography is flown for the preliminary design stage and it might well be incorporated into the reconnaissance stage. Mosaics can be established to show land use and to relate the alternative routes one against the other, and photo-interpretation carried out to find areas of poor drainage, mining subsidence, land slides etc. Potential sources of construction materials can be located and any control points for the road alignment, such as river crossings, marked. In addition, heights and gradients can be estimated by parallax bar observations. Ground reconnaissance is also carried out, particularly at critical places. At the conclusion of (i) the highway engineer should be aware of costs and problems involved in the construction.

At this juncture it is appropriate to mention a further method of remote sensing, defined here as the detection of electromagnetic radiation and its registration at a remote station. We refer to the French initiated SPOT (Systeme Pour l'Observation de la Terre) which embodies two similar scanners within a satellite orbiting at an altitude of 800 km. These scanners, or digital sensors, can collect either three separate bands of radiation in a multi-spectral role or a single band panchromatically. The image is made up of pixels each representing a certain area of the terrain below, the size of which depends upon the scanning mode.

The scanners can be directed vertically downwards or be placed in an inclined attitude so that the same section of terrain can be covered from two different satellite stations. Photographs of a desired area can be selected from a catalogue of SPOT images or taken on demand. These can be set-up in certain analytical plotters, e.g. Zeiss P-Series Planicomps, to establish stereomodels and hence measurement can be effected in a similar manner to conventional air-photographs.

Resolution to 10 m is achievable in the panchromatic mode and mapping has been carried out at scales of 1:50 000 and 1:100 000 and hence from the civil engineer's point of view the data available from this method is likely to be restricted to use at the reconnaissance stage. However, this satellite system, together with the similar US Landsat satellites, has extensive use in the realm of computer based Geographical Information Systems.

In order to carry out the detailed final alignment for a project like the highway scheme mentioned previously a conventional air survey would be necessary to achieve the accuracy. In so far as heighting by photogrammetry is concerned, a standard error between $\pm H/10\ 000$ and $\pm 1.5H/10\ 000$ may be attained for well-defined points, whilst for less clearly defined points it could be $\pm 2H/10\ 000$: contour-line accuracy of the order of $\pm 3H/10\ 000$ can be contemplated.

In developed areas the survey can be carried out at scales of 1:12 000 or 1:10 000 to produce maps at 1:2500 with contours marked at 2 m intervals and with spot heights to 0.5 m: existing 1:2500 sheets could be revised and contours superimposed. These maps will cover the bands of interest, which can be up to 2 km wide, relating to the alternatives obtained in (i) and both horizontal and vertical alignments for each will be studied. Factors considered include the fit of the highway into the topography, structures to be built, gradients and their influence on traffic and emphasis on climatic conditions, costs of rights-of-way, construction and maintenance. In undeveloped areas the photography can be carried out at scales of 1:20 000 to 1:40 000 for plotting at 1:10 000 with contours at 5 m intervals. Also for low-cost roads the 1:2500 maps could well serve at the design stage, with some supplementary survey at critical places.

For the design stage of high-cost roads such as motorways, plans at scales of 1:500 are produced from 1:3000 photography and they are supplied with contours at 0.5 m intervals and spot heights to 0.1 m. They must cover the horizontal alignment chosen at the end of the location stage and the photography can well consist of a series of longitudinal strips, since a 230 mm square photograph covers about 700 m at 1:3000 scale: accordingly, only a narrow portion of that will be needed for the design survey. Ground control can be put down to a fairly dense network or alternatively a geodetically-supported aerial triangulation can be effected, pre-marked bench marks being provided at frequent intervals near the proposed alignment. These provide control over lateral tilt errors which might arise.

Orthophotomaps

An *orthophotograph* is a photographic product from which the distortions due to ground relief and tilt have been removed and as a result the orthophotograph possesses the geometrical characteristics of a map. Such a photograph cannot be taken from the air and it needs to be produced using a special instrument which may be attached to a conventional plotting machine in an on-line mode. An orthonegative film, free from relief and tilt displacement and at uniform scale, is produced in that instrument as the operator systematically scans the stereo model. The orthonegative, therefore, retains all the detail present on the original photograph unlike a conventional line map where a selection of detail present has to be made by the operator.

In the above procedure, the orthonegative is produced at the same time as the scanning takes place, by separating the measuring procedure in the stereo-plotter and the projection in the orthophoto instrument. It is, however, possible to use many orthophoto systems in an off-line mode. This requires the use of a storage device in which the profile data or contours can be digitally stored during scanning and then used later to control the exposure of the orthophoto. The off-line mode of operation has many advantages over the on-line mode.

When additional information, such as contours, place names and a grid are added, the orthophotograph becomes an orthophotomap (see Fig. 13.28). The contours may be produced in many ways as by-products of the scanning process, but perhaps the most accurate and simplest method is to contour, in the normal way, the stereomodel set up in the plotting machine. This additional information is added to the orthophotograph during the printing stage.

Orthophotomapping shows a considerable saving in time and some saving in cost, as compared with conventional mapping from air photographs, and orthophotomaps also show a great deal of surface information which is absent from a conventional map. Although the technique is probably best suited for use at scales which are smaller than those desired by the engineer, there are a number of examples where orthophotomaps have been produced for him as alternatives to the conventional map.

In highway work, orthophotomaps, at 1:5000 and 1:10 000 scales with 2 m contours, have been used during the preliminary design stage and at 1:1000 and 1:2000 scales in the final design. All details of the road, such as centre line and side widths, can be superimposed onto the orthophotomap to good effect.

In addition, large scale orthophotomaps have been employed as highway inventory and maintenance maps. Most items of street furniture, such as traffic signs, road markings, lamps and barriers, can be interpreted and annotated directly on the orthophotomap.

Digital maps

The Ordnance Survey has been producing digital maps since 1970. At first they were just used as a convenient means of reproducing the hard copies of the map but they are now available as computer data. In this form they can be used in Geographical Information Systems, computer aided design packages and as a base for a customer's command and control system, e.g. recording the location of

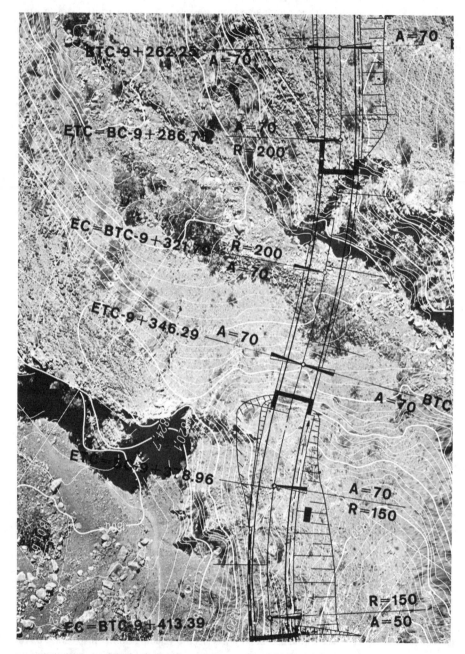

Fig. 13.28 Part of Orthophoto combined with contour lines, grid lines and road design data. Scale 1:1000. Note that $A = \sqrt{RL}$ in the design of the transition curves

underground assets owned by a utility. The digital map consists of strings of co-ordinates of all salient features stored digitally on magnetic media which can either be viewed and edited on the computer screen or hard copied at a plotter. The information is copyright and can, of course, only be reproduced by legitimate users under licence. At present there is appreciable coverage of the industrial regions and major conurbations in the UK and complete coverage of the UK can be expected during the life of this book.

As an example of the use of digital maps, the Ordnance Survey map provides an ideal base for a Geographical Information System operated by a local authority. The basic data base would cover the topography and land usage for the metropolitan area which could be used to provide line information plus contour and spot height data along the lines of a conventional plan. However, it could also be used to produce thematic plans for town planning purposes and each department could have an overlay plan or data base containing specific details. For example, the finance department may wish to record the owner of the land and valuation of the property, whilst the planning department may record details of all appropriate planning applications. This information would be recalled by displaying the map on the computer screen and then pointing and 'clicking' at the appropriate property using a mouse. Information so recorded has other uses, in the case of the above examples the owner of property and any proposed development are of interest to the civil engineer who is planning a highway realignment scheme. Applications of the technology seem endless and are only limited by financial restraints and legal requirements relating to the confidentiality of data.

Within an engineering context it is technically possible to have a single data base containing all details about a street, i.e. overlay maps showing the location of services, data base details of street furniture, further data base details about previous excavations, traffic accidents, etc. This would be of immense benefit in the event of excavations or highway realignment since a composite knowledge of underground pipes and surface features would be obtained on command. At the time of writing this edition, a trial system similar to that described above was under way in Dudley involving the co-operation of the local authority and the utilities and it is probable that other more extensive schemes will emerge during the life of this book.

| Digital Terrain Model (DTM) | A Digital Terrain Model can be defined as a statistical sampling of the X, Y and Z co-ordinates of the terrain, i.e. ground surface, so that by interpolation the Z co-ordinate of any point can be accurately estimated from its X, Y co-ordinates. The morphological shapes must therefore be properly represented by the co-ordinated points chosen, and not only should those Z co-ordinates be established to the required accuracy, but the points should be sufficiently numerous and adequately arranged. Data for a DTM may be gathered by land survey, photogrammetry or from an existing map. Of these, photogrammetric techniques are the most widely used. |

Spatial co-ordinates of these characteristic points may be stored in the computer to allow the interpolation of other points, and the DTM can be classified according to method of point registration into one of the following groups: (a) ordered points, (b) semi-ordered points and (c) random points.

Ordered points imply that the position of every one of the characteristic points is fixed relative to its surrounding points, and this suggests square, rectangular or triangular grid layouts: uniformly shaped ground is the most appropriate for these grids. Linear interpolation is adopted to give the level at T in Fig. 13.29(b) as

$$Z_T = \frac{Z_p}{L^2}(L-X)(L-Y) + \frac{Z_Q}{L^2}X(L-Y) + \frac{Z_R}{L^2}XY + \frac{Z_S}{L^2}Y(L-X)$$

Fig. 13.29

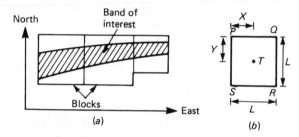

North

Band of interest

Blocks

East

(a)

(b)

Photogrammetric measurement is eminently suited to gathering of data for this type of DTM particularly if the grid lines can be made to coincide with the X and Y directions of the plotting machine axes.

In rolling terrain the contour lines can be digitized so that X, Y co-ordinates change for constant Z co-ordinates. This form of DTM lies within the group designated as semi-ordered points, in which a relationship exists between some of the points forming the model. Also included in this group is the terrain line, a feature such as a well-defined change of slope, for which the X, Y, Z co-ordinates are measured and stored.

Random-point models demand that a computer be available: an interpolation surface has to be formed from the points observed which can be measured and stored by an ordered or semi-ordered method over the band of interest. The relationships mentioned for those groups are not used, but instead the interpolation surface is based on a number of points nearest to the point whose level is required. A surface fit based on a 'least-squares' procedure is adopted to establish a parabolic surface.

When forming a DTM, the accuracy of measurement of individual points obtained by photogrammetry is lower than that which can be obtained by ground survey methods. Photogrammetric measurement is more economic and many more points may be measured and comparative tests indicate similar overall accuracies after interpolation. However the potential of the DTM can only be fully exploited when it is used in conjunction with a computer and associated software.

Probably the most well-used software is MOSS, which is marketed by Moss Systems Ltd. in a modular form including packages for inputting both ground and air survey data, and editing and plotting the results. The ground surface model can then be viewed on a computer screen in plan or perspective, contoured, and used to draw cross-sections or enable a volumetric analysis to be carried out for the survey area. A design module allows proposed civil engineering projects such as roads, railways and airfields, or other applications like open-cast mining, to be integrated into the model. These can also be viewed perspectively, plotted and fully incorporated into the sectioning and volumetric calculations. The speed and versatility of such software allows the engineer to investigate different alignments or layouts for his proposed scheme with the minimum of effort and this goes some way to achieving optimum design.

Applications of terrestrial photogrammetry

Terrestrial photogrammetry was commonly used before the advent of the aeroplane, but nowadays its use is restricted to specialist applications where aerial photogrammetry and more conventional ground surveys are difficult, e.g. in steep gorges or mountainous sites. Even here modern electronic instruments, including reflectorless EDM and GPS, are starting to be used in preference because they do not need complex software and specialist plotters to interpret the survey. It is probable that in the near future terrestrial photogrammetry will only be used in close range situations as discussed below.

Applications of close range photogrammetry

The number and variety of applications of close-range photogrammetry to the solution of measurement problems in engineering is growing. The technique has been applied in many diverse fields of civil engineering and at various stages of an overall project ranging from the preliminary design stage, where model simulation is often used, to the checking of the completed structure and the monitoring of its behaviour with time. Close-range photogrammetry is also used as a research tool in civil engineering such as the testing of models of box girders.

Other possibilities include:

(1) geological work on vertical faces and application in rock mechanics and slope stability problems;
(2) recording of architectural features and conditions including the plotting of facades: representation of proposed structures on photographs in conjunction with existing nearby buildings;
(3) recording of traffic accidents or construction accidents;
(4) measurement of verticality of towers etc. during construction;
(5) measurement of river flows.

Exercises 13

1 An area of 220 km^2 is to be photographed at a scale of 1 in 8000 from the air using a camera of focal length 150 mm, the photographs being 230 mm square. A longitudinal overlap of 60 per cent and a lateral overlap of 25 per cent must be provided.

If the operating speed of the aircraft is 225 km/h find

(a) the flying height of the aircraft and the interval between exposures;

(b) the number of prints required if the flying strips are 16 km long.

Answer: (a) 1200 m, 11.8 sec; (b) 253.

2 In a pair of overlapping vertical air photographs, the mean distance between the two principal points, both of which are at datum level, is 89.6 mm. At the time of photography the aircraft was 2500 m above datum and the camera had a focal length of 150 mm. In the common overlap a tall chimney, whose base is at datum level, is observed, and the difference in parallax (measured with a direct reading bar) of the top and base is found to be 2.40 mm. Estimate the height of the chimney.

Answer: 65.2 m.

3 Overlapping images *ac* and *bd* of a car of length 4.60 m, travelling in a straight path, were obtained on the same photograph by exposures made at a time interval of 0.10 s. Measurements were then taken from the photograph as follows:

$$ab = \ \ 6.5 \text{ mm} \qquad bd = 23.2 \text{ mm}$$
$$ac = 24.6 \text{ mm} \qquad cd = \ \ 6.1 \text{ mm}$$

a and *b* refer to a feature at one end of the car, and *c* and *d* to a feature at the other end. Estimate the speed of the car.

(*Salford*)

Answer: 12.1 m/s.

4 The image co-ordinates of three points, *A*, *B* and *C*, and of the principal points *P* and *Q* on two overlapping vertical aerial photographs were as follows:

Point	Left photo		Right photo	
	x (mm)	y (mm)	x (mm)	y (mm)
P	0.0	0.0	−89.2	0.0
Q	+89.4	0.0	0.0	0.0
A	+12.8	+44.6	−76.6	+44.2
B	+16.4	+6.3	−72.8	+5.9
C	+20.2	−30.7	−69.6	−31.2

Given that the ground co-ordinates of *A* and *C* were 60 000 mE, 72 000 mN and 61 260 mE, 71 200 mN respectively, estimate those of *B*.

(*Salford*)

Answer: 60 629 mE, 71 579 mN.

5 (a) (i) Explain the three main geometrical differences between a map and an air survey photograph of the same area.

(ii) Outline the circumstances when the two documents (map and photograph) would be geometrically equivalent.

(iii) Discuss the characteristics of an orthophotograph which mean that it is geometrically equivalent to a map.

(b) (i) Derive an expression which would enable you to assess the displacement of a point, imaged on an air survey photograph, due to the variations in ground relief.

(ii) The photographic image of an Ordnance Survey pillar is 80 mm from the centre of an air survey photograph. The pillar is 160 m above Ordnance Datum and the aircraft flying height is 1600 m above sea level. How much is the image displaced because of the elevation of the OS pillar? (*London*)

Answer: 8 mm.

6 The image of a ground point has co-ordinates of $x = +75.30$ mm and $y = +54.00$ mm on a photograph taken at a height of 1000 m with a camera of focal length 152 mm. Assuming that rotations of $\omega = +\frac{1}{2}°$, $\phi = -1°$, $\kappa = -\frac{1}{2}°$ actually occurred at exposure determine the corrected co-ordinates and hence ground co-ordinates with respect to the camera position.

Answer: 72.27 mm, 51.41 mm, 475.46 mE, 338.24 mN.

7 A survey was carried out using a phototheodolite set up at stations situated on a proposed centre line. At station A, the camera axis was directed on to a levelling staff held vertically at station B (reduced level 56.4 m AOD) which was also on the centre line. On the subsequent photograph, taken with the plate vertical, the images of the 1.00 m and 3.00 m graduations on the staff were found to be 2.0 mm and 8.6 mm above the principal horizontal line.

The image of a point C lying on the cross-section at B was then established to be 45.0 mm to the right of the principal vertical line and 12.8 mm above the principal horizontal line. If the focal length of the camera was 165 mm, determine (i) the co-ordinates of C with respect to A, and (ii) the reduced level of C.

Answer: (ii) 60.7 m.

8 Two photographs were taken by a phototheodolite of focal length 165 mm from stations A and B, 100 m apart, the camera axes being normal to AB in each case. A control point C, whose co-ordinates with respect to A were 33.98 mE, 450.50 mN, was imaged on the photograph from A at 12.45 mm to the right of the principal vertical line and 10.52 mm above the principal horizontal line. The corresponding co-ordinates on the photograph from B were 24.13 mm to the left of the principal vertical line and 20.95 mm above the principal horizontal line.

Determine the inclination from the vertical of the plate at B, given that the effective camera level at *B* was 3.24 m higher than at *A*, where the plate was known to be vertical.

Answer: 4°.

9 On a certain air-photograph the images of three ground stations, *P*, *Q* and *R*, were identified. The following co-ordinates were measured and compared to the known ground co-ordinates.

Position	Photo co-ordinates		Ground co-ordinates	
	x (mm)	*y* (mm)	*x* (mE)	*y* (mN)
P	9.1	87.2	382 092.6	399 001.0
Q	−6.4	2.8	382 375.5	398 926.5
R	57.5	−17.5	382 458.9	399 136.8

Determine the ground co-ordinates for a point that has photo co-ordinates $x = +18.6$ mm, $y = +60.1$ mm.

Answer: 382 187.2 mE, 399 025.9 mN

10 (*a*) Explain the meaning of the following terms in the context of digital mapping:

(i) vector data;
(ii) raster data;
(iii) topology;
(iv) spaghetti data;
(v) structured data.

(*b*) Describe the processes which are necessary to map 'spaghetti' data into a 'structured' format.

(*c*) What are the advantages in holding map data in this more complex 'structured' format?

(*London*)

Answer: The reader will appreciate that in a general book such as this, the above terms cannot be covered in detail. He is thus encouraged to review current literature appertaining to Geographical Information Systems before attempting a comprehensive answer to the question.

Index